深度学习私房菜
跟着案例学TensorFlow

程世东 ◎ 编著

电子工业出版社
Publishing House of Electronics Industry
北京·BEIJING

内容简介

本书通过案例讲解如何使用 TensorFlow 解决深度学习的实际任务，每章都包含 TensorFlow 1.x 和 2.0 的代码实现。全书共分 10 章，主要讲解卷积神经网络、LSTM、Seq2Seq、Transformer、BERT、文本卷积、GBDT、FM、FFM、Dlib、MTCNN、VGG、AlphaGo / AlphaZero、BiLSTM、DQN、Gym、GAN 等技术，包含的项目有 CIFAR-100 图像分类、彩票预测、古诗生成、推荐系统、广告点击率预测、人脸识别、中国象棋、汉字 OCR、FlappyBird 和超级马里奥、人脸生成。

本书假设读者具有适当的 Python 编程基础和深度学习基础（比如梯度下降、反向传播等知识）。全书以案例的方式讲解涉及的知识，包括理论、算法和解决思路，适合相关专业的大学生或研究生，以及对深度学习感兴趣的读者参考阅读。

未经许可，不得以任何方式复制或抄袭本书之部分或全部内容。

版权所有，侵权必究。

图书在版编目（CIP）数据

深度学习私房菜：跟着案例学 TensorFlow / 程世东编著．—北京：电子工业出版社，2019.8
ISBN 978-7-121-36499-0

I. ①深… II. ①程… III. ①人工智能—算法—研究 IV. ①TP18

中国版本图书馆 CIP 数据核字 (2019) 第 089264 号

策划编辑：郑柳洁
责任编辑：葛　娜
印　　刷：三河市良远印务有限公司
装　　订：三河市良远印务有限公司
出版发行：电子工业出版社
　　　　　北京市海淀区万寿路 173 信箱　　邮编：100036
开　　本：787×980　1/16　印张：30.25　字数：691 千字
版　　次：2019 年 8 月第 1 版
印　　次：2020 年 3 月第 2 次印刷
定　　价：128.00 元

凡所购买电子工业出版社图书有缺损问题，请向购买书店调换。若书店售缺，请与本社发行部联系，联系及邮购电话：（010）88254888，88258888。

质量投诉请发邮件至 zlts@phei.com.cn，盗版侵权举报请发邮件至 dbqq@phei.com.cn。

本书咨询联系方式：（010）51260888-819，faq@phei.com.cn。

前言

在过去的几年中，人工智能和深度学习是一个不断被提及的话题，最令大众熟知的恐怕就是 AlphaGo 与柯洁、李世乭的围棋大战了。而最近，人工智能 OpenAI Five 在 *DOTA 2* 的比赛中击败了世界冠军 OG，人工智能的发展总会给人带来惊喜。除了上述"大事件"，人工智能和深度学习早就深入我们的生活当中，比如无人驾驶汽车、人脸识别，或者订外卖时 App 给我们做的推荐。在经历了互联网、移动互联网的浪潮之后，可以说现在我们已经步入了人工智能的时代。

我第一次接触机器学习和深度学习时就被深深地吸引了，感觉自己就像一块海绵被投入到大海里，看书、看视频、看源码，学习新的知识使我感到非常兴奋。相信此刻的你也跟当年学习的我的状态是一样的吧，本书将我对深度学习的理解、从开源社区学到的知识分享给大家，希望能为你的学习提供一些帮助。

本书以案例的形式，讲解各种深度学习理论和相应场景的实践，包含 TensorFlow 1.x 和 TensorFlow 2.0 的代码实现。全书共分 10 章。第 1 章讲解了卷积神经网络理论知识，第 2 章讲解了如何进行 CIFAR-100 图像分类实践。第 3 章介绍了循环神经网络（RNN 和 LSTM），以及在彩票预测和古诗生成上的实践。在第 3 章的最后，介绍了 Seq2Seq、Transformer 和 BERT 模型。第 4 章以电影推荐系统为例，分享了推荐系统的实现。第 5 章介绍了广告点击率预测。第 6 章讲解了人脸识别的实践，包含使用 OpenCV、dlib 和 MTCNN 进行人脸检测，使用 dlib、FaceNet 和 VGG16 等方式提取人脸特征，然后讨论了比较人脸特征的几种方式。最后，使用上述技术实现了一个在视频中找人的应用。第 7 章分析了 AlphaGo 和 AlphaZero 的论文原文，讲解了蒙特卡罗树搜索（MCTS）和神经网络的结构，并且通过实现中国象棋复现了 AlphaZero。第 8 章介绍了 OCR 在汉字识别上的应用，并且讲解了 BiLSTM 的多种类型，最后实现了一个端到端的汉字识别网络。第 9 章介绍了 DQN 算法，用于玩 *Flappy Bird*。然后介绍了 OpenAI Baselines 和 Gym 的用法，并用于玩"超级马里奥"。最后介绍了 OpenAI 提出的具有好奇心的强化学习算法。第 10 章讲解了生成对抗网络（GAN）及衍生的变种 DCGAN、WGAN 和 WGAN-GP 算法，用于人脸生成的实践。最后介绍了 PG-GAN 和 TL-GAN 的理论。

限于篇幅，以及作者能力有限，书中难免有错漏之处，本书仅仅是将作者掌握的知识做了总结与分享。当然，这些知识不属于我个人，首先要感谢那些工作在人工智能第一线的科学家们，是他们将研究成果公布出来，让大家可以阅读论文和博客。更要感谢开源社区的贡献者们，使我们可以阅读源码参考学习。还要感谢同样喜欢技术分享的人们，我所能做的跟他们一样，就是将分享的"火炬"继续传递下去。书中在引用时都会给出引用的出处，在这里一并表示感谢。

写作本书的过程，是对自己所学知识的一次梳理，回过头来重新审视自己对某些知识的理解，又是一次成长。同时，写作的过程又是孤独的、寂寞的，有时觉得自己就像在山洞里练剑一样。感谢我的妻子对我的理解和支持，在写作期间，她在国内带着孩子，听从了我"在交稿前不要来打扰"的安排。最要感谢的是我的妈妈，是她培养了我学习与钻研的习惯。最后还要感谢电子工业出版社的郑柳洁和葛娜老师，她们对本书的出版和编辑提供了很多专业性的指导和帮助，没有她们的付出，本书无法与大家见面。

希望本书的内容能够为你提供帮助，权当抛砖引玉，为你的深度学习知识打下基础。

程世东

2019 年 4 月于日本

读者服务

轻松注册成为博文视点社区用户（www.broadview.com.cn），扫码直达本书页面。

- 提交勘误：您对书中内容的修改意见可在【提交勘误】处提交，若被采纳，将获赠博文视点社区积分（在您购买电子书时，积分可用来抵扣相应金额）。
- 与读者交流：在页面下方【读者评论】处留下您的疑问或观点，与其他读者一同学习交流。

页面入口：http://www.broadview.com.cn/36499

目录

1 卷积神经网络与环境搭建   1
  1.1 概述 1
  1.2 卷积神经网络 2
      1.2.1 卷积层 3
      1.2.2 修正线性单元 6
      1.2.3 池化层 8
      1.2.4 全连接层 8
      1.2.5 softmax 层 9
      1.2.6 LeNet-5 网络 9
  1.3 准备开发环境 10
      1.3.1 Anaconda 环境搭建 10
      1.3.2 安装 TensorFlow 1.x 11
      1.3.3 FloydHub 使用介绍 13
      1.3.4 AWS 使用介绍 18
  1.4 本章小结 26

2 卷积神经网络实践：图像分类   27
  2.1 概述 27
  2.2 卷积神经网络项目实践：基于 TensorFlow 1.x 27
      2.2.1 数据预处理 28
      2.2.2 网络模型 33
      2.2.3 训练网络 39
  2.3 卷积神经网络项目实践：基于 TensorFlow 2.0 41
      2.3.1 TensorFlow 2.0 介绍 41
      2.3.2 CIFAR-100 分类网络的 TensorFlow 2.0 实现 44
  2.4 本章小结 60

3 彩票预测和生成古诗 — 61

- 3.1 概述 — 61
- 3.2 RNN — 61
- 3.3 LSTM — 63
- 3.4 嵌入矩阵 — 66
- 3.5 实现彩票预测 — 69
  - 3.5.1 数据预处理 — 70
  - 3.5.2 构建神经网络 — 71
  - 3.5.3 训练神经网络 — 75
  - 3.5.4 分析网络训练情况 — 83
  - 3.5.5 生成预测号码 — 88
- 3.6 文本生成 — 93
- 3.7 生成古诗：基于 TensorFlow 2.0 — 96
  - 3.7.1 数据预处理 — 96
  - 3.7.2 构建网络 — 99
  - 3.7.3 开始训练 — 102
  - 3.7.4 生成古诗 — 102
- 3.8 自然语言处理 — 106
  - 3.8.1 序列到序列 — 106
  - 3.8.2 Transformer — 108
  - 3.8.3 BERT — 112
- 3.9 本章小结 — 118

4 个性化推荐系统 — 119

- 4.1 概述 — 119
- 4.2 MovieLens 1M 数据集分析 — 120
  - 4.2.1 下载数据集 — 120
  - 4.2.2 用户数据 — 120
  - 4.2.3 电影数据 — 122
  - 4.2.4 评分数据 — 123
- 4.3 数据预处理 — 123
  - 4.3.1 代码实现 — 124

	4.3.2	加载数据并保存到本地	127
	4.3.3	从本地读取数据	128
4.4	神经网络模型设计		128
4.5	文本卷积神经网络		130
4.6	实现电影推荐：基于 TensorFlow 1.x		131
	4.6.1	构建计算图	131
	4.6.2	训练网络	139
	4.6.3	实现个性化推荐	144
4.7	实现电影推荐：基于 TensorFlow 2.0		154
	4.7.1	构建模型	154
	4.7.2	训练网络	166
	4.7.3	实现个性化推荐	166
4.8	本章小结		169

5 广告点击率预估：Kaggle 实战　　　　170

5.1	概述		170
5.2	下载数据集		170
5.3	数据字段的含义		171
5.4	点击率预估的实现思路		172
	5.4.1	梯度提升决策树	172
	5.4.2	因子分解机	172
	5.4.3	场感知分解机	174
	5.4.4	网络模型	175
5.5	数据预处理		176
	5.5.1	GBDT 的输入数据处理	177
	5.5.2	FFM 的输入数据处理	177
	5.5.3	DNN 的输入数据处理	179
	5.5.4	数据预处理的实现	180
5.6	训练 FFM		188
5.7	训练 GBDT		197
5.8	用 LightGBM 的输出生成 FM 数据		203
5.9	训练 FM		207

目录

5.10 实现点击率预估：基于 TensorFlow 1.x 218
    5.10.1 构建神经网络 219
    5.10.2 训练网络 225
    5.10.3 点击率预估 231
5.11 实现点击率预估：基于 TensorFlow 2.0 237
5.12 本章小结 245

6 人脸识别   246

6.1 概述 246
6.2 人脸检测 247
    6.2.1 OpenCV 人脸检测 247
    6.2.2 dlib 人脸检测 251
    6.2.3 MTCNN 人脸检测 254
6.3 提取人脸特征 264
    6.3.1 使用 FaceNet 提取人脸特征 264
    6.3.2 使用 VGG 网络提取人脸特征 265
    6.3.3 使用 dlib 提取人脸特征 272
6.4 人脸特征的比较 276
6.5 从视频中找人的实现 282
6.6 视频找人的案例实践 284
6.7 人脸识别：基于 TensorFlow 2.0 302
6.8 本章小结 303

7 AlphaZero / AlphaGo 实践：中国象棋   304

7.1 概述 304
7.2 论文解析 305
    7.2.1 蒙特卡罗树搜索算法 307
    7.2.2 神经网络 312
    7.2.3 AlphaZero 论文解析 314
7.3 实现中国象棋：基于 TensorFlow 1.x 317
    7.3.1 中国象棋着法表示和 FEN 格式 317
    7.3.2 输入特征的设计 321
    7.3.3 实现神经网络 323

		7.3.4 神经网络训练和预测 327
		7.3.5 通过自我对弈训练神经网络 330
		7.3.6 自我对弈 334
		7.3.7 实现蒙特卡罗树搜索：异步方式 340
		7.3.8 训练和运行 353
	7.4	实现中国象棋：基于 TensorFlow 2.0，多 GPU 版 354
	7.5	本章小结 364

8 汉字 OCR 365

	8.1	概述 365
	8.2	分类网络实现汉字 OCR 365
		8.2.1 图片矫正 366
		8.2.2 文本切割 368
		8.2.3 汉字分类网络 369
	8.3	端到端的汉字 OCR：基于 TensorFlow 1.x 371
		8.3.1 CNN 设计 372
		8.3.2 双向 LSTM 设计 374
		8.3.3 CTC 损失 385
		8.3.4 端到端汉字 OCR 的网络训练 388
	8.4	汉字 OCR：基于 TensorFlow 2.0 395
		8.4.1 CNN 的实现 395
		8.4.2 双向 LSTM 的实现 396
		8.4.3 OCR 网络的训练 403
	8.5	本章小结 406

9 强化学习：玩转 Flappy Bird 和超级马里奥 407

	9.1	概述 407
	9.2	DQN 算法 407
	9.3	实现 DQN 玩 Flappy Bird：基于 TensorFlow 1.x 412
	9.4	实现 DQN 玩 Flappy Bird：基于 TensorFlow 2.0 417
	9.5	使用 OpenAI Baselines 玩超级马里奥 424
		9.5.1 Gym 424
		9.5.2 自定义 Gym 环境 426

| 9.5.3 使用 Baselines 训练 431
       9.5.4 使用训练好的智能体玩游戏 437
       9.5.5 开始训练马里奥游戏智能体 438
9.6 具有好奇心的强化学习算法 443
9.7 本章小结 444

10 生成对抗网络实践：人脸生成　　445

10.1 概述 445
10.2 GAN 446
10.3 DCGAN 447
       10.3.1 生成器 448
       10.3.2 判别器 449
10.4 WGAN 449
10.5 WGAN-GP 451
       10.5.1 WGAN-GP 算法 451
       10.5.2 训练 WGAN-GP 生成人脸：基于 TensorFlow 1.x 452
       10.5.3 训练 WGAN-GP 生成人脸：基于 TensorFlow 2.0 462
10.6 PG-GAN 和 TL-GAN 469
10.7 本章小结 473

1 卷积神经网络与环境搭建

1.1 概述

在计算机视觉和人工智能领域,人们希望智能体能够像人一样,不仅能看到图像,更要看懂图像,比如目标检测和物体识别。这就需要训练智能体对物体进行分类,并能够识别出这些分类物体,甚至能够进行人脸识别(区分出是否是同一人)。当能够识别出图像中的物体之后,智能体还要能够理解图像所包含的含义,比如图像中一辆车行驶在马路上、一只猫在捉老鼠等。更进一步,人们还要求智能体能够预测图像中物体下一步要做什么事情,这就涉及人体骨骼关键点检测等技术。2019 年 2 月,由李飞飞等研究者发表的论文 *Peeking into the Future: Predicting Future Person Activities and Locations in Videos* 中提出了一种端到端的模型,可以用来预测行人的未来活动和行动轨迹,论文地址为 https://arxiv.org/abs/1902.03748。

总的来说,计算机视觉应用的最终目标是使智能体能够像人一样看待、看懂这个世界。视觉是人类从世界获取信息的主要途径,也是智能体了解世界的途径。我们希望自动驾驶汽车能够识别出人、道路等其他物体,这一切都可以使用卷积神经网络实现。

20 世纪 90 年代末,Yann LeCun 研究的卷积神经网络开始流行,LeNet-5 网络能够用来识别手写数字。自 2010 年以来,每年的 ImageNet 大规模视觉识别挑战赛(ILSVRC),很多研究团队都在 ImageNet 的超大图像数据集(含有 1000 个标签类别、超过 100 万张的训练图像)上评估验证他们的最新算法。2012 年,基于卷积神经网络的 AlexNet 在 ImageNet 上实现了 15% 的错误率,这一成绩轻松击败了第二名 26% 的错误率。此后在历年的比赛中,更多的基于卷积神经网络的优秀网络接踵而来,比如 VGG(2014 年)、GoogLeNet(2014 年)、ResNet(2015 年)等。

本章将讲解这一重要的网络——卷积神经网络的原理。

1.2 卷积神经网络

卷积神经网络（Convolutional Neural Network, CNN）跟传统的神经网络有些类似，它们都由可学习的神经元节点组成，包含权重和偏置。传统神经网络的各种计算在卷积神经网络中都一样，比如前向传播、反向传播等。

传统神经网络的结构如图 1.1 所示。可以看出，网络的每一层都是以全连接的形式连接起来的。而卷积神经网络与传统神经网络相比，在结构上做了改进，卷积神经网络通常包含以下几层：

- 卷积层（Convolutional Layer）
- 修正线性单元（Rectified Linear Unit, ReLU）
- 池化层（Pooling Layer）
- 全连接层（Fully-Connected Layer）
- softmax 层

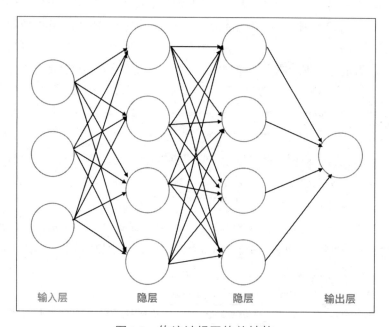

图 1.1　传统神经网络的结构

1.2.1 卷积层

卷积层是整个卷积神经网络的核心，这里就不介绍卷积的计算了，我们来直观感受卷积是如何工作的，如图 1.2 所示。

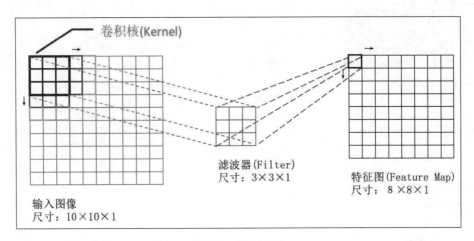

图 1.2 卷积过程

假设输入是一张灰度图图像，那么输入的深度就是 1，如图 1.2 所示输入图像尺寸是 $10 \times 10 \times 1$。现在选择一个尺寸（通常选择 3×3 或 5×5，如图 1.2 左侧所示），从输入图像左上角开始水平移动（水平方向到头的话，垂直移动一步后再次水平移动），直到滑过整个输入图像，最终到达图像右下角。这个在输入图像上选择的尺寸（Kernel Size），叫作卷积核（Kernel）。在这个过程中，每次移动都将该尺寸的输入图像与相同尺寸（3×3 或 5×5）的卷积参数做运算，这个卷积参数叫作滤波器（Filter，如图 1.2 中间所示），并得到一个输出（如图 1.2 右侧所示）。每移动一步，都会得到一个输出，当滤波器与整个输入图像运算完之后，最终得到的输出就是对输入图像提取出的特征，叫作特征图（Feature Map）。这就是卷积层的功能，在整个移动（卷积）过程中，卷积层的权重（滤波器，Filter Size）是不变的，即权值共享。

卷积核在输入图像上移动时，每移动一步的长度叫作步幅（Stride，如图 1.3 左侧所示），这个长度表示移动的像素数，最后输出的特征图尺寸跟移动的步幅相关。如果步幅是 1 的话，则输出图像的尺寸和输入图像基本相同。为什么是基本相同呢？这取决于卷积核在图像边界上的处理方式。有两种方式：一种是卷积核在图像边界范围内移动，称作有效填充（Valid Padding），如图 1.4 所示；另一种是在图像边界外用 0 填充，卷积核在填充 0 的位置开始移动，这样输出图像的尺寸和输入图像是一致的，称作相同填充（Same Padding），如图 1.5 所示。当步幅为 2 时，输出图像的尺寸是输入图像的一半，相当于降采样。

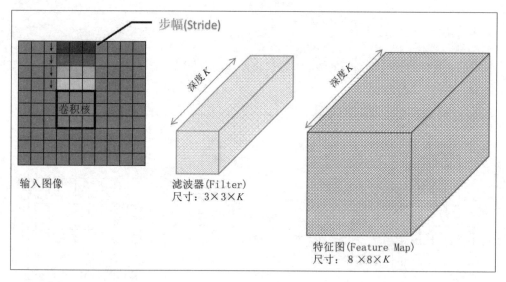

图 1.3 滤波器的深度

图 1.4 有效填充

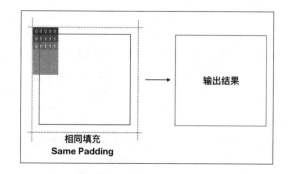

图 1.5 相同填充

现在来看看更常见的情况，通常我们会对输入图像提取多个特征，这代表我们要使用多个滤波器。图 1.3 中间所示的滤波器尺寸是 $3 \times 3 \times K$，表示卷积层使用 K 个滤波器，输出深度为 K（K 维）的特征图（也可以理解成是一个具有 K 个颜色通道的图像，如图 1.3 右侧所示），每一维特征图都是由输入和训练的卷积参数（滤波器）做点积，然后加上偏置得到的，训练的每一维滤波器都可以用来提取图像的不同特征。

卷积神经网络就是使用权值共享的卷积层来代替全连接层的。图 1.2 所示的演示只是经过了一层卷积，考虑传统的神经网络通常会使用多个隐层，在使用多个卷积层之后，整个网络看起来如图 1.6 所示。

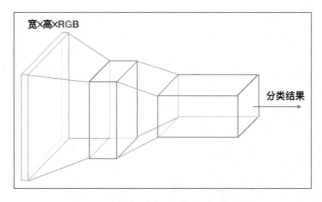

图 1.6　卷积神经网络金字塔结构

图 1.6 左侧是输入图像,经过卷积之后,会逐步改变图像的维度,宽度和高度逐渐变小,并且深度不断变大,这样深度的信息通常可以表示复杂的抽象特征。如果在网络的最后使用分类器,那么根据卷积层学到的图像特征最终会得到该图像的分类结果。

对于卷积层,我们只需要设计好卷积核的尺寸、滤波器的个数(输出的深度)、移动的步幅以及填充方式,然后将输入图像交给卷积层,输出的特征图即是提取到的特征。第一层卷积层提取到输入图像的特征作为输入,再由第二层卷积层继续提取特征,接着后面的卷积层继续提取,直到网络的最后一层。这样越往后的卷积层提取到的特征越复杂。最开始的卷积层学习到的是最基本的特征,例如直线、斜线、弧线。可以将这个过程理解成放大镜,最开始看到的是最高倍放大的结果,即图中的某个边缘。第二层卷积层在基本特征的基础上,就好比稍微缩小了放大镜的倍率,看到的图像稍微清晰一些,是由直线、斜线和弧线组成的某种图像,例如矩形、圆形、多边形。再继续,下一层卷积层提取到的特征更高级一些,好比再次缩小了放大镜的倍率,好像看到了某种物体的一部分,例如眼睛、鼻子、耳朵……直到最后一层卷积层,提取到更抽象的特征,能够通过识别到的眼睛、鼻子、耳朵确认输入图像是否是某种人脸或者动物或者任意我们要识别的东西,从而得到分类结果。

卷积层的好处首先就是权值共享。由全连接的方式变成了小范围的局部连接,并且用局部连接的卷积核滑过整个输入图像的权值是共享的。这大大减少了要学习的参数数量,也减小了网络的复杂度,降低了过拟合风险。

另一个好处就是带来了平移不变性。比如我们要在图像中识别一只狗,不管这只狗是大是小、是黄色的还是黑色的、是站着的还是躺着的,也不管在图像的哪个位置,只要卷积神经网络学习到狗的特征之后,无论在什么条件下都能够识别出图像中的狗,而不用训练网络单独学习狗在左边的情况、狗在右边躺下的情况……否则,工作量就大了。一旦出现没有想到的情况,就识别失败了。实际上,这就是人们手工制定特征和神经网络学习特征的差别,泛化能力不同。所以,卷积神经网络的平移不变性带来了很好的泛化能力,提高了网络的鲁棒性。

1.2.2 修正线性单元

在卷积层之后,通常要加入非线性激活函数,目的是通过引入非线性,使得神经网络能够逼近任意函数,解决复杂的现实问题。

关于激活函数,有很多种选择,比如 Sigmoid 函数。

公式:$f(z) = \dfrac{1}{1+e^{-z}}$

导数:$f'(z) = f(z)(1 - f(z))$

Sigmoid 函数的曲线如图 1.7 所示(实线)。

Sigmoid 函数也叫作 Logistic 函数,特点是能够将输入映射到 (0,1)。输出值在 0~1 之间,符合概率输出的意义,适合用于解决二分类问题。但是 Sigmoid 函数有缺陷,我们来看看反向传播时 Sigmoid 函数的导数曲线,如图 1.7 所示(虚线)。

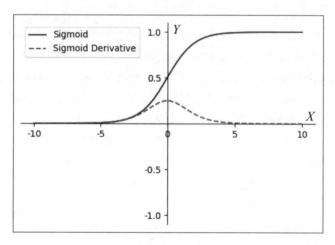

图 1.7　Sigmoid 函数的曲线和 Sigmoid 函数的导数曲线

Y 轴是梯度值,从 Sigmoid 函数的导数曲线可以看出,梯度在饱和区域非常平缓,趋近于 0。这意味着,当网络层数较多时,Sigmoid 函数在不断的反向传播中梯度会逐渐变小,这导致更新网络参数异常缓慢,网络难以收敛,这就是梯度消失问题。这可以理解成梯度趋近于 0 的神经元都"死"了,不再起任何作用,一旦梯度消失蔓延下去,整个网络也不再有效。梯度趋近于 0,会造成大量信息丢失,是训练深层网络必须要解决的问题。另外,Sigmoid 函数和反向传播计算量大,包含指数和除法等运算。

现在介绍另一个激活函数,即线性整流函数,又称修正线性单元(ReLU)。

公式:$f(x) = \max(0, x)$

在神经网络中，ReLU 作为激活函数，是对来自上一层网络的输出 $\boldsymbol{x}$，经过本层网络的线性变换 $\boldsymbol{w}^\mathrm{T}\boldsymbol{x}+b$ 之后做非线性变换的结果。

公式：$f(\boldsymbol{x}) = \max(0, \boldsymbol{w}^\mathrm{T}\boldsymbol{x})$

ReLU 函数的曲线如图 1.8 所示。

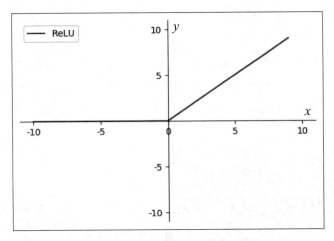

图 1.8　ReLU 函数的曲线

从曲线图可以看出，ReLU 是分段线性函数：

$$\mathrm{ReLU}(x) = \begin{cases} x, & \text{如果 } x > 0 \\ 0, & \text{如果 } x \leqslant 0 \end{cases}$$

输入是负值，输出都是 0；正值不变，输出等于输入。因为 ReLU 的分段线性特性，前向传播、后向传播和求导都是分段线性的，计算足够简单，使得训练网络更容易，加快了网络的收敛。更重要的是，解决了梯度消失问题，可以用于训练更深层次的网络。

当然，ReLU 也并不完美，使用 ReLU 训练网络有时很脆弱。假如一个非常大的梯度经过 ReLU 神经元，参数更新之后，神经元就"死"了，导致这个神经元的梯度永远都是 0。感兴趣的读者可以去了解"dying ReLU"问题。

有很多 ReLU 的变体可供我们使用，比如 Leaky ReLU、PReLU、Randomized Leaky ReLU、ELU 和 SELU 等激活函数，这里就不展开介绍了，感兴趣的读者可以继续研究。本章我们使用 ReLU 即可。

1.2.3 池化层

池化层（Pooling Layer）的作用是对卷积层输出的特征图做降采样，深度不变，减小特征图的尺寸，这样会丢弃一部分提取的特征，从而减少要学习的参数，避免过拟合。可以将池化层理解成对输入图像的压缩，或者说特征降维，不会引入新的参数。比如图像中是一张人脸，我们将图像尺寸缩小一倍或几倍，仍然能从图像中看出是一张人脸，这说明对图像做降维，抛弃的是冗余信息，不会改变图像的主要特征。

池化层常用的方法是最大池化（Max Pooling）和平均池化（Average Pooling），最大池化取滑动窗口内的最大值；平均池化取滑动窗口内的平均值。通常使用 2×2 的尺寸，步幅常使用 2，我们以此为例来看看最大池化是怎么计算的，如图 1.9 所示。

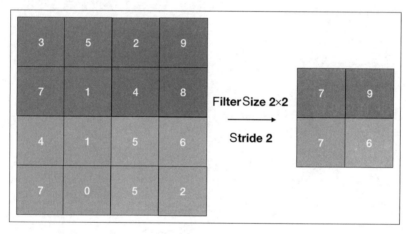

图 1.9　最大池化

1.2.4 全连接层

通常全连接层在网络的尾部，跟传统神经网络的连接方式是一样的，可以参考图 1.1 所示的网络结构。定义一个卷积神经网络，就是通过设计卷积核和池化层的尺寸、数量，然后不断堆叠卷积层（激活函数）+ 池化层，最后把多个特征图连接成一个扁平的特征向量，叫作 Flatten 层，如图 1.10 所示。

通过 Flatten 层的处理，将提取到的多个特征压缩成一个特征向量，然后就可以把这个向量作为全连接层的输入了。卷积网络的尾部通常使用 1~2 个全连接层。

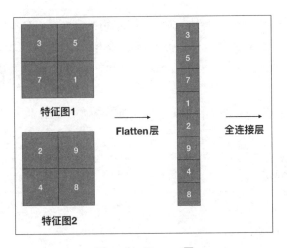

图 1.10 Flatten 层

1.2.5 softmax 层

对于分类网络，在全连接层的最后，通常使用 softmax 层来做分类。根据前几层提取到的特征，通过 softmax 层得到分类的概率。

1.2.6 LeNet-5 网络

LeNet-5 是最早将卷积神经网络用于解决分类问题的网络，同时也是非常经典和著名的网络。因为 LeNet 网络足够简单，我们就以 LeNet 网络为例，将本节讲述的各层概念统一起来。LeNet-5 网络结构如图 1.11 所示（图片来自 Yann LeCun 的论文 *Gradient-Based Learning Applied to Document Recognition*）。

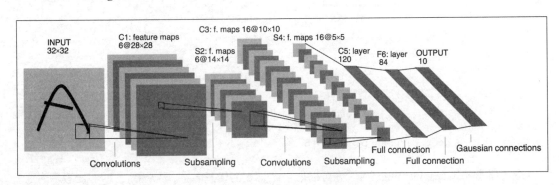

图 1.11 LeNet-5 网络结构

输入层是 32×32 的灰度图，不算输入层的话，网络共有 7 层。输入层之后的 C1 层就是卷积层，然后是池化层 S2，紧接着又是一对卷积层（C3）和池化层（S4）。此时一共有 16 个 5×5 的特征图，后面连接两个全连接层 C5 和 F6，最后是输出层，得到 10 个输出，对应 10 个数字分类的概率。

以上就是卷积神经网络的介绍。接下来，我们准备开发环境。

1.3 准备开发环境

1.3.1 Anaconda 环境搭建

首先搭建开发环境。我们要安装的是 Anaconda，打开浏览器，访问地址：https://www.anaconda.com/download/，然后选择对应的操作系统（Windows、macOS 或 Linux）。这里以 macOS 为例，如图 1.12 所示，选择 Python 3.7 version 一侧，点击 Download 按钮进行下载，下载以后正常安装即可。

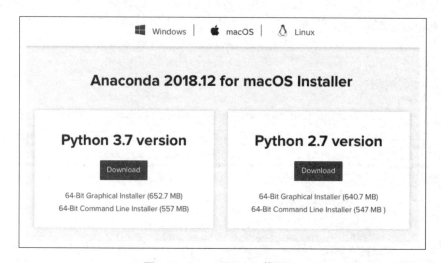

图 1.12　Anaconda 下载页面

安装完成之后，打开终端，输入如下命令创建一个本地环境：

conda create -n deeplearning python=3.5

指定环境名为 deeplearning，当然你可以取自己喜欢的任意名字，然后指定 Python 版本为 3.5，你也可以指定自己想要的版本。

输入命令之后，会提示安装几个软件包，并提示确认安装输入 y，这里我们输入 y 继续安装，如图 1.13 所示。

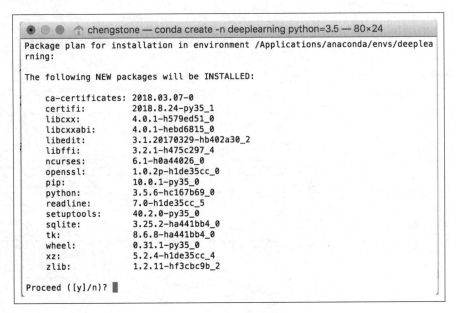

图 1.13　新建环境安装软件包

输入 y 之后，会安装各种所需要的软件包，在安装过程中需要全程联网。安装好之后，输入如下命令激活环境：

```
source activate deeplearning
```

以后本书中的所有项目都是在这个环境下进行开发和运行的。

1.3.2　安装 TensorFlow 1.x

接下来安装 TensorFlow，在安装前请确保开发环境已经激活。在终端输入如下命令，搜索可以安装的 TensorFlow 源和版本：

```
anaconda search -t conda tensorflow
```

运行命令之后，终端会打印出大量的信息，包含 TensorFlow 源的名字、版本、包的类型和支持的系统（Windows、macOS 或 Linux），如图 1.14 所示，这里只截取了一部分。

```
Using Anaconda API: https://api.anaconda.org
Packages:
     Name                             |  Version | Package Types | Platforms               | Builds
     ------------------------- |  -------- | ------------- | ----------------------- | ----------
     GlaxoSmithKline/tensorflow        | 0.12.0   | conda         | linux-64                | py27hb0d0e74_0
                                       : TensorFlow is a machine learning library.
     HCC/tensorflow                    | 1.9.0    | conda         | linux-64                | py27_1, py27_1, py27_0, py27h5b78b684f_1_0, py5h78b684f_0, np113py35_0, py27h78b684f_0, py27h7
     8b684f_1, py36h78b684f_1, np113py27_0, np113py36_0, py35_0, py35_1
                                       : Computation using data flow graphs for scalable machine learning.
     HCC/tensorflow-cpucompat          | 1.7.0    | conda         | linux-64                | py27_0, py36_0, py34_0, np113py35_0, np113py27_0, np113py36_0, py35_0
                                       : Computation using data flow graphs for scalable machine learning.
     HCC/tensorflow-fma                | 1.5.0    | conda         | linux-64                | py27_1, py27_1, py27_0, py34_0, np113py35_0, np113py27_0, np113py36_0, py35_0, py35_1
                                       : Computation using data flow graphs for scalable machine learning.
     HCC/tensorflow-gpu                | 1.7.0    | conda         | linux-64                | np113py35_1, np113py35_2, np113py27_1, np113py27_2, py35hc2b91c5c_3, np113py36_2, np113py36_1, py36h
     2b91c5c_3, py27h2b91c5c_3
                                       : Computation using data flow graphs for scalable machine learning.
     Qwant/onnx-tensorflow             | 1.2.0    | conda         | linux-64                | py36_0
                                       : Tensorflow Backend and Frontend for ONNX
     RMG/tensorflow                    | 1.0.0    | conda         | linux-64, osx-64        | py27_0
     SentientPrime/tensorflow          | 0.6.0    | conda         | osx-64                  | py27_0
                                       : TensorFlow helps the tensors flow
     aaronzs/tensorflow                | 1.10.0   | conda         | linux-64, osx-64, win-64| py36h39705f4_0, py36h8a03d48_0, py35hc784f49_0, py36hdb853c_0, py35h2d7a08b_0, py36he4e0f
     4f_0, py36_1, py35hc0f5839_0, py36hebc11a6_0, py35h89e3332_0, py35h6467dd0_0, py35heb185b1_0, py35hf9a0815_0, py36h2003710_0, py36_0, py36h4df9c7b_0, py35_0, py35_
     1
                                       : TensorFlow helps the tensors flow
     aaronzs/tensorflow-gpu            | 1.10.0   | conda         | linux-64, win-64        | py35h8ac8084_0, py35_1, py36_0, py36hbec5d8f_0, py36h7b1156b_0, py35h14e71af_0, py35_0, py35_1
                                       : TensorFlow helps the tensors flow
     acaprez/tensorflow-gpu            | 1.4.0    | conda         | linux-64                | np113py35_0
                                       : Computation using data flow graphs for scalable machine learning.
     acellera/tensorflow-cuda          | 0.12.1   | conda         | linux-64                | 0
     anaconda/tensorflow               | 1.12.0   | conda         | linux-ppc64le, linux-64, win-64, osx-64 | gpu_py27h99ab47f_0, gpu_py27hd8bfc1a_0, mkl_py36ha6f0bda_0, eigen_py27h6ba8
     707_0, gpu_py36he68c306_0, mkl_py35h34e052c_1, gpu_py27hd7ad7f5_0, gpu_py27hd3a7791e_1_0, gpu_py35hb39db67_1, mkl_py36h2b2bbaf_0, eigen_py35h40edd97_1, eigen_py27hf386fcc_1, eigen
     _py36hc7785b_0, eigen_py35hac2770_0, eigen_py36hc7ac661_0, gpu_py36hfc8490_1, gpu_py36h4fc8490_1, gpu_py27h39Sd9f40_1, gpu_py36h3l1df88_1, eigen_py27h4fd0958_0, gpu_py36h3l3df88_1, gpu_py36hbec23
     59_1, mkl_py27h857755f_0, eigen_py36h995bb4_0, py36_0, eigen_py36h906d37_0, eigen_py27h45ac830_1, gpu_py35h5ed898b_0, gpu_py35h0075c17_1, gpu_py36h8dbd23f_0, h7b2774c_0, gpu_py2
     7h6f941b3_0, mkl_py36h44b7a51_0, eigen_py36he2359_0, mkl_py36h5dc63e2_0, gpu_py35h5d5ef7_0, eigen_py36h8c89287_0, hb1b1514_0, gpu_py36h1514_0, np111py27_0, hc2d9325_0, h6451
     07b_0, mkl_py36h69b6ba0_0, mkl_py27h6c571c4_0, mkl_py36h0cb61a4_0, np112py27_0, eigen_py36hae858b4_0, gpu_py36h9c9050a_0, gpu_py27hc55d17a_0, np111py36_0, gpu_py35h42
     d5ad8_1, hb381393_0, py35_0, eigen_py36h0b764b7_0, gpu_py36h0b764b7_1, eigen_py36hf07811a_1, np35h41bbc20_0, mkl_py35h41bbc20_0, mkl_py35h377fd_0, gpu_py36h02c5d5e_1, eige
     n_py27hf93ee88_0, mkl_py36h5be851a_0, mkl_py35h5be851a_1, gpu_py36ha5f9131_0, eigen_py27hdc0099c_0, eigen_py36h3f7ef1_0, gpu_py36h4459f94_0, gpu_py36hcebf108_0, mkl_py35heddcb22_
     0, gpu_py35ha6119f3_0, gpu_py36h2b0e158_1, eigen_py35h38c8211_0, h2742514_0, mkl_py27h5e87bc0_0, eigen_py27h0dbb4d0_1, np111py35_0, gpu_py35h5b66a776_0, np112py36_0, h469b60b_0, gp
     u_py35h4e76799b_0, eigen_py36hdce78_1, mkl_py36hd6ce78_0, gpu_py35hd9c640d_0, mkl_py35h40f5c2_0, eigen_py36hbd5f568_0, np112py35_0, eigen_py36h346fd36_0, gpu_py36h97a2126_0, eige
     n_py35h8c89287_1, gpu_py27h233f449_1, mkl_py27h6c6124_1, eigen_py27h6eea64_0, py27_0, h01c6a4e_0, eigen_py36hf386fcc_0, h16da8f7_0, eigen_py27hfe19c55_0, h
     b1d968_0, mkl_py36h361250_0, gpu_py27h9580370_0, h57681fa_0, gpu_py36hfdee9c2_1, gpu_py27h956c076_0, h5c3c37f_0, mkl_py36h4f00359_0, gpu_py35h60c0932_1, eigen_py35h0be21f4_1
                                       : TensorFlow is a machine learning library.
     anaconda/tensorflow-base          | 1.12.0   | conda         | linux-64, osx-64, win-64| py35hdbcaa48_1, mkl_py27h2c3e929_0, py35h1a1b453_1, py35h1a1b453_0, gpu_py36h53903_0, py
     36_0, py27hee38f2d_0, gpu_py36h9f529ab_0, gpu_py36h9f529ab_1, gpu_py35h6ecc378_0, mkl_py35h2ca6a6a_0, eigen_py35hdfca3bf_0, py35hdbcaa48_2, gpu_py27h9f529ab_0, py35h4f133c_0, mkl
     _py36h4c2b49_2, py27hdbcaa40_1, py27hdbcaa48_1, py35hff88cb2_1, py35hff88cb2_0, eigen_py27hdfca3bf_0, mkl_py35h81393da_0, py36h1a1b453_0, eigen_py35h9f529ab_1, py27_0, py
     36hdbcaa48_0, py27hdbcaa40_1, py27hdbcaa48_1, py35hff88cb2_1, py35hff88cb2_0, mkl_py35h78e0e9a_0, eigen_py27h4f0eeca_0, py35h4df133c_0, mkl_py27h2ca6a6a_0, eigen_py35hdfca3bf_0, mkl_py35
     5h9f529ab_0, eigen_py36h4f0eeca_0, py35h000c003_2, mkl_py35h78e0e9a_0, eigen_py27h4f0eeca_0, py35h4df133c_0, mkl_py27h2ca6a6a_0, eigen_py35hdfca3bf_0, mkl_py35
     h3c3e929_0, eigen_py35h45dfd0_8, py35h3435052_0, eigen_py35h1a1b453_1, mkl_py36h70e0e9a_0, eigen_py35he28f2d_0, py35hf64866_0, gpu_py27h9f529ab_0, py35h6e53903_0, p
```

图 1.14　TensorFlow 软件包信息

我们要安装的是 Anaconda 发布的 TensorFlow，如果电脑支持 GPU，则可以选择安装 anaconda/tensorflow-gpu；如果电脑不支持 GPU，则可以安装 anaconda/tensorflow。注意，在安装前一定要确认安装源支持的系统版本，比如 anaconda/tensorflow-gpu 的信息是：

```
anaconda/tensorflow-gpu    | 1.12.0 | conda    | linux-ppc64le, linux-64, win-64 |
```

可以看到，在这个软件包支持的系统当中不包含 macOS，所以，如果是 Mac 电脑，并且要安装 GPU 版本的 TensorFlow 的话，就不能安装这个源的软件包了，可以选择其他支持的软件包来安装。这里以安装 anaconda/tensorflow 为例进行说明。

接下来在终端输入如下命令，进一步查看安装源的信息：

```
anaconda show anaconda/tensorflow
```

终端会打印出该安装源所支持的软件包的各种版本，如图 1.15 所示。

可以看到，笔者写本章节时，TensorFlow 的版本已经到了 1.12.0。我们输入如下命令安装 TensorFlow：

```
conda install --channel https://conda.anaconda.org/anaconda tensorflow=1.12.0
```

这样就可以指定安装的版本是 1.12.0，也可以去掉 "=1.12.0" 这部分，直接安装 TensorFlow。

```
Using Anaconda API: https://api.anaconda.org
Name:      tensorflow
Summary: TensorFlow is a machine learning library.
Access:    public
Package Types:  conda
Versions:
    + 0.10.0rc0
    + 1.0.1
    + 1.1.0
    + 1.2.1
    + 1.3.0
    + 1.4.1
    + 1.5.0
    + 1.6.0
    + 1.7.0
    + 1.7.1
    + 1.8.0
    + 1.9.0
    + 1.10.0
    + 1.11.0
    + 1.12.0

To install this package with conda run:
    conda install --channel https://conda.anaconda.org/anaconda tensorflow
```

图 1.15　软件包的各种版本

至此，主要的环境和软件包就安装完成了。运行以后的代码，如果提示缺少某个包，我们直接安装即可。记住，一定要在激活环境后的终端安装软件包。

安装软件包的命令格式是：

conda install 包名

pip install 包名

1.3.3　FloydHub 使用介绍

FloydHub 是一个很方便的深度学习平台，其主旨是让深度学习的网络构建和训练更快速和简单。只要开启一个 FloydHub 实例，我们把所有注意力集中到代码当中，把所有精力用在网络构建和训练上，而不用关心各种依赖软件包的安装，这是 FloydHub 非常便利的地方。跟 Amazon 的 AWS（Amazon Web Services）相比，确实方便了不少。

这里简单介绍 FloydHub 的使用，更详细的使用方法请参考官方的说明。

登录网址 https://www.floydhub.com，首先要注册一个账号，点击 Sign Up for Free 按钮进行注册。

注册成功并登录后，进入 Powerups 看看可以购买的 GPU 实例，地址：https://www.floydhub.com/settings/powerups。

如图 1.16 所示，可以看到官方提供的 GPU 实例的配置、价格和时长，以及笔者本人还有 4 小时 5 分钟的 GPU 时间可以使用。购买需要绑定信用卡，可以点击左侧的 Billing 选项进行绑定。

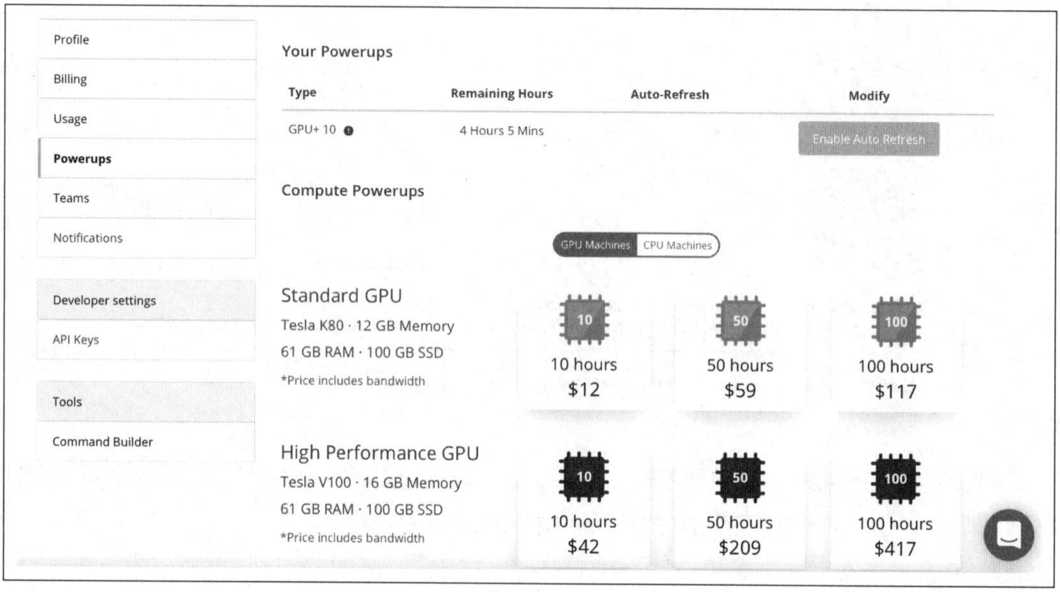

图 1.16　FloydHub 的 GPU 实例

现在讲解如何启动一个任务。前提是我们已经注册完账号了，现在打开终端，并激活之前创建的 Anaconda 环境。接下来要安装 Floyd CLI，在终端输入如下命令安装：

```
conda install -y -c conda-forge -c floydhub floyd-cli
```

或者

```
pip install -U floyd-cli
```

然后在终端登录 FloydHub，有两种登录方式，一种是使用 FloydHub 的用户名和密码登录，输入命令 floyd login：

```
$ floyd login
Login with your FloydHub username and password to run jobs.
Username [alice]: alice
Password:
Login Successful as alice
```

另一种是使用认证 Token 登录，输入命令 floyd login --token：

```
$ floyd login --token
Please paste the authentication token from https://www.floydhub.com/settings/
security.
This is an invisible field. Paste token and press ENTER:
Login Successful as alice
```

认证 Token，可以访问地址 https://www.floydhub.com/settings/security 来查看。

在终端登录后，接下来需要新建一个项目。访问地址 https://www.floydhub.com/projects，打开项目首页，点击右侧的 New project 按钮新建一个项目。也可以直接访问地址 https://www.floydhub.com/projects/create 新建项目。

如图 1.17 所示，输入项目名，然后点击下面的 Create project 按钮，创建项目。

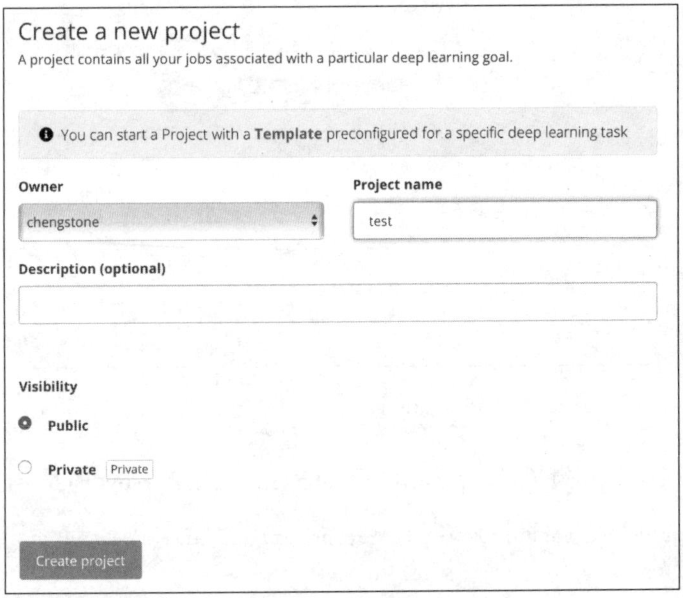

图 1.17　新建项目

项目创建好以后，需要将本地的代码和 FloydHub 上的项目关联起来。首先在终端进入代码的根目录，比如输入如下命令：

```
cd my_projects/my_training_code
```

然后，输入如下命令初始化项目，其中 test 是刚创建的项目名：

```
$ floyd init test

Project "test" initialized in current directory
```

最后，就可以在 FloydHub 云端运行这个程序了。可以访问地址 https://www.floydhub.com/tools/builder 来帮助构建运行的命令。

如图 1.18 所示，选择若干选项之后，在页面顶部生成了命令行。

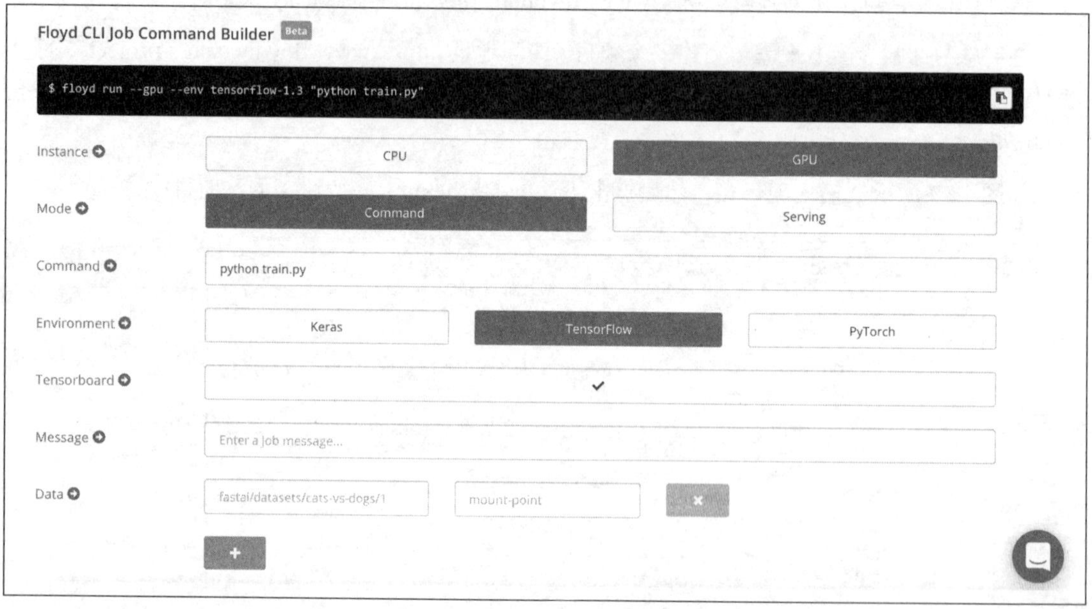

图 1.18 FloydHub 的命令行构建器

在终端输入命令行以运行项目，比如运行下面命令（引用自 FloydHub 文档）：

```
$ floyd run --gpu --env tensorflow-1.3 "python train_and_eval.py"

Creating project run. Total upload size: 25.4KiB
Syncing code ...
[================================] 27316/27316 - 00:00:00

JOB NAME
----------------------
mckay/projects/mnist-cnn/1
```

To view logs enter:
```
floyd logs mckay/projects/mnist-cnn/1
```

如果要运行的是 jupyter notebook，则可以加入参数--mode jupyter。但是现在已经不推荐使用该命令了，因为 FloydHub 支持在项目中创建工作区（Workspace），该工作区以 JupyterLab IDE 的方式运行 jupyter notebook，可以很轻松地在 CPU 和 GPU 之间切换，也能方便地挂载数据集。可以在项目中点击右侧的 Create Workspace 按钮创建工作区，如图 1.19 所示。

图 1.19　创建工作区

在运行 floyd run 命令之后，项目就在 FloydHub 云端运行了。此外，还有一些较常用的命令，下面简单说明。

初始化数据。在终端进入数据目录，然后输入命令：

```
$ floyd data init 项目名
```

上传数据：

```
$ floyd data upload
```

下载输出文件：

```
$ floyd data clone <username>/<project_name>/<run_number>/output
```

所有命令，请参考：http://docs.floydhub.com/commands/。

上传和下载数据集的官方指导，请参考：https://docs.floydhub.com/guides/data/mounting_data/。

官方帮助文档，请访问地址 https://docs.floydhub.com 查看。

1.3.4　AWS 使用介绍

AWS（Amazon Web Services）是亚马逊公司旗下的云计算服务平台，可以为我们提供高质量的云计算服务，对本书中的任务来说就是提供 GPU 计算资源。本书中的 AlphaZero 实践就是在 AWS 平台上进行的。另外，训练 AlphaZero 下中国象棋和汉字 OCR 两章的 TensorFlow 2.0 GPU 训练代码也是在 AWS 上实现的。

本书部分章节可以使用 CPU 来训练，部分内容因为数据集体量较大、网络结构较深，使用 CPU 训练会很慢，如果你不能接受 CPU 训练较慢的话，则可以考虑使用 GPU 来训练网络。如果你没有 GPU 的训练环境，则可以使用上一节介绍的 FloydHub 来完成训练。虽然使用 FloydHub 可以快速地上手训练网络、调试代码，但是在灵活性上笔者认为还是 AWS 更好。我们是在 AWS 上租用了一个虚拟主机实例，自己选择机器和 GPU 配置的，开发环境可以自定义搭建。

本节简单介绍如何使用 AWS。

首先创建一个 AWS 账号，访问网址 https://aws.amazon.com/cn/，点击右侧的 创建AWS账户 按钮创建账户，如图 1.20 所示。

图 1.20　创建 AWS 账户

创建账户需要绑定信用卡，此时不会产生任何费用，只有在实例运行之后才会计费。创建成功后，稍等一会儿，AWS 会给你发送一封确认邮件，确认之后即可使用 AWS 账户登录。

接下来需要选择要运行的实例（选择要使用的机器配置），可以访问地址 https://aws.amazon.com/cn/ec2/instance-types/ 查看实例类型，主要看 加速计算 的实例。当然，这里显示的并不全，比如 g2 的实例就没有显示出来。当选好心仪的实例后，建议你再看看实例价格是否可以接受，可以访问地址 https://aws.amazon.com/cn/ec2/pricing/on-demand/ 查看按需实例的定价，注意要选择相应的区域（就是机房），笔者使用的区域是亚太地区（东京）。

当价位能够接受之后，在使用相应实例之前，需要先确认你是否有使用该实例的权限。访问地址 https://console.aws.amazon.com/ec2/v2/home?#Limits 确认你的 EC2（Elastic Compute Cloud，弹性计算云）服务限制，如图 1.21 所示。

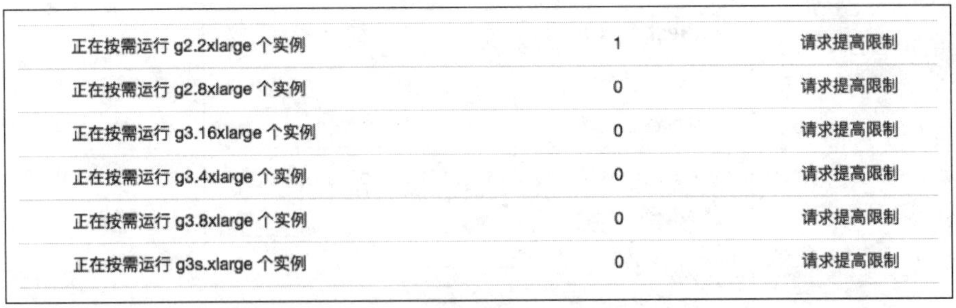

正在按需运行 g2.2xlarge 个实例	1	请求提高限制
正在按需运行 g2.8xlarge 个实例	0	请求提高限制
正在按需运行 g3.16xlarge 个实例	0	请求提高限制
正在按需运行 g3.4xlarge 个实例	0	请求提高限制
正在按需运行 g3.8xlarge 个实例	0	请求提高限制
正在按需运行 g3s.xlarge 个实例	0	请求提高限制

图 1.21　EC2 服务限制

假如你现在要用 g3.4xlarge 实例，可以看到当前的限制是 0，说明你没有使用该实例的权限，需要请求提高限制。

点击该实例右侧的 请求提高限制 链接，进入申请页面（创建工单），如图 1.22 所示。

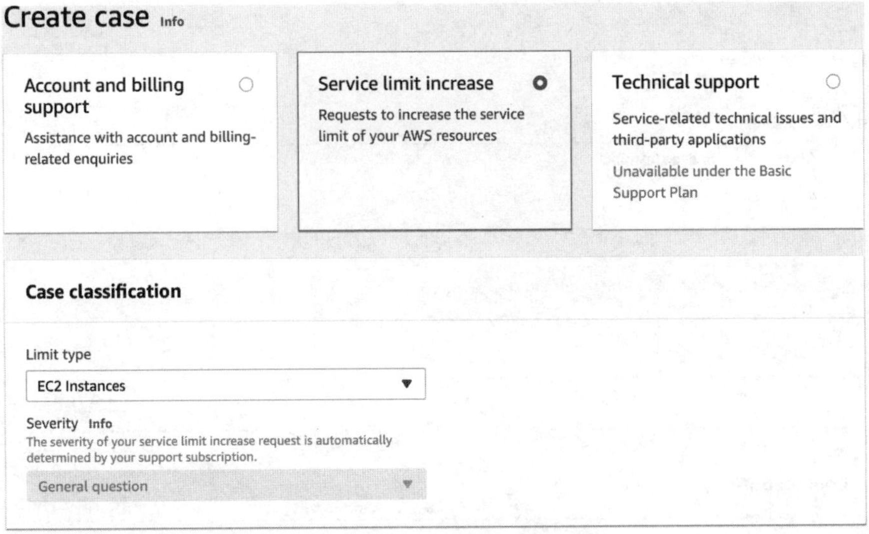

图 1.22　请求提高限制页面

选择 Service limit increase（服务限制增加）选项，下面的 Limit type（限制类型）选择 EC2 Instances（EC2 实例）。

继续往下看，如图 1.23 所示，在 Request（请求）中，选择 Region（地区），然后选择要申请增加限制的实例，这里选择 g3.4xlarge。选择这个实例只是为了举例，你可以根据自己的需求决定使用哪个实例。最后在 New limit value（新限制数）一栏选择要使用多少个实例，通常 1 个就够了。

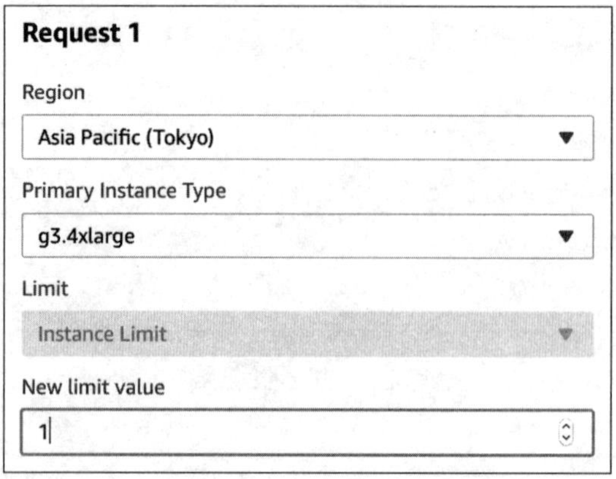

图 1.23　请求设置

最后还要填写理由，如图 1.24 所示，点击 Submit（提交）按钮完成请求。

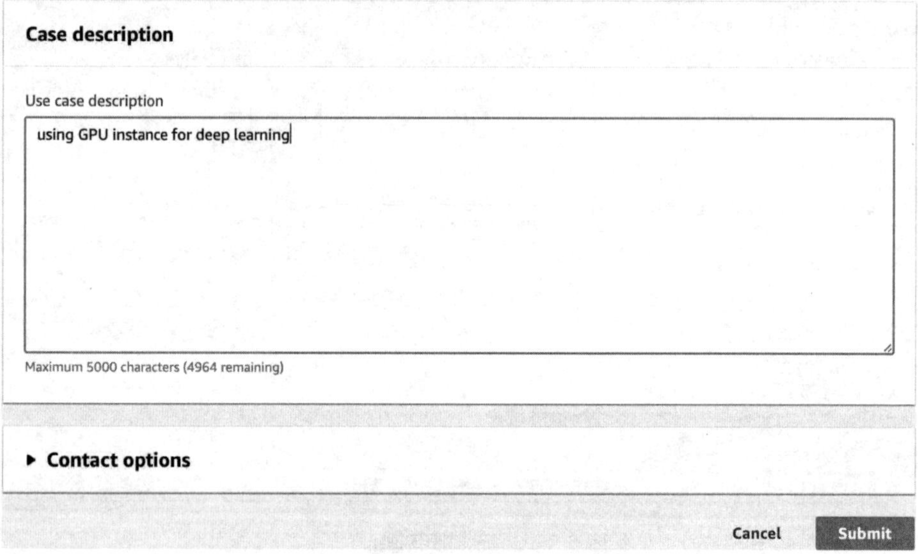

图 1.24　填写理由

请求完之后，通常要等到第二天才能收到确认邮件，告知请求的结果。AWS 处理完请求之后，再次查看自己的 EC2 服务限制，会发现申请实例的限制数已经变成 1 了，这样就可以运行实例了。

访问地址 https://console.aws.amazon.com/ec2/v2/home，打开 EC2 控制台页面，如图 1.25 所示，点击 启动实例 按钮。

1 卷积神经网络与环境搭建

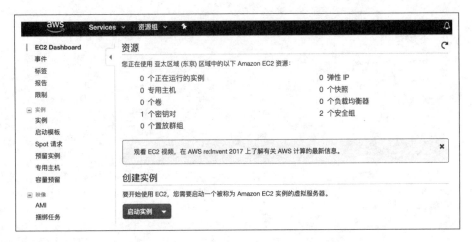

图 1.25 在 EC2 控制台启动实例

接下来开始一系列的启动配置，实例是远程主机，首先选择运行在该主机上的镜像文件 AMI（Amazon Machine Image，Amazon 系统映像），如图 1.26 所示。

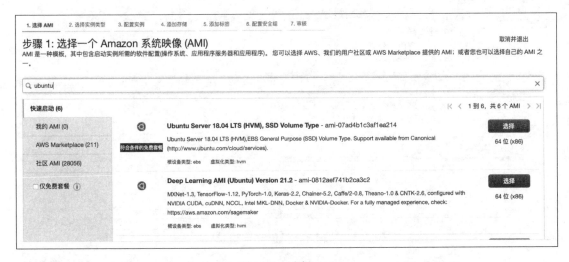

图 1.26 选择 AMI

我们可以在搜索栏中查找，比如这里搜索的关键词是 ubuntu，然后会出现跟 Ubuntu 有关的 AMI。你也可以搜索其他关键词，比如 deep learning 等。找到自己想要使用的 AMI，点击 选择 按钮进入 选择实例类型 页面，如图 1.27 所示。

选择要使用的实例，这里选择了 g2.2xlarge。此时已经可以启动实例了，你也可以继续进行其他配置，这里就不详细解释了。我们来看看配置安全组页面，如图 1.28 所示。

21

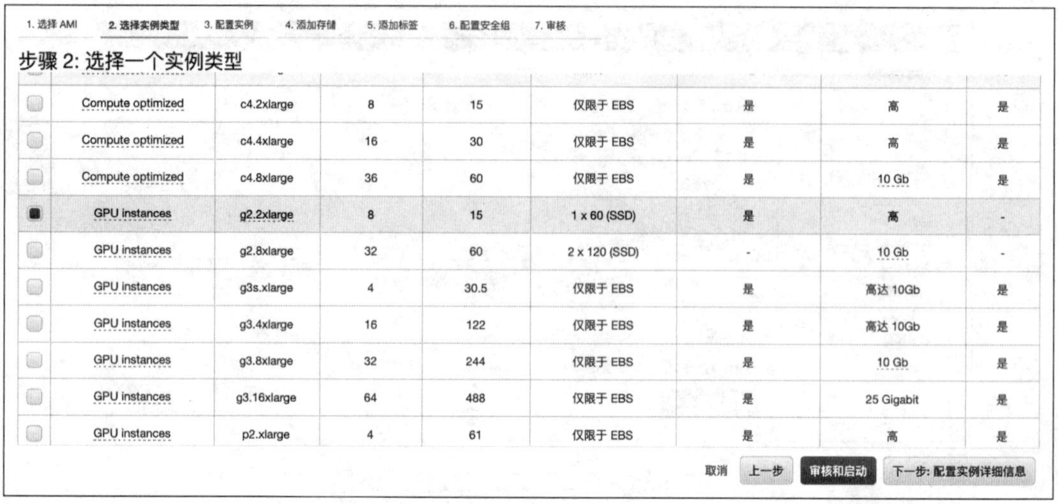

图 1.27　选择实例类型

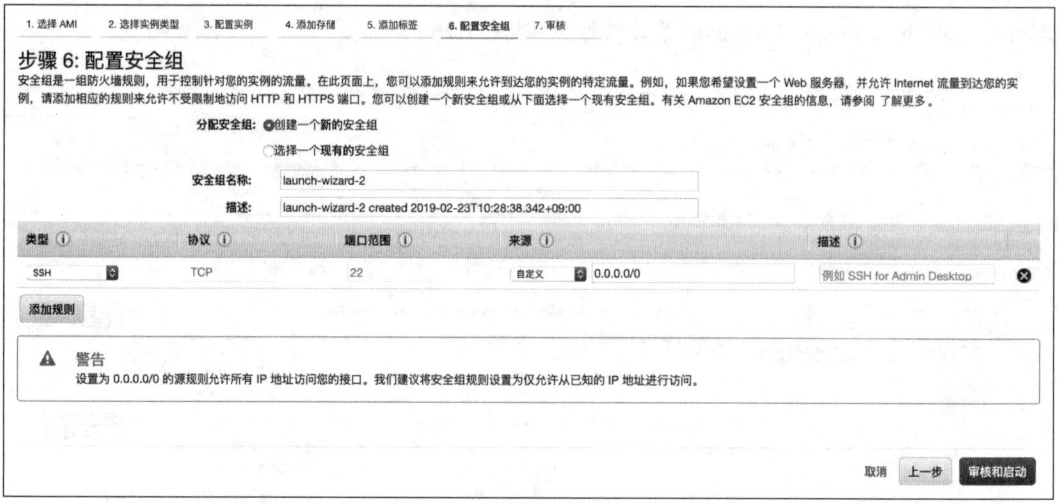

图 1.28　配置安全组

默认会有 SSH 连接的规则，我们可以在这里添加其他网络规则，方便访问其他端口。现在创建一个新规则，可以让我们在实例上运行 jupyter notebook，然后在本地主机上访问它。点击添加规则 按钮，配置一个新规则，如图 1.29 所示。

类型选择 自定义TCP规则，端口范围设成 8888，来源选择 任何位置。现在点击 审核和启动 按钮进入最后一个页面，之前所有的配置都会出现在这里，供你重新审查确认配置是否正确，当一切都没问题的话，点击 启动 按钮启动实例。此时会弹出 选择现有密钥对或创建新密钥对 的窗口，如图 1.30 所示。

类型	协议	端口范围	来源	描述
SSH	TCP	22	自定义 0.0.0.0/0	例如 SSH for Admin Desktop
自定义 TCP 规则	TCP	8888	任何位置 0.0.0.0/0, ::/0	例如 SSH for Admin Desktop

添加规则

图 1.29　自定义规则

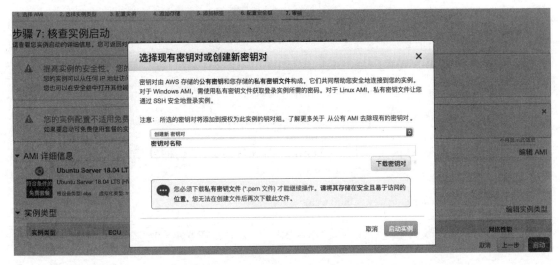

图 1.30　选择现有密钥对或创建新密钥对

此时有三种选择，如图 1.31 所示，可以创建新密钥对，可以选择现有密钥对，也可以在没有密钥对的情况下继续。这个你随意选择，笔者选择的是使用现有密钥对，当然最开始是没有的，需要创建一个新密钥对。

图 1.31　三种选择

选择 创建新密钥对，然后在 密钥对名称 中输入一串字符作为你的密钥对名称，然后点击

下载密钥对 按钮。比如笔者的密钥对名称是 secret，此时会下载一个名为 secret.pem 的文件，一定要保存好这个文件，以后用 SSH 连接实例时都需要这个密钥文件。

现在点击 启动实例 按钮来启动实例，注意，此时开始计费了，在 EC2 控制台的实例页面（https://console.aws.amazon.com/ec2/v2/home#Instances：）可以看到当前的实例状态，如图 1.32 所示。

Name	实例 ID	实例类型	可用区	实例状态	状态检查	警报状态	公有 DNS (IPv4)	IPv4 公有 IP
	i-0b0290f0775cc2ad0	g2.2xlarge	ap-northeast-1a	running	2/2 的检查已通过	无	ec2-52-69-64-30.ap-no...	52.69.64.30

图 1.32　实例状态

当 状态检查 一栏是 2/2的检查已通过 时，说明实例已经启动完成，现在可以使用 SSH 连接远程主机了。

连接命令是：

```
ssh -i <pem文件路径> EC2主机用户名@<EC2 公有 DNS或IPv4 公有 IP地址>
```

举个例子，比如 secret.pem 文件在 ~/Downloads/文件夹下，实例的公有 IP 地址是 52.69.64.30，实例的用户名是 ubuntu，那么连接该实例的命令就是：

```
ssh -i ~/Downloads/secret.pem ubuntu@52.69.64.30
```

注意，下载 pem 文件后，该文件的访问权限可能太大，在登录实例时会显示如图 1.33 所示的警告信息。

```
@@@@@@@@@@@@@@@@@@@@@@@@@@@@@@@@@@@@@@@@@@@@@@@@@@@
@       WARNING: UNPROTECTED PRIVATE KEY FILE!       @
@@@@@@@@@@@@@@@@@@@@@@@@@@@@@@@@@@@@@@@@@@@@@@@@@@@
Permissions 0644 for '/Users/chengstone/Downloads/secret.pem' are too open.
It is required that your private key files are NOT accessible by others.
This private key will be ignored.
Load key "/Users/chengstone/Downloads/secret.pem": bad permissions
ubuntu@54.64.206.179: Permission denied (publickey).
```

图 1.33　pem 文件权限问题

现在输入如下命令修改 pem 文件的权限为 0600：

```
chmod 0600 ~/Downloads/secret.pem
```

此时再次输入 SSH 命令连接远程主机即可成功。

注意，在使用 SSH 连接时，主机的用户名一定要正确，否则会连接失败，如图 1.34 所示。此时使用的用户名是 hahaha，这个用户名肯定是不对的，结果显示没有权限（Permission denied）。

1 卷积神经网络与环境搭建

```
ssh -i ~/Downloads/secret.pem hahaha@54.64.206.179
hahaha@54.64.206.179: Permission denied (publickey).
```

图 1.34　错误的用户名

有的主机名可能不是 ubuntu，而是 root，所以登录时一旦出现没有权限的提示信息，就有可能是用户名输入错了。

登录以后，就跟在自己的电脑中操作完全一样了。注意，当使用完实例之后，别忘记停止实例，一旦实例运行后就开始计费了。当想删除实例时，可以选择终止。停止和终止实例如图 1.35 所示，在实例上单击鼠标右键，弹出操作菜单，然后选择相应的菜单项即可。

图 1.35　停止和终止实例

我们需要在本地电脑和远程实例之间传送文件，使用 scp 命令可以做到。

（1）从实例主机下载文件到本地：

`scp EC2主机用户名@<EC2 公有 DNS或IPv4 公有 IP地址>:/文件路径/文件名 本地路径`

举例：

`scp ubuntu@52.69.64.30:/home/ubuntu/xxx.py ~/Downloads`

（2）上传本地文件到实例：

`scp 本地路径/文件名 EC2主机用户名@<EC2 公有 DNS或IPv4 公有 IP地址>:/远程路径`

跟第一个命令一样，只不过是远程路径跟本地路径交换位置。举例：

`scp ~/Downloads/xxx.py ubuntu@52.69.64.30:/home/ubuntu`

（3）从实例下载文件夹到本地：

```
scp -r EC2主机用户名@<EC2 公有 DNS或IPv4 公有 IP地址>:/文件路径/远程目录 本地路径
```

跟第一个命令相比，多了参数-r，表示将文件夹内的文件全部下载下来。举例：

```
scp -r ubuntu@52.69.64.30:/home/ubuntu/ ~/Downloads
```

（4）上传本地文件夹到实例：

```
scp -r 本地路径 EC2主机用户名@<EC2 公有 DNS或IPv4 公有 IP地址>:/文件路径/远程目录
```

跟第三个命令一样，同样有参数-r，并将实例路径和本地路径交换位置即可。举例：

```
scp -r ~/Downloads ubuntu@52.69.64.30:/home/ubuntu/
```

如果发现上传的文件没有出现在远程主机上，则有可能是远程主机的文件夹没有写入权限，可以使用如下命令放开权限：

```
chmod 0777 远端文件夹
```

（5）scp命令同样支持使用密钥对的连接方式，在上面的四个命令中添加参数 -i pem文件路径即可。举例：

```
scp -i ~/Downloads/secret.pem ~/Downloads/xxx.py ubuntu@52.69.64.30:/home/ubuntu
```

1.4 本章小结

通过本章的介绍，相信你已经了解了卷积神经网络的原理及它是如何工作的。接下来我们通过实践，构建了一个解决分类任务的神经网络，通过使用卷积神经网络来解决现实生活中的分类问题。

2

卷积神经网络实践：图像分类

2.1 概述

经过第 1 章的讲解，我们已经了解了卷积神经网络的理论，本章我们通过图像分类的实践项目，来理解卷积神经网络是如何工作的。

2.2 卷积神经网络项目实践：基于 TensorFlow 1.x

现在开始卷积神经网络的项目实践，相信读者的运行环境已经准备好了。

限于篇幅，本章只放重要代码，完整代码可以在 GitHub 上下载，地址：https://github.com/chengstone/CIFAR100_tutorial。代码文件是 Cifar100_Classifier.ipynb，在终端输入如下命令打开该文件：

```
jupyter notebook Cifar100_Classifier.ipynb
```

如果电脑中没有安装 jupyter notebook，则可以输入以下命令安装（注意，运行前要先使用 source activate 激活本地环境）：

```
conda install jupyter notebook
```

也可以安装 jupyterlab，输入以下命令安装：

```
conda install -c conda-forge jupyterlab
```

然后输入以下命令打开 jupyterlab：

```
jupyter lab
```

2.2.1 数据预处理

本章的项目是图像分类，使用 CIFAR-100 数据集，共包含 100 个类别，每个类别有 600 张图像，其中 500 张图像在训练集中，剩余 100 张图像在测试集中。先访问网址 http://www.cs.toronto.edu/~kriz/cifar.html 下载数据集，如图 2.1 所示。

Download		
Version	Size	md5sum
CIFAR-100 python version	161 MB	eb9058c3a382ffc7106e4002c42a8d85
CIFAR-100 Matlab version	175 MB	6a4bfa1dcd5c9453dda6bb54194911f4
CIFAR-100 binary version (suitable for C programs)	161 MB	03b5dce01913d631647c71ecec9e9cb8

图 2.1 数据集下载页面

我们下载第一个"CIFAR-100 python version"，将下载的数据集解压缩，放到项目根目录下。数据集解压缩后的目录如图 2.2 所示。

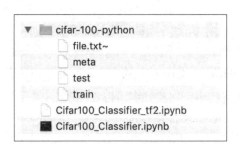

图 2.2 训练集目录形式

CIFAR 数据集官网上已经提供给我们读取这些数据的 Python 代码，如下所示。

```
def unpickle(file):
    import pickle
    with open(file, 'rb') as fo:
        dict = pickle.load(fo, encoding='bytes')
    return dict
```

我们调用 unpickle 函数，传入文件路径即可读取数据，数据是以字典形式保存的。

meta 文件内容是标签的名字字符串，运行以下代码查看 meta 字典的键值。

```
meta=unpickle("./cifar-100-python/meta")
meta.keys()
```

输出如下键值：

```
dict_keys([b'fine_label_names', b'coarse_label_names'])
```

能够看到 meta 字典中包含两种标签名，一种是 fine 标签，另一种是 coarse 标签。这跟 CIFAR-100 的数据分类有关，CIFAR-100 数据集包含 100 种精细（fine）分类，这 100 种精细分类还对应着 20 种粗略（coarse）分类，或者说是超类（superclass）。例如观赏鱼、比目鱼、鲨鱼、鳟鱼等都是不同种类的鱼，所以对应着 4 种精细分类，但它们都属于鱼这一种超类。

打印粗略分类的标签名：

```
meta[b'coarse_label_names']

[b'aquatic_mammals',
 b'fish',
 b'flowers',
 b'food_containers',
 b'fruit_and_vegetables',
 b'household_electrical_devices',
 b'household_furniture',
 b'insects',
 b'large_carnivores',
 b'large_man-made_outdoor_things',
 b'large_natural_outdoor_scenes',
 b'large_omnivores_and_herbivores',
 b'medium_mammals',
 b'non-insect_invertebrates',
 b'people',
 b'reptiles',
 b'small_mammals',
 b'trees',
 b'vehicles_1',
 b'vehicles_2']
```

接下来加载训练集：

```
trainset=unpickle("./cifar-100-python/train")
trainset.keys()
```

打印出的训练集字典键值如下：

```
dict_keys([b'filenames', b'batch_label', b'fine_labels', b'coarse_labels', b'data'])
```

其中，fine_labels 和 coarse_labels 就是训练集的目标 y（标签），data 是训练数据 x。我们看一下训练数据的形状。

```
trainset_y=trainset[b'coarse_labels']
trainset_x=trainset[b'data']
trainset_x.shape
```

打印出的形状是：

```
(50000, 3072)
```

本章的实践使用粗略分类进行训练，通过以下代码保存训练集数量和分类数量，得到 50000 和 20。

```
n_trainset=len(trainset_x)
n_class=len(meta[b'coarse_label_names'])
```

使用相同的方法加载测试集：

```
testset=unpickle("./cifar-100-python/test")
testset_x=testset[b'data']
testset_y=testset[b'coarse_labels']
```

继续分析数据图像，每张图像的维度是 3072，我们要把这样的图像转换成具有 RGB 三色通道的图像，以便查看。转换方式是先将图像形状从（3072）转成（3, 32, 32），再改变形状的轴次序，变成形状是（32, 32, 3）的图像（宽和高都是 32，并且含有 RGB 信息）。代码如下：

```
# 转换训练集图像
trainset_x=trainset_x.reshape(-1,3,32,32)
trainset_x=np.rollaxis(trainset_x, 1, 4)
# 转换测试集图像
testset_x=testset_x.reshape(-1,3,32,32)
testset_x=np.rollaxis(testset_x, 1, 4)
```

接下来可以做一些数据增强处理。一是将图像转成灰度图；二是通过灰度图得到直方图均衡化（Histogram Equalization）特征；三是通过灰度图得到对比度受限自适应直方图均衡化（Contrast Limited Adaptive Histogram Equalization，CLAHE）特征。直方图均衡化用于调节图像的对比度，使灰度直方图分布更加均匀，这样低对比度的图像有了更高的对比度，对图像中很暗或很亮的部分像素进行调节，得到更清晰的图像。CLAHE 是在 AHE（Adaptive Histgram Equalization，自适应直方图均衡化）的基础上对每个小块对比度做了限制，以控制 AHE 带来的噪声。经过以上三步处理，一张图像会变成三张，数据集会变成之前的 3 倍。

除了对图像数据做 3 种处理，还要对学习目标 y（分类 ID）做 one-hot 编码处理。因为现在的数据中分类 ID 是从 0 到 19 的，在定义网络时，网络的输出是 20 维的向量，每一维对应一个分类。

在处理数据集时，还要再拆分训练集。我们要把训练集拆分成训练集和验证集，拆分比例是 8∶2。这样在训练网络时，使用训练集训练网络，使用验证集得到训练的指标（损失和准确率），利用验证集的训练指标能够知道网络训练的情况和泛化能力。

先引入要用到的包：

```
import cv2
import matplotlib.pyplot as plt
import pandas as pd
import os
from scipy import misc
from sklearn.model_selection import train_test_split
from sklearn.utils import shuffle
import pickle
import numpy as np
```

我们来看看实现的代码：

```
import os
import cv2
import numpy as np
import shutil
from PIL import Image
import sys

current_num = 0
```

```python
def CLAHE(img):
    # 对比度受限自适应直方图均衡化
    clahe = cv2.createCLAHE(clipLimit=2.0, tileGridSize=(8,8))
    # 将CLAHE应用到输入图像上
    cl1 = clahe.apply(img)
    return cl1

def Histograms_Equalization(img):
    # 直方图均衡化
    equ = cv2.equalizeHist(img)
    return equ

# one-hot编码
def make_one_hot(data, num):
    return (np.arange(num)==data[:,None]).astype(np.integer)
```

对训练集做处理：

```python
from sklearn.utils import shuffle
new_X_train = []
new_y_train = []
# 循环遍历训练集
for index in range(n_trainset):
    sys.stdout.write(" {} / {}\r".format(index, n_trainset))
    # 训练集图像转灰度图
    img_gray = cv2.cvtColor(trainset_x[index], cv2.COLOR_RGB2GRAY)
    # 保存归一化图像和目标y（分类ID）
    new_X_train.append(img_gray.astype('float32') / 255.0)
    new_y_train.append(trainset_y[index])
    # 直方图归一化并保存归一化图像和目标y（分类ID）
    he_image = Histograms_Equalization(img_gray)
    new_X_train.append(he_image.astype('float32') / 255.0)
    new_y_train.append(trainset_y[index])
    # CLAHE并保存归一化图像和目标y（分类ID）
    clahe_img = CLAHE(img_gray)
    new_X_train.append(clahe_img.astype('float32') / 255.0)
    new_y_train.append(trainset_y[index])
```

```
print("All done!")
# 将输入数据扩展一个维度, (n, 32, 32)变成(n, 32, 32, 1)
all_xs = np.expand_dims(new_X_train, 3)
# 将目标y转成one-hot编码
all_ys = make_one_hot(np.array(new_y_train), n_class)
# 随机打散训练集,并按照比例将训练集拆分成训练集和验证集
train_xs, valid_xs, train_ys, valid_ys = train_test_split(
    all_xs, all_ys, test_size=0.2, random_state=0)
```

最后处理测试集,这回不用做数据增强了(当然做增强处理也可以),只是将图像转成灰度图,并且将目标 y 转成 one-hot 编码。

```
from sklearn.utils import shuffle
# 处理测试集
test_set = []
# 循环遍历测试集
for j in range(len(testset_x)):
    # 图像转灰度图
    img_gray = cv2.cvtColor(testset_x[j], cv2.COLOR_RGB2GRAY)
    # 扩展维度,(32, 32)转成(32, 32, 1)
    img_gray = np.expand_dims(img_gray, 2)
    # 归一化
    img_gray = img_gray / 255.0
    # 保存图像
    test_set.append(img_gray)
test_set = np.array(test_set)
# one-hot编码
y_test = make_one_hot(np.array(testset_y), n_class)
```

至此,数据预处理告一段落。

2.2.2 网络模型

网络模型参照 LeNet-5 网络结构,其包含的层如表 2.1 所示。

网络模型结构如图 2.3 所示。

表 2.1　网络模型包含的层

层	描述
输入层	32×32×1 灰度图
卷积层（5×5）	1×1 步幅（Stride），有效填充（Valid Padding），输出 28×28×100
ReLU	
最大池化层	2×2 步幅，输出 14×14×100
卷积层（3×3）	1×1 步幅，有效填充，输出 12×12×150
ReLU	
最大池化层	2×2 步幅，输出 6×6×150
卷积层（3×3）	1×1 步幅，相同填充（Same Padding），输出 6×6×250
ReLU	
最大池化层	2×2 步幅，输出 3×3×250
全连接层	输出长度 512
全连接层	输出长度 300
全连接层	输出长度 20

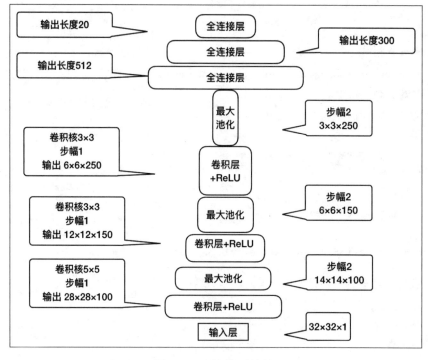

图 2.3　网络模型结构

下面用代码来实现这个模型。

```python
import tensorflow as tf
import datetime
# 定义网络模型类
class network(object):
    # 构造函数,定义网络结构
    def __init__(self):

        self.graph = tf.Graph()
        with self.graph.as_default():
            # 定义输入占位符
            self.x = tf.placeholder(tf.float32, shape=[None, 32, 32, 1],
                                    name='input_tensor')
            # 定义目标y占位符
            self.y = tf.placeholder(tf.float32, shape=[None, n_class],
                                    name='labels')
            # 卷积层:输入=32×32×1,输出=28×28×100,激活函数ReLU
            conv1 = tf.layers.conv2d(self.x, filters=100, kernel_size=5,
                                     activation=tf.nn.relu)
            print("conv1.shape = ", conv1.get_shape())   # (?, 28, 28, 100)

            # 最大池化层:输入=28×28×100,输出=14×14×100
            self.conv1 = tf.layers.max_pooling2d(conv1, pool_size=2, strides=2)

            # 卷积层:输出=12×12×150,激活函数ReLU
            conv2 = tf.layers.conv2d(self.conv1, filters=150, kernel_size=3,
                                     activation=tf.nn.relu)
            print("conv2.shape = ", conv2.get_shape())   # (?, 12, 12, 150)

            # 最大池化层:输入=12×12×150,输出=6×6×150
            self.conv2 = tf.layers.max_pooling2d(conv2, pool_size=2, strides=2)
            print("max_pool conv2.shape = ", self.conv2.get_shape())   # (?, 6, 6, 150)

            # 卷积层:输出=6×6×250,激活函数ReLU
            conv3 = tf.layers.conv2d(self.conv2, filters=250, kernel_size=3,
```

```python
                                  padding='same', activation=tf.nn.relu)
print("conv3.shape = ", conv3.get_shape())  # (?, 6, 6, 250)

# 最大池化层：输入=6×6×250，输出=3×3×250
self.conv3 = tf.layers.max_pooling2d(conv3, pool_size=2, strides=2)
print("max_pool conv3.shape = ", self.conv3.get_shape())  # (?, 3, 3, 250)

# Flatten，用来将多维输入压扁成一个向量。输入=3×3×250，输出=2250
self.fc0    = tf.layers.flatten(self.conv3)
print("fc0.shape = ", self.fc0.get_shape())  # (?, 2250)

# 全连接层：输入=2250，输出=512，激活函数ReLU
self.fc1 = tf.layers.dense(self.fc0, units=512, activation=tf.nn.relu)

# 全连接层：输入=512，输出=300，激活函数ReLU
self.fc2 = tf.layers.dense(self.fc1, units=300, activation=tf.nn.relu)

# 全连接层：输入=300，输出=20，将网络计算出的分类概率结果保存起来
self.logits = tf.layers.dense(self.fc2, units=n_class)

with tf.name_scope('loss'):
    # 将网络计算的分类概率和真实的学习目标y一起计算softmax分类交叉熵
    self.cross_entropy = tf.reduce_mean(
        tf.losses.softmax_cross_entropy(self.y,
                                        logits=self.logits))

with tf.name_scope('train_step'):
    # 使用Adam优化器，学习率设为1e-4
    self.train_step = tf.train.AdamOptimizer(1e-4).minimize(
                    self.cross_entropy)

with tf.name_scope('accuracy'):
    # 将网络计算的分类概率和真实的学习目标y一起计算准确率
    correct_prediction = tf.equal(tf.argmax(self.logits,1),
                                  tf.argmax(self.y,1))
```

```python
            self.accuracy = tf.reduce_mean(tf.cast(correct_prediction,
                                                    tf.float32))

        self.saver = tf.train.Saver()
        self.sess = tf.Session(graph=self.graph)
        self.sess.run(tf.global_variables_initializer())

        self.save_dir = './session'
        if not os.path.isdir(self.save_dir):
            os.mkdir(self.save_dir)
        self.ckpt = tf.train.get_checkpoint_state(self.save_dir)
        if self.ckpt and self.ckpt.model_checkpoint_path:
            self.saver.restore(self.sess, tf.train.latest_checkpoint(self.
            save_dir))
            print("Successfully loaded:", tf.train.latest_checkpoint(self.
            save_dir))
        else:
            print("Could not find old network weights")
# 网络训练函数
def training(self, xs, labels, ii, epoch_i, batch_i, batch_num):
    # 传入网络输入x和目标y
    feed_dict = {
        self.x: xs,
        self.y: labels
    }
    # 开始训练网络,得到准确率和交叉熵损失
    _, accuracy, loss = self.sess.run([self.train_step, self.accuracy, self.
    cross_entropy], feed_dict=feed_dict)
    if (ii % 20 == 0):
        time_str = datetime.datetime.now().isoformat()
        print('Training {}: Epoch {:>3} Batch {:>4}/{}   train_loss = {:.5f}'.
        format(
                time_str,
                epoch_i,
                batch_i,
                batch_num,
```

```python
                    loss))

        return accuracy, loss
    # 神经网络测试函数
    def testing(self, xs, labels, ii, epoch_i, batch_i, batch_num):
        # 传入测试数据x和目标y
        feed_dict = {
            self.x: xs,
            self.y: labels
        }
        # 将数据传入网络,通过推理(Inference)做前向传播,得到准确率和交叉熵损失
        accuracy, loss = self.sess.run([self.accuracy, self.cross_entropy],
                                       feed_dict=feed_dict)

        if (ii % 20 == 0):
            time_str = datetime.datetime.now().isoformat()
            print('#Testing# {}: Epoch {:>3} Batch {:>4}/{} accuracy = {:.3f}'
                  'test_loss = {:.5f}'.format(
                    time_str,
                    epoch_i,
                    batch_i,
                    batch_num,
                    accuracy,
                    loss))

        return accuracy, loss
    # 保存模型参数
    def save(self):
        save_path = self.saver.save(self.sess, os.path.join(self.save_dir, '
        best_model.ckpt'))
        print("Model saved in file: {}".format(save_path))
    # 定义前向传播函数
    def forward(self, xs):
        # 传入输入数据x
        feed_dict = {
            self.x: xs
        }
```

```python
        # 通过推理做前向传播，得到分类概率
        logits = self.sess.run([tf.nn.softmax(self.logits)], feed_dict=feed_dict)
        logits = np.reshape(np.array(logits), (n_classes))
        # 返回网络预测的分类概率
        return logits
```

2.2.3 训练网络

定义网络结构之后，就可以开始训练网络了。

```python
# 批量大小
total_batch_size = 64

# 定义获取批量数据函数
def get_batches(Xs, ys, batch_size):
    # 循环数据，每次返回一个批次的数据
    for start in range(0, len(Xs), batch_size):
        end = min(start + batch_size, len(Xs))
        yield Xs[start:end], ys[start:end]

# 循环训练5次
epochs = 5
net = network()

best_loss = 9999
# 开始训练循环
for ii in range(epochs):
    # 定义获取批量数据迭代器
    train_batches = get_batches(train_xs, train_ys, total_batch_size)
    test_batches = get_batches(valid_xs, valid_ys, total_batch_size)
    # 计算得到训练集数据的循环次数
    batch_num = (len(train_xs) // total_batch_size)
    # 循环遍历所有训练集数据
    for batch_i in range(batch_num):
        # 获取每个批次的训练数据
        x, y = next(train_batches)
```

```python
        # 传入数据，开始训练网络
        net.training(x, y, ii * (batch_num) + batch_i, ii, batch_i, batch_num)
# 计算得到验证集数据的循环次数
batch_num = (len(valid_xs) // total_batch_size)
test_loss = 0.0
test_acc = 0.0
# 循环遍历所有验证集数据
for batch_i  in range(batch_num):
    # 获取每个批次的验证集数据
    x, y = next(test_batches)
    # 传入数据，测试网络，得到验证集数据的准确率和损失
    acc, loss = net.testing(x, y, ii * (batch_num) + batch_i,
                            ii, batch_i, batch_num)
    test_loss = test_loss + loss
    test_acc = test_acc + acc

test_acc = test_acc / batch_num
test_loss = test_loss / batch_num
# 当出现更小的验证集损失时，更新保存的损失，并且保存网络参数
if test_loss < best_loss:
    best_loss = test_loss
    print("best loss = {}  acc = {}".format(best_loss, test_acc))
    net.save()
else:
    print("test loss = {}  acc = {}".format(test_loss, test_acc))
```

训练迭代 5 次之后的网络，在验证集数据上的损失是：best loss = 1.8720741564901466，准确率是：acc = 0.42715010683760685。

现在使用训练后的模型，看看在测试集数据上的表现。

```python
net = network()
# 定义获取批量数据迭代器
test_batches = get_batches(test_set, y_test, total_batch_size)
# 计算得到测试集数据的循环次数
batch_num = (len(test_set) // total_batch_size)
test_loss = 0.0
test_acc = 0.0
```

```python
# 循环遍历所有测试集数据
for batch_i in range(batch_num):
    # 获取每个批次的测试集数据
    x, y = next(test_batches)
    # 传入数据，测试网络，得到测试集数据的准确率和损失
    acc, loss = net.testing(x, y, 0 * (batch_num) + batch_i, 0, batch_i, batch_num)
    test_loss = test_loss + loss
    test_acc = test_acc + acc

test_acc = test_acc / batch_num
test_loss = test_loss / batch_num
print("test loss = {}  acc = {}".format(test_loss, test_acc))
```

在测试集数据上的损失是：test loss = 1.9895841059012291，准确率是：acc = 0.3975360576923077。

2.3 卷积神经网络项目实践：基于 TensorFlow 2.0

2.3.1 TensorFlow 2.0 介绍

2019 年 1 月初，Google 推出了 TensorFlow 2.0 预览版，在 3 月发布了 Alpha 版，又在 6 月发布了 Beta 版。我们现在就可以安装预览版或 Alpha 版，来感受 TensorFlow 2.0 带来的新特性和新的构建网络的方式。

本书不想写成一本讲解 API 的字典读物，那样不如大家去官方 API 文档中查一查就好了。所以，本节将简单介绍 TensorFlow 2.0 带来的一些变化，在每章的项目实践中会对用到的 TensorFlow 2.0 实现做讲解。简单来讲，就是以案例驱动，用到什么就讲解什么，相信各不相同的实践项目应该能够覆盖大多数应用场景。

Keras 作为 TensorFlow 的核心组件，是从 TensorFlow 1.4 版本开始加入进来的，而在 TensorFlow 2.0 版本中 Keras 将作为构建神经网络的主要组件。TensorFlow 1.x 版本经常用到的 tf.layers 没有了，取而代之的是 tf.keras.layers，这样构建网络的各类层都要使用 Keras 下的层。大部分时候，我们将使用 Keras 组件下的各类方法来构建和训练网络，比如损失函数、优化器、Estimator、评估指标 metrics 等。

另外一个大变化就是，tf.contrib 被彻底砍掉。这导致 TensorFlow 1.x 代码中使用的 tf.contrib 方法，需要以其他方式来实现。部分原 tf.contrib 的功能移到了 tensorflow_addons 库中，感兴趣的读者可以访问 https://github.com/tensorflow/addons 来了解该库的安装和使用。

TensorFlow 1.x 使用的 API 都被封装到 tf.compat.v1 中，因此，如果代码改成 TensorFlow 2.0 实现的话，部分 TensorFlow 1.x API 可以在 tf.compat.v1 中查找。

1. 动态图机制（Eager Execution）

最重要的改变就是动态图机制，在 TensorFlow 1.5 版本中已经加入了动态图机制，但不是默认模式，需要调用 tf.enable_eager_execution() 开启动态图。而在 TensorFlow 2.0 版本中，默认模式就是动态图，你可以调用 tf.executing_eagerly() 函数查看当前是否是动态图模式。

所谓动态图模式，就是立即执行的意思，跟 TensorFlow 1.x 的图模式有很大的不同。在 TensorFlow 1.x 中，想要得到一个张量的结果，需要先构建图，以说明某个张量的结果是从图的哪个输入到哪个输出的，这样数据从图中流过就得到了张量的结果。图中的某些节点暂时在没有值的情况下，是可以通过定义占位符表示的，当图开始运行起来之后，需要将数据传给某些输入占位符，数据流到输出占位符时，就得到结果了。所以，TensorFlow 1.x 的图模式要先把图定义好，然后需要在 Session 中运行图。总之，TensorFlow 1.x 版本的构建和调试网络要稍微麻烦一些，代码量很大。

动态图模式是函数式思想，马上就能得到结果，抛弃了 Session 和构建图。举个例子，假设现在有变量 x，值是 2，要想得到 $x+1$，在 TensorFlow 1.x 中实现可能是这样的：

```
x=2
sess=tf.Session()
print(tf.add(x,1).eval(session=sess))
```

打印的结果是 3，如果直接调用 tf.add(x,1)，得到的结果是一个张量 <tf.Tensor 'Add_3：0' shape=() dtype=int32>，而不是想要的结果 3。

在 TensorFlow 2.0 中实现就很简单了：

```
x=2
print(tf.add(x,1))
```

得到结果 <tf.Tensor: id=2, shape=(), dtype=int32, numpy=3>，此时的张量是有结果 3 的，在 Tensor 的 numpy 属性中。将上面代码改成下面这样：

```
x=2
y=tf.add(x,1)
print(y.numpy())
```

打印的结果是 3，不再是张量（Tensor），每个 Tensor 的 numpy 方法都会得到该张量 numpy 形式的值。

看到变化了吧！TensorFlow 2.0 中的实现不再需要 Session，也不需要 Graph，当然 TensorFlow 2.0 也是支持构建图的，TensorFlow 1.x 的构建方式仍然可以用。在效率上，动态图和图模式都差不多，但是对于工程人员来说，动态图模式更加友好，更方便调试，可以马上执行代码得到结果。另外，构建同样的网络，动态图模式的代码量更少。

关于动态图模式的函数式思想，在构建网络时会做说明。

2. AutoGraph 机制

TensorFlow 2.0 提供了 AutoGraph 机制，可以将 Python 代码自动转换成图模式，只需要在函数前加上 @tf.function 即可。但是要注意，转换成图模式的这部分代码将不再是动态图模式。

举例如下：

```
@tf.function
def test():
    x=2
    print(tf.executing_eagerly())
    y=x*x
    print(y)

def test2():
    print(tf.executing_eagerly())
    test()
    print(tf.executing_eagerly())

test2()
```

打印的结果是：

```
True
False
4
True
```

第一行和最后一行是 True，这是 test2 函数打印的，说明在 test2 函数中是动态图模式。第二行是 False，是由 @tf.function 修饰的图模式 test 函数打印的，说明在 test 函数中不再是动态图模式了。

关于 TensorFlow 2.0 的其他特性和使用方法，会在每章的 TensorFlow 2.0 实现中逐一讲解，更多关于 TensorFlow 2.0 API 的介绍请访问官方地址：https://www.tensorflow.org/versions/r2.0/api_docs/python/tf?hl=zh-cn。接下来开始本章的 TensorFlow 2.0 项目实践。

2.3.2　CIFAR-100 分类网络的 TensorFlow 2.0 实现

1. 安装 TensorFlow 2.0

首先创建一个 TensorFlow 2.0 的本地环境，输入如下命令：

```
conda create -n tf2 python=3.6
```

以后每章的 TensorFlow 2.0 实现都在这个环境下运行。

然后激活该环境：

```
source activate tf2
```

接下来开始安装 TensorFlow 2.0，现在的预览版是夜间版，每天都会更新：

```
pip install tf-nightly-2.0-preview
```

如果要安装 GPU 版本，则输入如下命令：

```
pip install tf-nightly-gpu-2.0-preview
```

安装 GPU 版本请注意，要求 GPU 算力（Compute Capability）至少是 3.5，需要安装 CUDA 10。

安装 Alpha 版输入如下命令：

```
pip install tensorflow==2.0.0-alpha0
```

安装 Alpha 版的 GPU 版本，输入如下命令：

```
pip install tensorflow-gpu==2.0.0-alpha0
```

安装 Beta 版输入如下命令：

```
pip install tensorflow==2.0.0-beta0
```

其他软件包缺少哪个装哪个即可，本书不另做说明。

本节代码的地址是：https://github.com/chengstone/CIFAR100_tutorial，文件名是 Cifar100_Classifier_tf2.ipynb，在本地使用 jupyter notebook Cifar100_Classifier_tf2.ipynb 命令打开。

大部分内容跟上一节的 TensorFlow 1.x 实现差不多，包括数据预处理、网络模型结构等，不同的是本节使用 TensorFlow 2.0 来实现。本节会对多种实现方式进行说明，以后章节不会每种方式都实现一遍。

2. 将 TensorFlow 1.x 代码升级到 TensorFlow 2.0

TensorFlow 2.0 提供了一个升级命令，可以将 TensorFlow 1.x 代码升级到 TensorFlow 2.0。但是这个命令只是简单的 API 映射而已，并不会按照 TensorFlow 2.0 该有的编程风格转换。比如它会将 tf.variable_scope 转换成 tf.compat.v1.variable_scope。之前说过，TensorFlow 1.x 的部分 API 被挪到 tf.compat.v1 中了。诸如此类的转换，只是将代码转换成可以在 TensorFlow 2.0 下成功运行的程度，而实质上仍然是 TensorFlow 1.x 的风格。而且部分 API 转换不了，比如砍掉的 tf.contrib。但是这个升级命令还是有用的，部分 API 的参数有变化，它会帮你添加或删除参数，并且会提供转换报告文档。官方的命令说明文档地址为：https://github.com/tensorflow/docs/blob/master/site/en/r2/guide/upgrade.md，以下该命令的介绍出自官方文档。

转换单个 Python 文件，命令是：

```
tf_upgrade_v2 --infile 源文件.py --outfile 输出文件-upgraded.py
```

批量转换 Python 文件，命令是（默认将其他文件复制到目标文件夹中）：

```
tf_upgrade_v2 --intree 源文件夹 --outtree 输出文件夹-upgraded
```

只批量转换 Python 文件，不复制其他文件，命令是：

```
tf_upgrade_v2 --intree 源文件夹 --outtree 输出文件夹-upgraded --copyotherfiles False
```

3. Keras 实现

这里使用基本的 Keras 方式来实现。

先引入软件包：

```
from tensorflow import keras
import tensorflow as tf
import datetime
from tensorflow.python.ops import summary_ops_v2
```

其中 summary_ops_v2 用来记录训练日志，供 TensorBoard 使用。目前这个功能有一些问题，记录日志有时程序会崩溃，官方的例子也将 summary_ops_v2 的使用注释掉了，可能以后会解决吧。如果在运行记录日志的代码时出现问题，则将该部分代码注释掉即可。

网络模型结构不变，大家可以对照 2.2.2 节关于网络模型结构的描述和实现来看本节的实现。

本章的分类网络足够简单，使用 Keras 实现可以很快速地构建和训练网络。首先实现 get_model 函数来构建网络模型。使用 Keras 构建模型需要将所有的层包装在 tf.keras.Sequential 里面，实现代码如下：

```python
# 注意，InputLayer并不是网络的第一层，只是用来说明输入尺寸的
def get_model():
    # keras.Sequential返回的即是模型model
    model = keras.Sequential([
            # 在keras.Sequential中按照顺序定义每一层，首先定义输入层来说明输入形状
            # 当然，InputLayer可以省略，这样就需要在第一个Conv2D层中指定输入形状
            # 网络前两层（InputLayer和Conv2D）可以用这句代码代替
# keras.layers.Conv2D (kernel_size = (5, 5), filters = 100, activation='relu',
# input_shape=(32, 32, 1)),
            keras.layers.InputLayer(input_shape=(32, 32, 1)),
            keras.layers.Conv2D (kernel_size = (5, 5), filters = 100,
                        activation='relu'),
            keras.layers.MaxPool2D(),
            keras.layers.Conv2D (kernel_size = (3, 3), filters = 150,
                        activation='relu'),
            keras.layers.MaxPool2D(),
            keras.layers.Conv2D (kernel_size = (3, 3), filters = 250,
                        padding='same', activation='relu'),
            keras.layers.MaxPool2D(),
            keras.layers.Flatten(),
            keras.layers.Dense(512, activation='relu'),
            keras.layers.Dense(300, activation='relu'),
            keras.layers.Dense(n_class, activation='softmax')
        ])
    return model

# 每一层的含义就不详细解释了，Conv2D是卷积层，MaxPool2D是最大池化层，Dense是全连接
# 层，虽然写法跟TensorFlow 1.x不太一致，但是每个参数的概念是一致的
```

```
# 如果在之前的图像预处理部分没有使用img_gray = np.expand_dims(img_gray, 2)
# 网络一开始的InputLayer可以使用这句代码代替：keras.layers.Reshape(target_shape
# =[32, 32, 1], input_shape=(32, 32,))。也就是说，可以通过一个reshape层将输入重新
# 调整形状，作用都是将32×32变成32×32×1
```

现在我们可以调用 get_model 函数得到构建的模型，接下来调用模型的 compile 函数来配置学习过程，就是设置损失函数、优化器和训练指标。

```
# 构建模型
model = get_model()
# 设置优化器是Adam，学习率是1e-4，损失是分类交叉熵，指标是准确率
# 如果在数据预处理部分标签没有做one-hot编码
# 这里的损失函数可以使用sparse_categorical_crossentropy
model.compile(optimizer=tf.keras.optimizers.Adam(1e-4),
              loss='categorical_crossentropy',
              metrics=['accuracy'])
```

此时我们可以调用模型的 summary 函数打印模型概述信息。

```
model.summary()
```

打印的结果如下：

```
Model: "sequential"
_____
Layer (type)                 Output Shape              Param #
=================================================================
conv2d_6 (Conv2D)            (None, 28, 28, 100)       2600
_____
max_pooling2d_6 (MaxPooling2 (None, 14, 14, 100)       0
_____
conv2d_7 (Conv2D)            (None, 12, 12, 150)       135150
_____
max_pooling2d_7 (MaxPooling2 (None, 6, 6, 150)         0
_____
conv2d_8 (Conv2D)            (None, 6, 6, 250)         337750
_____
max_pooling2d_8 (MaxPooling2 (None, 3, 3, 250)         0
```

```
------------------------------------------------------------------
flatten_2 (Flatten)           (None, 2250)              0
------------------------------------------------------------------
dense_4 (Dense)               (None, 512)               1152512
------------------------------------------------------------------
dense_5 (Dense)               (None, 300)               153900
------------------------------------------------------------------
dense_6 (Dense)               (None, 20)                6020
==================================================================
Total params: 1,787,932
Trainable params: 1,787,932
Non-trainable params: 0
------------------------------------------------------------------
```

这样能够看到模型每层的信息以及参数总量。

接下来，我们来定义一些回调函数，Keras 训练时会调用这些回调函数。我们定义三个回调函数，第一个是保存模型的回调函数。

```
# 模型的保存路径
checkpoint_path = "training_models/cp-{epoch:04d}.ckpt"
# 定义保存模型的回调函数
# 参数是保存路径、log等级、只保存最好的模型、只保存权重以及保存频率(每一代保存一次)
cp_callback = tf.keras.callbacks.ModelCheckpoint(
    checkpoint_path, verbose=1, save_best_only=True, save_weights_only=True,
    period=1)
```

第二个回调函数用于提前停止（Early Stop）训练，就是在验证集损失没有任何进展的情况下停止训练。第三个回调函数用于记录 TensorBoard 日志。

```
callbacks = [
    cp_callback,
    # 当验证集损失没有提高时中断训练，观察2代。也就是说，在2代内损失没有任何提高，
    # 则停止训练
    tf.keras.callbacks.EarlyStopping(patience=2, monitor='val_loss'),
    # TensorBoard日志回调函数，保存地址是 \verb|./logs|
    tf.keras.callbacks.TensorBoard(log_dir='./logs')
]
```

现在可以调用模型的 fit 函数，传入训练数据开始训练了。

```python
# 保存初始权重
model.save_weights(checkpoint_path.format(epoch=0))
# 开始训练，一共训练1代，批量大小是64
model.fit(train_xs, train_ys, epochs=1, batch_size=64,
          validation_data=(valid_xs, valid_ys),
          callbacks=callbacks)
```

训练完之后，可以调用模型的evaluate函数，传入测试集评估网络，通过指标准确率查看模型训练的情况。

```python
test_loss, test_acc = model.evaluate(test_set, y_test)

print('\nTest accuracy:', test_acc)
```

在测试集上准确率约为25%：

```
10000/10000============================] - 11s 1ms/sample - loss: 2.4464 - acc: 0.2537

Test accuracy: 0.2537
```

使用Keras方式构建和训练模型很方便，代码量也很少。

4. 以函数式方式构建模型

本节只介绍函数式构建模型的方法，模型训练等代码请参考上一节的实现代码。

除基本的Keras实现方式之外，我们还可以使用函数式方式构建模型，这种方法以后会经常用到。那么，什么是函数式方式呢？我们通过下面代码来了解。

```python
# 先定义输入占位符，当然这不是必需的，上一节说过，可以在第一个卷积层里通过
# input_shape来指定输入的形状
inputs = keras.Input(shape=(32, 32, 1))

# 现在我们还是定义每一层，参数同上一节，没有变化，但是实现方式有一点变化
# 就是定义的每一层都可以被看成一个函数，将上一层作为参数传给本层，函数的结果是返回
# 本层的张量。每一层都这样做，其实跟TensorFlow 1.x差不多
x = keras.layers.Conv2D (kernel_size = (5, 5), filters = 100,
                        activation='relu')(inputs)
x = keras.layers.MaxPool2D()(x)
```

```python
x = keras.layers.Conv2D (kernel_size = (3, 3), filters = 150, activation='relu')(x)
x = keras.layers.MaxPool2D()(x)
x = keras.layers.Conv2D (kernel_size = (3, 3), filters = 250,
                        padding='same', activation='relu')(x)
x = keras.layers.MaxPool2D()(x)
x = keras.layers.Flatten()(x)
x = keras.layers.Dense(512, activation='relu')(x)
x = keras.layers.Dense(300, activation='relu')(x)
predictions = keras.layers.Dense(n_class, activation='softmax')(x)
# 最后通过tf.keras.Model来封装模型，构建的模型需要知道输入和输出是谁，传入输入和
# 输出的张量作为参数
model = keras.Model(inputs=inputs, outputs=predictions)
```

5. 继承 keras.Model 构建模型

我们还可以继承 keras.Model 来创建一个模型类。

```python
# 定义模型类MyModel，继承自keras.Model
class MyModel(keras.Model):
    # 在初始化函数中，先把每一层都定义好，此时这些层还没有堆叠起来
    def __init__(self):
        super(MyModel, self).__init__(name='my_model')
        self.layer1 = keras.layers.Conv2D (kernel_size = (5, 5), filters = 100,
                                          activation='relu')
        self.layer2 = keras.layers.MaxPool2D()
        self.layer3 = keras.layers.Conv2D (kernel_size = (3, 3), filters = 150,
                                          activation='relu')
        self.layer4 = keras.layers.MaxPool2D()
        self.layer5 = keras.layers.Conv2D (kernel_size = (3, 3), filters = 250,
                                          padding='same', activation='relu')
        self.layer6 = keras.layers.MaxPool2D()
        self.layer7 = keras.layers.Flatten()
        self.layer8 = keras.layers.Dense(512, activation='relu')
        self.layer9 = keras.layers.Dense(300, activation='relu')
        self.predictions = keras.layers.Dense(n_class, activation='softmax')
    # call函数是真正将这些层堆叠起来的函数，当我们定义一个MyModel实例时，call函数
    # 会被调用，通过函数式方式堆叠所有层，最终输出最后一层的张量即可，这样返回的
```

```python
    # 就是模型了
    def call(self, inputs):
        x = self.layer1(inputs)
        x = self.layer2(x)
        x = self.layer3(x)
        x = self.layer4(x)
        x = self.layer5(x)
        x = self.layer6(x)
        x = self.layer7(x)
        x = self.layer8(x)
        x = self.layer9(x)
        predictions = self.predictions(x)

        return predictions
model=MyModel()
```

本节换一种方式来训练模型，Keras 模型的 fit 函数接收的数据集除之前使用的数组形式外，还支持 tf.data 的数据，我们一起来看看怎么用。

tf.data 使用起来很简单，只需要调用 tf.data.Dataset.from_tensor_slices 函数，传入数据集即可。

```python
# 传入训练数据集
dataset = tf.data.Dataset.from_tensor_slices((train_xs, train_ys))
# 然后就可以调用shuffle函数随机打散数据，设置批量大小是64
# 最后的repeat是到数据的结尾后再从头开始遍历数据，保证数据可以无限遍历下去
# 也可以在repeat中传入参数，比如repeat(5)，则相当于变成了5份数据集的量，这样数
# 据集是有终点的，一旦训练过程遍历到达数据的结尾，而训练又没有结束，就会抛出异常

dataset = dataset.shuffle(128).batch(64).repeat()
```

使用 tf.data 处理的数据集训练模型的代码如下：

```python
# 准备数据集
val_dataset = tf.data.Dataset.from_tensor_slices((valid_xs, valid_ys))
val_dataset = val_dataset.batch(64).repeat()

dataset = tf.data.Dataset.from_tensor_slices((train_xs, train_ys))
dataset = dataset.shuffle(128).batch(64).repeat()
```

```python
model.compile(optimizer=tf.keras.optimizers.Adam(1e-4),
              loss='categorical_crossentropy',
              metrics=['accuracy'])
# 传入tf.data处理的数据集时,fit函数需要指定参数steps_per_epoch,表示每一代(epoch)
# 的训练次数(step数),这个值是数据集样本数/批次大小,validation_steps也是一样的
model.fit(dataset, epochs=1, steps_per_epoch=(len(train_xs) // total_batch_size),
          validation_data=val_dataset, validation_steps=(len(valid_xs) //
          total_batch_size), callbacks=callbacks)
```

模型训练好之后,就可以加载参数进行预测了。

```python
checkpoint_path = "training_models/cp-{epoch:04d}.ckpt"
checkpoint_dir = os.path.dirname(checkpoint_path)
# 得到保存的参数
latest = tf.train.latest_checkpoint(checkpoint_dir)
# 构建模型(这里使用之前介绍的任意一种方式构建模型都可以)
model = get_model()
# 调用load_weights函数加载保存的参数
model.load_weights(latest)
```

通过调用模型的 predict 函数来进行预测,其中 generate_x 函数用来处理读入的图像生成输入数据,跟之前的图像预处理是一样的。

```python
# 图像预处理
def generate_x(img):
    img = cv2.cvtColor(img, cv2.COLOR_RGB2GRAY)

    img_gray = Histograms_Equalization(img)
    img_gray = np.expand_dims(img_gray, 2)
    img_gray = img_gray / 255.0
    img_gray = np.expand_dims(img_gray, 0)
    return img_gray
# 随机得到一个索引
index = random.randint(0, len(testset_x))
# 读取测试集中的图像,经过预处理,生成输入数据
input_img = generate_x(testset_x[index]).astype('float32')
# 将数据传给模型,得到模型输出(分类概率向量)
logits = model.predict(input_img)
```

```python
# 打印概率值最大的，就是预测的图像分类
print(np.argmax(logits[0]))
# 打印实际的图像分类
print(testset_y[index])
```

6. 动态图模式实现

动态图模式的实现方式以后会经常使用，这里的实现可以作为基线。首先定义一个类network，在初始化函数中先定义好模型。

```python
MODEL_DIR = "./models"

class network(object):
    def __init__(self):
        # 定义模型，注意最后一层全连接没有softmax
        self.model = keras.Sequential([
            keras.layers.InputLayer(input_shape=(32, 32, 1)),
            keras.layers.Conv2D (kernel_size = (5, 5), filters = 100,
                                activation='relu'),
            keras.layers.MaxPool2D(),
            keras.layers.Conv2D (kernel_size = (3, 3), filters = 150,
                                activation='relu'),
            keras.layers.MaxPool2D(),
            keras.layers.Conv2D (kernel_size = (3, 3), filters = 250,
                                padding='same', activation='relu'),
            keras.layers.MaxPool2D(),
            keras.layers.Flatten(),
            keras.layers.Dense(512, activation='relu'),
            keras.layers.Dense(300, activation='relu'),
            keras.layers.Dense(n_class)
        ])
```

再定义优化器。

```python
        self.optimizer = tf.keras.optimizers.Adam(learning_rate=1e-4)
```

然后定义日志记录器，这部分代码目前有问题，有时在记录日志时会崩溃，所以一旦出现这种情况，就先把与日志相关的代码注释掉，后面章节的这部分代码也一样。

```python
        # 如果文件夹不存在，则创建
        if tf.io.gfile.exists(MODEL_DIR):
            pass
        else:
            tf.io.gfile.makedirs(MODEL_DIR)
        # 定义好保存的文件夹
        train_dir = os.path.join(MODEL_DIR, 'summaries', 'train')
        test_dir = os.path.join(MODEL_DIR, 'summaries', 'eval')
        # 创建日志记录器
        self.train_summary_writer = summary_ops_v2.create_file_writer(train_dir, flush_millis=10000)
        self.test_summary_writer = summary_ops_v2.create_file_writer(test_dir, flush_millis=10000, name='test')
```

接下来定义参数保存地址，以及加载存在的参数。

```python
        # 参数的保存地址
        checkpoint_dir = os.path.join(MODEL_DIR, 'checkpoints')
        self.checkpoint_prefix = os.path.join(checkpoint_dir, 'ckpt')
        # 定义保存模型参数检查点，传入模型和优化器
        self.checkpoint = tf.train.Checkpoint(model=self.model, optimizer=self.optimizer)

        # 加载已经存在的参数
        self.checkpoint.restore(tf.train.latest_checkpoint(checkpoint_dir))
```

现在定义计算损失的函数，使用分类交叉熵损失：

```python
    def compute_loss(self, logits, labels):
        # 如果目标y不是one-hot编码的话
        # 则可以使用sparse_categorical_crossentropy作为损失函数
        return tf.reduce_mean(tf.keras.losses.categorical_crossentropy(labels, logits, from_logits=True))
```

定义计算准确率指标的函数，可以用在后面的训练函数 train_step 中。

```python
    def compute_accuracy(self, logits, labels):
        # 注意，本次实现代码中并没有用本函数计算准确率
        # 在后面我们用tf.keras.metrics.CategoricalAccuracy计算准确率
```

```
# 在这里两种实现方式都提供给大家，用哪种计算都可以
# 如果目标y不是one-hot编码的话
# 则可以使用tf.keras.metrics.sparse_categorical_accuracy计算分类准确率
return tf.keras.metrics.categorical_accuracy(labels, logits)
```

定义每一步的训练函数 train_step，该函数通过模型的前向传播得到网络的输出 logits，然后计算损失。也可以在这里计算指标，最后使用优化器优化损失。实现代码如下：

```
# 根据官方的例子，本函数通过@tf.function修饰成图模式，这里不再是动态图模式
# 当然，注释掉这句也是没问题的，而且训练速度会更快
@tf.function
def train_step(self, images, labels):

    # TensorFlow 2.0通过tf.GradientTape的磁带来记录计算损失的操作
    # 以便计算相对于参数变量的损失梯度
    with tf.GradientTape() as tape:
        # 这里跟TensorFlow 1.x完全不同，不需要session，也不需要run
        # 将模型看作函数
        # 直接将数据作为参数传给模型即可，模型返回的就是logits，体现了
        # TensorFlow 2.0的函数式思想
        # 此时参数training要设成True
        logits = self.model(images, training=True)
        # 使用logits和标签计算损失
        loss = self.compute_loss(logits, labels)
        # 这里也可以计算指标，但是本章的实现不在这里计算了
        # accuracy = self.compute_accuracy(logits, labels)
    # 利用磁带计算相对于参数变量的损失梯度
    grads = tape.gradient(loss, self.model.trainable_variables)
    # 使用优化器做优化
    self.optimizer.apply_gradients(zip(grads, self.model.trainable_variables))
    return loss, logits  #, accuracy
```

现在来定义训练函数。

```
# 接受tf.data处理的数据集，默认训练1代，log打印频率是50步
def training(self, train_dataset, test_dataset, epochs=1, log_freq=50):

    for i in range(epochs):
```

```python
            train_start = time.time()
            with self.train_summary_writer.as_default():
                start = time.time()
                # 这里定义一个平均指标用来计算平均损失
                # tf.keras.metrics的指标通过result函数返回积累值
                # reset_states函数用于清除积累值
                # avg_loss = tf.keras.metrics.Mean('loss', dtype=tf.float32)
                # 如果在train_step中计算准确率,则可以使用平均指标来计算平均准确率
                # 本次实现没有这么做
                # avg_accuracy=tf.keras.metrics.Mean('accuracy', dtype=tf.float32)
                # 这里通过tf.keras.metrics定义了分类准确率指标,两种计算指标的方法
                # 使用哪种都可以
                avg_accuracy = tf.keras.metrics.CategoricalAccuracy()

                # 遍历数据集
                for images, labels in train_dataset:
                    # 调用train_step训练一步,得到损失和模型的输出logits
                    # 如果在train_step中计算了准确率,则可以在这里返回
                    loss, logits = self.train_step(images, labels)  #, accuracy
                    # 将损失累积到平均指标中,它会自动计算平均值
                    avg_loss(loss)
                    # 如果train_step返回了准确率,则可以使用平均指标计算平均准确率
                    # avg_accuracy(accuracy)
                    # 这里是通过分类准确率指标来计算准确率的
                    # 参数是标签和模型输出logits
                    avg_accuracy.update_state(labels, logits)
                    # 当训练步数达到日志输出频率log_freq时,打印日志
                    if tf.equal(self.optimizer.iterations % log_freq, 0):
                        # 将指标记录到日志中
                        summary_ops_v2.scalar('loss', avg_loss.result(),
                                              step=self.optimizer.iterations)
                        summary_ops_v2.scalar('accuracy', avg_accuracy.result(),
                                              step=self.optimizer.iterations)

                        rate = log_freq / (time.time() - start)
                        # 将各项指标打印出来
```

```
                    print('Step #{}\tLoss: {:0.6f} accuracy: {:0.2f}% ({}
steps/sec)'.format(self.optimizer.iterations.numpy(), loss, (avg_accuracy.result()
* 100), rate))
                    avg_loss.reset_states()
                    avg_accuracy.reset_states()
                    start = time.time()

        train_end = time.time()
        print('\nTrain time for epoch #{} ({} total steps): {}'.format(i + 1,
self.optimizer.iterations.numpy(), train_end - train_start))

        # 使用验证集测试模型训练得怎么样
        with self.test_summary_writer.as_default():
            self.testing(test_dataset, self.optimizer.iterations)

        # 保存模型参数
        self.checkpoint.save(self.checkpoint_prefix)

# 训练完成后导出整个模型
self.export_path = os.path.join(MODEL_DIR, 'export')
tf.saved_model.save(self.model, self.export_path)
```

接下来是测试函数，跟训练函数差不多。

```
def testing(self, test_dataset, step_num):
    # 用平均指标计算平均损失
    avg_loss = tf.keras.metrics.Mean('loss', dtype=tf.float32)
    # avg_accuracy = tf.keras.metrics.Mean('accuracy', dtype=tf.float32)
    # 分类准确率指标
    avg_accuracy = tf.keras.metrics.CategoricalAccuracy()
    # 遍历验证集
    for (images, labels) in test_dataset:
        # 将数据传入模型，得到模型输出logits，此时参数training要设成False
        # 非常方便，就是函数调用，完全不需要session、run、feed_dict等
        logits = self.model(images, training=False)
        # 调用计算损失函数保存到平均损失指标中
        avg_loss(self.compute_loss(logits, labels))
```

```python
        # avg_accuracy(self.compute_accuracy(logits, labels))
        # 传入标签和模型输出计算准确率
        avg_accuracy.update_state(labels, logits)
    # 验证集完整遍历之后，打印指标
    print('Model test set loss: {:0.4f} accuracy: {:0.2f}%'.format(avg_loss.
result(), avg_accuracy.result() * 100))
    # 将指标保存到日志中
    summary_ops_v2.scalar('loss', avg_loss.result(), step=step_num)
    summary_ops_v2.scalar('accuracy', avg_accuracy.result(), step=step_num)
```

然后是传入测试集的评估函数，跟上面的函数实现差不多，这里没有计算损失。

```python
def evaluating(self, test_dataset):
    # 分类准确率指标
    avg_accuracy = tf.keras.metrics.CategoricalAccuracy()
    # 遍历测试集
    for (images, labels) in test_dataset:
        # 传入数据得到模型输出logits
        logits = self.model(images, training=False)
        # avg_accuracy(self.compute_accuracy(logits, labels))
        # 计算准确率
        avg_accuracy.update_state(labels, logits)
    # 测试集遍历完之后，打印指标
    print('Model accuracy: {:0.2f}%'.format(avg_accuracy.result() * 100))
```

最后定义预测函数，经过前向传播得到网络预测结果。

```python
def forward(self, xs):
    # 调用模型，得到模型输出
    predictions = self.model(xs)
    # 还记得构建模型时，最后一层全连接没有使用softmax吧
    # 这里在模型的输出上使用softmax
    logits = tf.nn.softmax(predictions)
    # 返回softmax结果，分类概率向量
    return logits
```

这样，network 类就实现好了，下面来看看网络的训练和预测。

首先准备好数据集。

```
# 验证集
val_dataset = tf.data.Dataset.from_tensor_slices((valid_xs.astype(np.float32),
                                                  valid)_ys))
val_dataset = val_dataset.batch(64)
# 训练集
dataset=tf.data.Dataset.from_tensor_slices((train_xs.astype(np.float32), train_ys))
dataset = dataset.batch(64)
```

定义类实例，训练网络。

```
net = network()

net.training(dataset, val_dataset)
```

最终的训练结果是：

```
Train time for epoch #1 (1875 total steps): 453.7773127555847
Model test set loss: 2.4213 accuracy: 26.07%
```

验证集准确率为 26.07%，现在看看网络在测试集上的准确率。

```
test_dataset = tf.data.Dataset.from_tensor_slices((test_set.astype(np.float32),
                                                   y_test))
test_dataset = test_dataset.batch(64)
net = network()
net.evaluating(test_dataset)
```

```
Model accuracy: 25.48%
```

测试集准确率是 25.48%。

最后要使用模型做预测的话，调用函数 forward 即可。

通过动态图模式的实现，能够看出其跟 TensorFlow 1.x 截然不同的实现思路。不再需要将模型、损失、准确率指标等提前构建成图，然后利用 Session 运行图。而是一切都以函数的方式，想要什么结果，直接调用相应的函数就好；要得到模型的输出，直接调用模型即可；调用损失函数，可以马上得到损失值。

```
# 随机得到一个索引
index = random.randint(0, len(testset_x))
# 读取测试集中的图像，经过预处理，生成输入数据
input_img = generate_x(testset_x[index]).astype('float32')
```

```
# 将数据传给模型，得到模型输出(分类概率向量)
logits = net.forward(input_img)
# 打印概率值最大的，就是预测的图像分类
print(np.argmax(logits[0]))
# 打印实际的图像分类
print(testset_y[index])
```

2.4 本章小结

通过本章的实践，我们构建了一个对 CIFAR-100 图像进行分类的神经网络。同时，你应该也学会了如何使用 TensorFlow 1.x 和 TensorFlow 2.0 构建和训练神经网络。现在，你已经有能力使用卷积神经网络来解决现实生活中的分类问题了。

关于 TensorFlow 2.0 的官方例子，请访问地址：https://github.com/tensorflow/docs/tree/master/site/en/r2。

3 彩票预测和生成古诗

3.1 概述

前两章介绍了卷积神经网络（CNN），该网络对于图像特征可以很好地学习，但是对于某些场景可能就不太适用了，比如基于时间序列的语音处理、机器翻译、股票趋势预测等应用场景，就需要使用循环神经网络（Recurrent Neural Network，RNN）来处理。本章首先介绍 RNN 和 LSTM，以及如何使用 RNN 来进行彩票预测和古诗生成，然后介绍 Seq2Seq、Transformer 和 BERT 模型。

3.2 RNN

现在假设有一个单词 sheep，我们要使用神经网络预测这个单词中每个字母的下一个字母，比如 s 的下一个字母是 h，h 的下一个字母是 e，依此类推。如果使用 CNN 或者 DNN 这类前馈网络的话，传入字母 s，期望的输出是 h，传入 h 希望得到 e，再传入 e，那么此时的输出就既有可能是 e，也有可能是 p。假设字母表一共只有 "s"、"h"、"e" 和 "p" 四个字母，这样网络的输出是字母表的概率向量，概率最大的就是预测的输出字母。那么，当输入是字母 e 时，前馈网络很难预测出下一个字母应该是 e 还是 p，因为网络没有处理时间序列的能力，它不能从单词的第一个字母开始一直读每一个字母，从而推断出合理的输出。这个过程如图 3.1 所示。

借用 Christopher Olah 的话说，"当读文章时，你会根据对之前单词的理解来推断每个单词的含义，而不会抛弃所有学会的知识，用空白的大脑来重新思考。人类的思想具有持久性，传统的神经网络不能做到这一点（比如根据电影中先前的事件推理后续的事件）"。详见 Christopher Olah 的文章 *Understanding LSTM Networks*。说真的，对于人来说，根据电影中先前的事件推理后续的事件也挺难的吧？除非看过一遍电影记住了情节，然后才可以根据先前的剧情推理后续的情节，但这不是推理，是记忆。

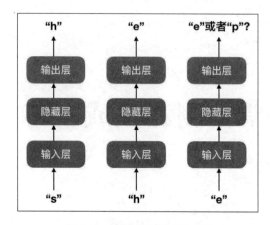

图 3.1　前馈网络推断过程

此时网络需要做出一些改变，继续上面 sheep 字母的预测。此时的输入不仅应该是当前字母，还应该输入前面字母的序列信息，或者说历史信息，这样网络就可以从历史的字母序列"sh"中推断出当前字母 e 的下一个字母应该是 e 而不是 p。换句话说，网络需要记忆。

如何做到这一点呢？我们需要将在历史时间步当中看到的字母序列输入给网络，具体的操作是改变隐藏层的输出，将隐藏层的输出重新连接到该隐藏层上。结构如图 3.2 所示（出自 Christopher Olah 的文章 *Understanding LSTM Networks*），像这种包含有循环的网络结构被称为循环神经网络（RNN），允许信息持久化——即记忆。这样，当网络看到字母 e 时，因为它看到了前面的 s 和 h，所以推断出下一个字母一定是 e。

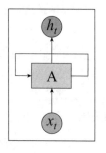

图 3.2　RNN 结构

根据图 3.2 所示，在当前时刻 t，隐藏层的输出 h_t（也叫作隐层状态，Hidden State），是由前一时刻 $t-1$ 的隐藏层输出 h_{t-1} 和时刻 t 的输入 x_t 同时作为网络的输入计算得到的。这样循环下去，可以使信息从前一时刻传递到当前时刻，再继续传递到下一时刻。我们也可以将图 3.2 所示的循环按照时间步展开，得到如图 3.3 所示的网络结构。

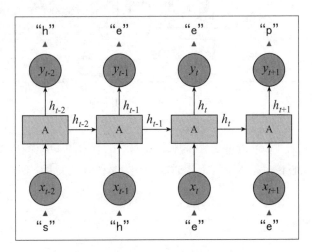

图 3.3 展开后的 RNN 网络结构

输入节点 x_t 和 RNN 单元上一时间步的输出 h_{t-1} 作为单元时间步 t 的输入，这样的结构很适合处理序列特征的数据。假设 W_{hh} 是 RNN 单元上的权重，单元的输出 h_t 是输入 X 和前一时间步的输出 h_{t-1} 与权重 W_{hh} 的乘积的线性组合：$h_t = f(W_{hh}h_{t-1} + WX)$。这里输入 x 的部分概括写成 WX 了，它实际上是相应时间步的 x 与权重 W 的乘积。在 RNN 中每一个时间步 h_t 都是可以展开的，比如输出 h_{t+1} 仍然是前一层的输出 h_t 再次乘以权重 W_{hh}，忽略 WX 和映射函数 f 的部分：$W_{hh}h_t = W_{hh}(W_{hh}h_{t-1})$。如果网络序列够长，那么对于网络的每一个时间步，都要一次次乘以权重，从而带来一个问题，就是 RNN 的梯度要么会因为越来越小而消失，要么会因为越来越大而爆炸。这取决于权重的值是小于 1（乘积趋于 0，梯度消失），还是大于 1（乘积趋于无穷大，梯度爆炸）。所以，RNN 虽然具有记忆，但是却不擅长处理信息距离过长的情况。将 RNN 比作人的话，就是说网络的记性不太好，短期记忆还行，对时间过长的信息就记不住了。

为了解决长期记忆的问题，可以使用 LSTM（Long Short Term Memory）来构建 RNN。

3.3 LSTM

LSTM 单元的结构如图 3.4 所示（出自 Christopher Olah 的文章 *Understanding LSTM Networks*）。

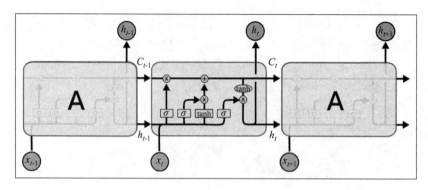

图 3.4　LSTM 单元的结构

C 是 LSTM 单元状态（Cell State），代表长期记忆，随着时间步不断在隐藏层中传递；状态 h 代表短期记忆。单元状态在一条水平线上流通，只有少量的线性交互，使得信息很容易保持，如图 3.5 所示。

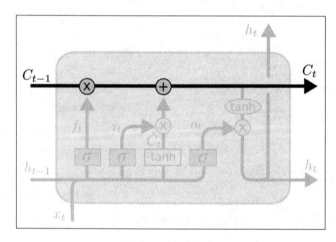

图 3.5　单元状态

LSTM 通过设计好的称作"门"的结构在单元状态上添加或去除信息，通常包括遗忘门、输入门和输出门。或者说通过"门"来决定记住某些信息，或者忘记某些信息，隐藏层的状态从单元状态计算得到，然后继续传递。

第一个门是遗忘门（Forget Gate），如图 3.6 所示。

遗忘门用来决定从单元状态中丢弃什么信息，σ 表示 Sigmoid 函数，输出一个 0~1 范围内的数值决定丢弃（遗忘）多少信息，相当于一个百分比。1 表示"全部保留"，0 表示"全部丢弃"。计算读取上一个时间步的状态 h_{t-1} 和 x_t，其中 h_{t-1} 表示历史信息，x_t 表示输入的新信息。输出

的 f_t（Sigmoid 输出值）与单元状态进行计算，接近 1 的值会直接在状态 C 的通道上通过，而接近 0 的值就是遗忘的信息，不会向前传递。这样就会使有用的长距离信息得到保留，并能够在网络上传递下去，并且忘记某些会导致预测错误的信息。可以将遗忘门当作 Dropout 来理解，只不过 Dropout 的百分比是通过 h_{t-1} 和 x_t 计算得到的。

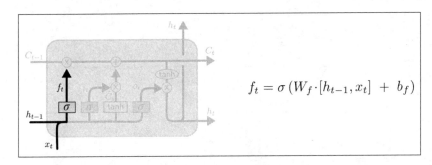

$$f_t = \sigma\left(W_f \cdot [h_{t-1}, x_t] \; + \; b_f\right)$$

图 3.6　遗忘门

第二个门是输入门，如图 3.7 所示。根据输入 x_t 和上一个时间步的状态 h_{t-1} 更新单元状态。遗忘门相当于 Dropout，丢弃部分状态信息，而输入门则要在状态上添加新信息。

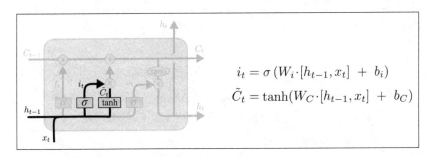

$$i_t = \sigma\left(W_i \cdot [h_{t-1}, x_t] \; + \; b_i\right)$$
$$\tilde{C}_t = \tanh(W_C \cdot [h_{t-1}, x_t] \; + \; b_C)$$

图 3.7　输入门

输入门分为两个部分：tanh 部分用来生成更新值 $\tilde{C}_t$，输出范围是 $[-1,1]$；σ 部分用来决定接受更新值的百分比，通过 $\tilde{C}_t$ 和 i_t 的计算最终决定把哪些新信息添加到单元状态 C 上。

当遗忘门和输入门计算好之后，就可以更新单元状态 C 了，过程如图 3.8 所示。

更新过程就是对旧的单元状态通过遗忘来丢弃部分信息 $f_t C_{t-1}$，然后加上新信息 $i_t \tilde{C}_t$，最终得到新的单元状态 C_t，这样当前的单元状态就被更新了。

最后，我们要确定 LSTM 单元的输出值 h，这是通过输出门来完成的，如图 3.9 所示。

单元的输出是基于上一步计算好的新单元状态 C_t 来计算的，先把新单元状态 C_t 通过 tanh 处理得到 $[-1,1]$ 的值，然后根据 Sigmoid 函数来确定保留单元状态的哪些部分作为单元的输出。

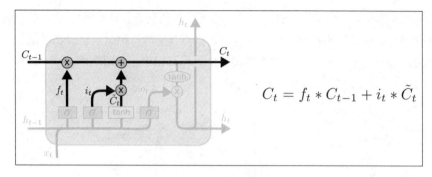

图 3.8　单元状态更新

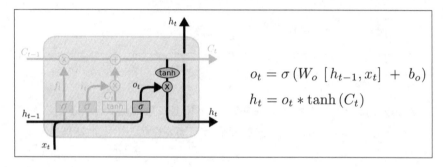

图 3.9　输出门

这样通过几个控制门组成的 LSTM 单元，在整个梯度的传递过程中是非常通畅的。因为在单元状态的传递过程中只有较少的一些线性求和运算，不再是大量的多层嵌套乘法，梯度在网络间的传递不会衰减，这样就很好地解决了 RNN 当中梯度消失或爆炸的问题。

当然，除 LSTM 以外，还有一些变种，比如 GRU（Gated Recurrent Unit）、SRU（Simple Recurrent Unit）、BiLSTM（Bidirectional LSTM，这个会在 OCR 章节中讲到），以及在 RNN 中加入注意力机制。

接下来，通过两个案例来看看如何通过 TensorFlow 使用 LSTM 解决实际问题。

3.4　嵌入矩阵

本节的案例是使用 LSTM 来预测彩票，那么预测哪种彩票呢？这里选择简单一些的，就是排列组合少一些的彩票也预测，如果证明我们的模型工作得很好，则可以再扩展到其他彩票上。预测的彩票是"排列 3"，从 000～999 的数字中选取一个 3 位数，一共有 1000 个，中奖概率就是千分之一，够简单吧。关于历史数据请访问地址：https://datachart.500.com/。

数据是按照每期一组数的顺序排列的，从第一期到最新的一期，是基于时间序列的数据。与回归预测有很大的区别，因为在特征上没有特殊的意义，不具备一组特征 x 映射到目标 y 的条件。但是按照时间序列来训练的话就不一样了，输入的 x 是某一期的开奖结果，要学习的 y 是下一期的开奖结果。我们需要从历史数据中寻找规律，这样基于时间序列的问题使用 LSTM 来解决再适合不过了。

限于篇幅，本章节重要代码放在书里，完整代码可以在 GitHub 上下载：https://github.com/chengstone/LotteryPredict，TensorFlow 1.x 的代码文件是 `cp_generation.ipynb`，TensorFlow 2.0 的代码文件是 `poetry_generation_tf2.ipynb`。

数据也为大家准备好了，就在 data 文件夹下，保存的形式是每行一期开奖号码，按照时间降序排列，也就是说，文件的最后一行是第一期开奖号码。

```
import os
import numpy as np

# 从本地读取数据文件
def load_data(path):
    input_file = os.path.join(path)
    with open(input_file, "r") as f:
        data = f.read()

    return data

# 加载数据集
data_dir = './data/cp.txt'
text = load_data(data_dir)
```

数据中的 10 条开奖记录如下：

```
202
243
580
300
598
900
761
262
891
```

623

本数据一共有 4656 条记录，4656 期，共出现 988 个不重复的结果，也就是说，有（1000 – 988）= 12 组号码还没有开出来过。我们可以把三个数组合成一组数，并且把一组数当作一个数或者一个字符串。这样在预处理数据集时会简单一些，从索引到字符串（0 → '000'）和从字符串到索引（'012' → 12）都是同一个数。

我们要构建的网络，其输入是每一期的开奖结果，就如本章开头的例子一样，可以将输入理解成单词中的一个字母，或者句子中的一个单词。所有可能的开奖结果一共有 1000 组数，000~999，通常的想法是将这个 1000 维的输入数据进行 one-hot 编码成 1000 维的稀疏向量，如图 3.10 所示。

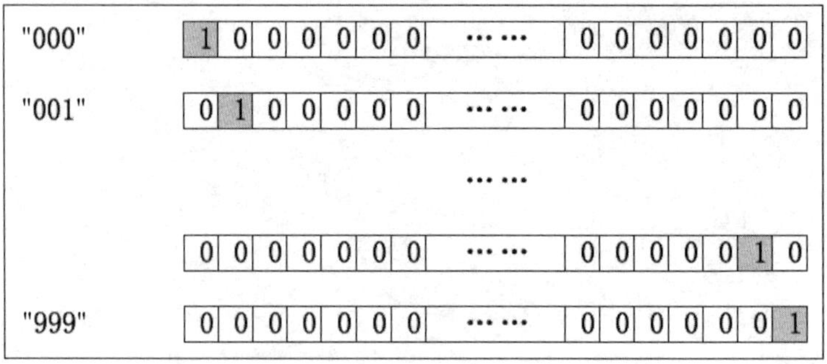

图 3.10 one-hot 编码

每输入一个号码，就需要号码对应的 one-hot 编码，然后与权重做计算。但是这样在输入层使用 one-hot 稀疏向量与网络第一层做矩阵乘法时会很没有效率，因为向量里面大部分都是 0，矩阵乘法浪费了大量的计算，最终矩阵运算得出的结果是稀疏向量中值为 1 的列所对应的权重矩阵中的行向量。这么说有点绕口，看图就清楚了，如图 3.11 所示。

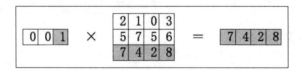

图 3.11 矩阵查表

这看起来很像用索引查表一样，将 one-hot 向量中值为 1 的位置作为下标，去索引权重矩阵中的行向量。

为了代替矩阵乘法，我们将权重矩阵当作一个查找表（Lookup Table），或者叫作嵌入矩阵（Embedding Matrix），将每个开奖号码所对应的数值作为索引，比如 "958" 对应的索引就是 958，然后在查找表中找第 958 行，如图 3.12 所示。

68

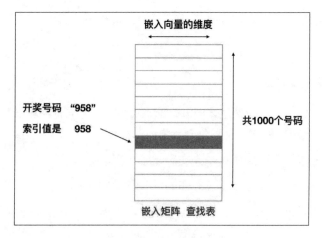

图 3.12　嵌入矩阵查找表

索引值就是输入 X，嵌入矩阵就是权重 W，而嵌入向量的维度，就是你想用多少维度的参数来表示特征。这其实跟之前的模型没有什么不同，嵌入层仍然是隐藏层。查找表只是矩阵乘法的一种便捷方式，它会像权重矩阵一样被训练，是要学习的参数。

3.5　实现彩票预测

网络的输入是开奖号码对应的数值，其作为嵌入矩阵的索引输入给嵌入层（Embed Layer），嵌入层输出对应号码的特征参数，再输入给 LSTM 层进行时间序列的学习，然后经过 softmax 预测出下一期的开奖结果，学习的目标就是下一期的开奖结果。网络结构如图 3.13 所示。

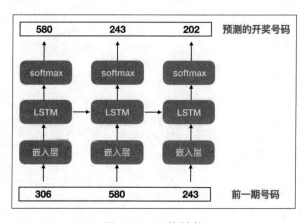

图 3.13　网络结构

接下来，编码实现这个网络，并演示如何通过 TensorFlow 使用嵌入层和 LSTM。

3.5.1 数据预处理

create_lookup_tables，在使用词向量之前，我们需要先准备好彩票开奖号码和 ID 之间的转换关系。在这个函数中，创建并返回两个字典：

- 开奖号码到 ID 的转换字典：vocab_to_int，用于输入的转换。
- ID 到开奖号码的转换字典：int_to_vocab，用于输出的转换。

```
import tensorflow as tf
import os
import pickle
import numpy as np
from collections import Counter


# 创建开奖号码（字符串）到数值、数值到开奖号码（字符串）两个字典
# 所有可能的开奖号码总共1000个，000~999。
def create_lookup_tables():
    # 开奖号码（字符串）到数值字典
    vocab_to_int = {str(ii).zfill(3) : ii for ii in range(1000)}
    # 数值到开奖号码（字符串）字典
    int_to_vocab = {ii : str(ii).zfill(3) for ii in range(1000)}
    return vocab_to_int, int_to_vocab


# 数据预处理
# 先将本地数据文件读取出来
text = load_data(data_dir)
# 将所有的开奖号码组织成list
words = [word for word in text.split()]

# 升序排列所有开奖号码，组成list，第一个元素是第一期开奖号码，依此类推
reverse_words = [text.split()[idx] for idx in (range(len(words)-1, 0, -1))]
# 创建两个字典
vocab_to_int, int_to_vocab = create_lookup_tables()
# 将升序排列的所有号码转成对应的索引ID，作为输入数据
int_text = [vocab_to_int[word] for word in reverse_words]
```

```
# 将预处理后的数据保存到本地
pickle.dump((int_text, vocab_to_int, int_to_vocab), open('preprocess.p', 'wb'))
```

3.5.2 构建神经网络

现在根据图 3.13 所示的网络结构构建神经网络，我们需要构建嵌入层、LSTM 层，然后将输入层和 softmax 层组合起来。

先定义输入 x 和学习目标 y 的占位符：

```
input_text = tf.placeholder(tf.int32, [None, None], name="input")
targets = tf.placeholder(tf.int32, [None, None], name="targets")
```

接下来构建嵌入层。input_text 是输入数据，作为嵌入矩阵的索引。vocab_size 是嵌入矩阵的行数，就是所有可能的号码个数，共 1000 个。embed_dim 是嵌入向量的维度，就是嵌入矩阵的列数。这样 vocab_size × embed_dim 就是嵌入矩阵的维度了。然后通过 tf.nn.embedding_lookup 传入嵌入矩阵和索引，就会得到索引值对应的嵌入向量，相当于在参数矩阵中查表得到特征向量。

```
# 定义嵌入矩阵
embed_matrix = tf.Variable(tf.random_uniform([vocab_size, embed_dim], -1, 1))
# embed_layer是从嵌入矩阵（查找表）中索引到的向量
embed_layer = tf.nn.embedding_lookup(embed_matrix, input_text)
```

现在来实现 LSTM。在 TensorFlow 中，可以使用 tf.contrib.rnn 中定义的各种 cell 创建 RNN 的节点，例如 ConvLSTMCell、GRUCell、LSTMCell、BasicLSTMCell 等，这里使用 BasicLSTMCell 作为 LSTM 单元，参数 num_units 用来定义 LSTM 单元中的隐藏节点数，也就是 LSTM 的输出维度。然后通过 MultiRNNCell 将多个 LSTM 层堆叠起来，输入的是多个 LSTM 层的列表。

```
# 循环定义两个LSTM单元构成LSTM列表，相当于定义了两层LSTM，输出的维度是rnn_size
lstm_cell = [tf.contrib.rnn.BasicLSTMCell(num_units=rnn_size) for _ in range(2)]
# 将两个LSTM层堆叠起来，构成RNN单元
cell = tf.contrib.rnn.MultiRNNCell(lstm_cell)

# 生成初始状态
initial_state = cell.zero_state(batch_size, tf.float32)
initial_state = tf.identity(initial_state, name="initial_state")
```

上面我们创建了 LSTM 的 RNN 单元，接下来通过 dynamic_rnn 使用这些单元创建 RNN。

```
# 将相应的开奖号码对应的参数向量传入RNN单元中，得到RNN的输出和状态
outputs, state = tf.nn.dynamic_rnn(cell, embed_layer, dtype=tf.float32)
final_state = tf.identity(state, name="final_state")
```

现在构建神经网络，其过程如图 3.13 所示。input_text 是定义的输入占位符，是开奖号码对应的 ID。首先将输入传给嵌入层，然后将嵌入层的输出传给 RNN，最后将 RNN 的输出传入全连接层，得到长度为 1000 的 logits。

超参数如下：

```
# 训练迭代次数
num_epochs = 50
# 批次大小
batch_size = 32
# RNN的尺寸（LSTM隐藏节点的维度）
rnn_size = 512
# 嵌入层的维度
embed_dim = 512
# 序列的长度，始终为1，每个批次都传入[批次×seq_length]个数据
# 这里就是[批次×1]，稍后会在获取批量数据处进行讲解
seq_length = 1
# 学习率
learning_rate = 0.01

# 表示经过多少批次以后打印训练信息
show_every_n_batches = 10


save_dir = './save'
```

1. 构建计算图

使用上面实现的神经网络来构建计算图。

损失函数没有使用 softmax 交叉熵，而是使用了 sequence_loss，原因是 seq2seq.sequence_loss 的 targets 参数不需要做 one-hot 编码，而且效果跟对目标 y 做 one-hot 编码，然后使用 softmax 交叉熵是一样的。

```
import tensorflow as tf
```

```python
tf.reset_default_graph()
train_graph = tf.Graph()
with train_graph.as_default():
    # 单词总数，即所有号码的个数，1000
    vocab_size = len(int_to_vocab)
    # 定义输入、目标和学习率占位符
    input_text = tf.placeholder(tf.int32, [None, None], name="input")
    targets = tf.placeholder(tf.int32, [None, None], name="targets")
    lr = tf.placeholder(tf.float32)

    input_data_shape = tf.shape(input_text)
    # 构建RNN单元和初始状态
    # 将一个或多个BasicLSTMCells叠加在MultiRNNCell中，这里我们使用两层LSTM cell
    cell = tf.contrib.rnn.MultiRNNCell([tf.contrib.rnn.BasicLSTMCell(num_units=rnn_size) for _ in range(2)])
    initial_state = cell.zero_state(input_data_shape[0], tf.float32)
    initial_state = tf.identity(initial_state, name="initial_state")

    # embed_matrix是嵌入矩阵，后面计算相似度（距离）的时候会用到
    embed_matrix = tf.Variable(tf.random_uniform([vocab_size, embed_dim], -1, 1))
    # embed_layer是从嵌入矩阵（查找表）中索引到的向量
    embed_layer = tf.nn.embedding_lookup(embed_matrix, input_text)

    # 使用RNN单元构建RNN
    outputs, state = tf.nn.dynamic_rnn(cell, embed_layer, dtype=tf.float32)
    # RNN输出状态
    final_state = tf.identity(state, name="final_state")
    # 网络的输出logits
    # 全连接层，将vocab_size（1000个开奖结果）作为输出的维度
    logits = tf.layers.dense(outputs, vocab_size)

    # 对网络的输出做softmax，得到1000个号码的概率向量
    probs = tf.nn.softmax(logits, name='probs')

    # 损失函数，输入网络的输出logits和学习目标（就是下一期中奖号码）
    cost = tf.contrib.seq2seq.sequence_loss(
```

```
        logits,
        targets,
        tf.ones([input_data_shape[0], input_data_shape[1]]))

# 使用余弦距离计算相似度
norm = tf.sqrt(tf.reduce_sum(tf.square(embed_matrix), 1, keep_dims=True))
normalized_embedding = embed_matrix / norm

# 使用Adam优化器
optimizer = tf.train.AdamOptimizer(lr)

# 梯度裁剪
gradients = optimizer.compute_gradients(cost)
capped_gradients = [(tf.clip_by_value(grad, -1., 1.), var) for grad, var in
gradients if grad is not None]
train_op = optimizer.apply_gradients(capped_gradients)

# 准确率
correct_pred = tf.equal(tf.argmax(probs, 2), tf.cast(targets, tf.int64))
accuracy = tf.reduce_mean(tf.cast(correct_pred, tf.float32), name='accuracy')
```

2. 获取批量数据

定义 get_batches 函数用来取得批量数据。

每一个批次都包含两个部分：

- 第一个部分是输入，形状为 [batch size, sequence length]。
- 第二个部分是目标，形状为 [batch size, sequence length]。

我们的序列长度始终是1，每个批次都是一列输入和一列目标，形状是[batch_size,1]。

当最后的数据量不足一个批次时，则抛弃这些数据。

```
# 对于本章彩票预测的实现，seq_length始终是1
def get_batches(int_text, batch_size, seq_length):
    """
    返回含有输入和目标的批量数据
    :参数 int_text: 单词（开奖号码）对应的ID
```

:参数 batch_size: 批次大小
:参数 seq_length: 序列长度，始终是1
:返回: 批量数据数组
"""
# 计算批次总数，不够组成一个batch的话就舍掉
batchCnt = len(int_text) // (batch_size * seq_length)
# 输入好理解，就是每一期的开奖结果，下标从0开始(冒号前的0没有写，默认从0开始)
int_text_inputs = int_text[:batchCnt * (batch_size * seq_length)]
# 目标是下一期的开奖结果，所以向右移动一个位置即可，下标从1开始
int_text_targets = int_text[1:batchCnt * (batch_size * seq_length)+1]

# 下面就是对数据切块，得到[batch size, sequence length]的形状
result_list = []
x = np.array(int_text_inputs).reshape(1, batch_size, -1)
y = np.array(int_text_targets).reshape(1, batch_size, -1)

x_new = np.dsplit(x, batchCnt)
y_new = np.dsplit(y, batchCnt)

for ii in range(batchCnt):
    x_list = []
    x_list.append(x_new[ii][0])
    x_list.append(y_new[ii][0])
    result_list.append(x_list)

返回批量数据
return np.array(result_list)
```

### 3.5.3 训练神经网络

在预处理过的数据上训练神经网络。

这里除保存预测准确率之外，还保存了三类准确率。

- Top $K$ 准确率：在预测结果中，前 $K$ 个结果的预测准确率。
- 与预测结果距离最近的 Top $K$ 准确率：先得到预测结果，然后使用嵌入矩阵计算与预测结果 Top 1 距离最近的相似度向量，取这个相似度向量中前 $K$ 个结果的预测准确率。
```

- 浮动距离中位数范围 K 准确率：在得到预测结果之后，计算正确结果在预测结果中的距离中位数，这个距离实际上是元素在向量中的位置与第一个元素位置的距离。这个距离数据告诉我们真正的结果在预测向量中的位置。每次训练之后，距离中位数都会有变化，所以是浮动的，当然也可以考虑使用众数或均值。使用中位数，是为了指出真正的结果在预测向量中的位置（平均位置或者说更具代表性的位置）在哪里。所以这个准确率就是以中位数为中心、以范围 K 为半径预测准确的概率。

这里距离中位数的准确率，我们分别在预测结果向量和与预测结果 Top 1 距离最近的相似度向量中做了统计。

浮动距离中位数的概率越高，说明模型训练得越不好。在理想情况下，应该是 Top K 准确率越来越高，说明模型预测得越来越准确。一旦模型预测得很差，那么在预测向量中就一定会有一部分区域是热点区域，也就是距离中位数指示的区域，这样就可以通过距离中位数来进行预测了。我们使用距离中位数来帮助预测，相当于为预测做了第二套方案，一旦模型预测得不准确，就可以尝试使用距离中位数来预测。

这三类准确率都是有范围的，我们只能知道在某个范围内猜中的概率会高一些，但是到底范围内的哪一个是准确值则很难说。

下面对几个变量进行解释。

- batches：训练批量数据。
- test_batches：测试批量数据。
- topk_acc：预测结果的 Top K 准确率。
- sim_topk_acc：与预测结果距离最近的 Top K 准确率。
- range_k：表示 k 值是一个范围，不像 Top K 是最开始的 K 个。
- floating_median_acc_range_k：以每次训练得出的距离中位数为中心、以范围 K 为半径的准确率，使用预测结果向量。
- floating_median_sim_acc_range_k：同上，使用的是相似度向量。
- losses：保存训练损失和测试损失。
- accuracies：保存各类准确率。

如果对上面的解释不是很理解，那么看看代码会清楚一些。

```
%matplotlib inline
%config InlineBackend.figure_format = 'retina'
import seaborn as sns
import matplotlib.pyplot as plt
```

```python
# 先准备好批量数据
batches = get_batches(int_text[:-(batch_size+1)], batch_size, seq_length)
test_batches = get_batches(int_text[-(batch_size+1):], batch_size, seq_length)
top_k = 10
# 预测结果的Top K准确率
topk_acc_list = []
topk_acc = 0
# 与预测结果距离最近的Top K准确率
sim_topk_acc_list = []
sim_topk_acc = 0
# 表示k值是一个范围,不像Top K是最开始的K个
range_k = 5
# 以每次训练得出的距离中位数为中心,以范围K为半径的准确率,使用预测结果向量
floating_median_idx = 0
floating_median_acc_range_k = 0
floating_median_acc_range_k_list = []
# 同上,使用的是相似度向量
floating_median_sim_idx = 0
floating_median_sim_acc_range_k = 0
floating_median_sim_acc_range_k_list = []
# 保存训练损失和测试损失
losses = {'train':[], 'test':[]}
# 保存各类准确率
accuracies = {'accuracy':[], 'topk':[], 'sim_topk':[], 'floating_median_acc_range_k
':[], 'floating_median_sim_acc_range_k':[]}

with tf.Session(graph=train_graph) as sess:
    sess.run(tf.global_variables_initializer())
    saver = tf.train.Saver()
    for epoch_i in range(num_epochs):
        state = sess.run(initial_state, {input_text: batches[0][0]})

        # 训练的迭代,保存训练损失
        for batch_i, (x, y) in enumerate(batches):
            feed = {
```

```python
            input_text: x,
            targets: y,
            initial_state: state,
            lr: learning_rate}
        # 训练网络
        train_loss, state, _ = sess.run([cost, final_state, train_op], feed)
        losses['train'].append(train_loss)

        # 每经过show_every_n_batches的训练次数后打印训练信息
        if (epoch_i * len(batches) + batch_i) % show_every_n_batches == 0:
            print('Epoch {:>3} Batch {:>4}/{}   train_loss = {:.3f}'.format(
                epoch_i,
                batch_i,
                len(batches),
                train_loss))

    # 使用测试数据的迭代
    acc_list = []
    prev_state = sess.run(initial_state, {input_text: np.array([[1]])})
    for batch_i, (x, y) in enumerate(test_batches):
        # 测试网络，得到测试集损失、准确率、预测的结果(softmax输出的概率向量)和
        # RNN状态
        test_loss, acc, probabilities, prev_state = sess.run(
            [cost, accuracy, probs, final_state],
            {input_text: x,
             targets: y,
             initial_state: prev_state})

        # 保存测试损失和准确率
        acc_list.append(acc)
        losses['test'].append(test_loss)
        accuracies['accuracy'].append(acc)

        print('Epoch {:>3} Batch {:>4}/{}   test_loss = {:.3f}'.format(
            epoch_i,
            batch_i,
```

```python
                len(test_batches),
                test_loss))

        # 利用嵌入矩阵和输出的预测结果计算得到预测相似度矩阵sim
        # 这个预测相似度矩阵的作用是，一旦预测结果不对，那么真正的
        # 结果可能在跟预测结果距离较近的范围内，也就是在相似度较高
        # 的top_k中
        valid_embedding = tf.nn.embedding_lookup(normalized_embedding,
np.squeeze(probabilities.argmax(2)))
        similarity = tf.matmul(valid_embedding, tf.transpose(
normalized_embedding))
        sim = similarity.eval()

        # 保存预测结果的Top K准确率和与预测结果距离最近的Top K准确率
        topk_acc = 0
        sim_topk_acc = 0
        for ii in range(len(probabilities)):
            # 取负是为了方便从大到小排序，因为argsort返回数组中从小到大的索引值
            # 取负以后最大的概率值就变成最小的了，索引值就是号码
            nearest = (-sim[ii, :]).argsort()[0:top_k]
            # 如果真正的开奖结果在预测相似度矩阵sim的top_k范围内，
            # 则说明预测结果跟距离较近的号码有相关性
            if y[ii] in nearest:
                sim_topk_acc += 1
            # 如果真正的开奖结果在预测结果的top_k范围内，则说明预测的质量还行
            if y[ii] in (-probabilities[ii]).argsort()[0][0:top_k]:
                topk_acc += 1

        # 计算出topk_acc和sim_topk_acc的准确率
        topk_acc = topk_acc / len(y)
        topk_acc_list.append(topk_acc)
        accuracies['topk'].append(topk_acc)

        sim_topk_acc = sim_topk_acc / len(y)
        sim_topk_acc_list.append(sim_topk_acc)
        accuracies['sim_topk'].append(sim_topk_acc)
```

```python
# 计算真实值在预测值中的距离数据
realInSim_distance_list = []
realInPredict_distance_list = []
for ii in range(len(probabilities)):
    # 对预测相似度矩阵按照概率从大到小排序
    sim_nearest = (-sim[ii, :]).argsort()
    # 得到真实开奖号码在预测相似度矩阵中的位置
    idx = list(sim_nearest).index(y[ii])
    realInSim_distance_list.append(idx)

    # 对预测结果按照从大到小排序
    nearest = (-probabilities[ii]).argsort()[0]
    # 得到真实开奖号码在预测结果中的位置
    idx = list(nearest).index(y[ii])
    realInPredict_distance_list.append(idx)

print('真实值在预测值中的距离数据：')
print('max distance : {}'.format(max(realInPredict_distance_list)))
print('min distance : {}'.format(min(realInPredict_distance_list)))
print('平均距离 : {}'.format(np.mean(realInPredict_distance_list)))
print('距离中位数 : {}'.format(np.median(realInPredict_distance_list)))
print('距离标准差 : {}'.format(np.std(realInPredict_distance_list)))

print('真实值在预测值相似向量中的距离数据：')
print('max distance : {}'.format(max(realInSim_distance_list)))
print('min distance : {}'.format(min(realInSim_distance_list)))
print('平均距离 : {}'.format(np.mean(realInSim_distance_list)))
print('距离中位数 : {}'.format(np.median(realInSim_distance_list)))
print('距离标准差 : {}'.format(np.std(realInSim_distance_list)))

# 这个是真实开奖结果在预测相似度矩阵中的位置中位数
floating_median_sim_idx = int(np.median(realInSim_distance_list))
floating_median_sim_acc_range_k = 0

# 这个是真实开奖结果在预测结果中的位置中位数
```

```python
            floating_median_idx = int(np.median(realInPredict_distance_list))
            floating_median_acc_range_k = 0
            # 计算以位置中位数为中心、以范围K为半径的准确率
            for ii in range(len(probabilities)):
                # 在预测结果中得到以位置中位数为中心、以范围K为半径的号码
                nearest_floating_median = (-probabilities[ii]).argsort()[0][
floating_median_idx - range_k:floating_median_idx + range_k]
                # 如果真正的开奖结果在上面所获得的号码当中
                if y[ii] in nearest_floating_median:
                    floating_median_acc_range_k += 1

                # 在预测相似度矩阵中得到以位置中位数为中心、以范围K为半径的号码
                nearest_floating_median_sim = (-sim[ii, :]).argsort()[
floating_median_sim_idx - range_k:floating_median_sim_idx + range_k]
                # 如果真正的开奖结果在上面所获得的号码当中
                if y[ii] in nearest_floating_median_sim:
                    floating_median_sim_acc_range_k += 1

            # 计算相应的准确率
            floating_median_acc_range_k = floating_median_acc_range_k / len(y)
            floating_median_acc_range_k_list.append(floating_median_acc_range_k)
            accuracies['floating_median_acc_range_k'].append(
                floating_median_acc_range_k)

            floating_median_sim_acc_range_k=floating_median_sim_acc_range_k/len(y)
            floating_median_sim_acc_range_k_list.append(
                floating_median_sim_acc_range_k)

            accuracies['floating_median_sim_acc_range_k'].append(
                floating_median_sim_acc_range_k)

        print('Epoch {:>3} floating median sim range k accuracy {}'.format(epoch_i,
np.mean(floating_median_sim_acc_range_k_list)))
        print('Epoch {:>3} floating median range k accuracy {} '.format(epoch_i,
np.mean(floating_median_acc_range_k_list)))
        print('Epoch {:>3} similar top k accuracy {} '.format(epoch_i, np.mean(
```

```
sim_topk_acc_list)))
        print('Epoch {:>3} top k accuracy {} '.format(epoch_i, np.mean(
topk_acc_list)))
        print('Epoch {:>3} accuracy {} '.format(epoch_i, np.mean(acc_list)))

    # 保存模型
    saver.save(sess, save_dir)
    print('Model Trained and Saved')
    embed_mat = sess.run(normalized_embedding)
```

最后一次迭代输出的结果如下:

真实值在预测值中的距离数据:
max distance : 946
min distance : 7
平均距离 : 556.65625
距离中位数 : 570.0
距离标准差 : 274.8418419490335
真实值在预测值相似向量中的距离数据:
max distance : 965
min distance : 20
平均距离 : 472.71875
距离中位数 : 562.0
距离标准差 : 311.20714909596387
Epoch 49 floating median sim range k accuracy 0.01125
Epoch 49 floating median range k accuracy 0.02875
Epoch 49 similar top k accuracy 0.0
Epoch 49 top k accuracy 0.004375
Epoch 49 accuracy 0.0

正常的开奖概率是 1‰,上面训练最后打印出的准确率和相似度向量 Top K 一样都是 0,一个都没猜中。

Top K 是 4.3‰ 左右,但因为是 Top 10,所以实际上是 0.43‰ 左右。

浮动中位数的准确率在 11.25‰~28.75‰ 之间,但由于这个范围是 Range 10,所以实际上在 1.125‰~2.875‰ 之间。

没有比正常开奖概率高多少。下面分析各项数据,看看网络训练的情况。

3.5.4 分析网络训练情况

1. 距离图表

真实值在预测值相似向量中的距离数据,如图 3.14 所示。

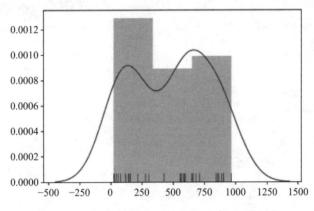

图 3.14 真实值在预测值相似向量中的距离

距离中位数:562.0

距离标准差:311.20714909596387

```
sns.distplot(realInSim_distance_list, rug=True)
```

真实值在预测值中的距离数据,如图 3.15 所示。

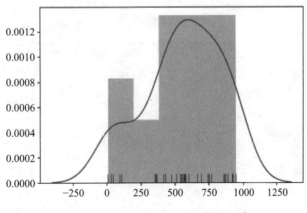

图 3.15 真实值在预测值中的距离

距离中位数：570.0

距离标准差：274.8418419490335

```
sns.distplot(realInPredict_distance_list, rug=True)
```

从最后一次的距离图表来看，模型训练得还不够好，从 0 到 1000 的距离都有可能，也就是没有学到规律。距离中位数都在 570 附近，是否说明一直买 570 这个号码中奖概率会高呢？可以考虑用中位数来弥补预测的不足。

2. 显示训练损失

```
plt.plot(losses['train'], label='Training loss')
plt.legend()
_ = plt.ylim()
```

训练损失如图 3.16 所示。

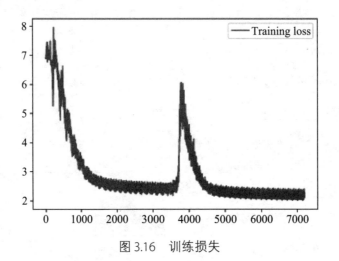

图 3.16　训练损失

3. 显示测试损失

```
plt.plot(losses['test'], label='Test loss')
plt.legend()
_ = plt.ylim()
```

测试损失始终没有降下来，这说明经过 50 次迭代训练网络没有学习到彩票预测的规律。epochs 高一点增加训练次数的话，会出现下降—上升—下降的波浪形曲线。测试损失如图 3.17 所示。

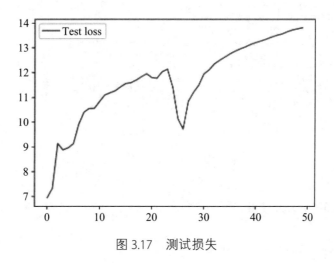

图 3.17　测试损失

4. **显示准确率**

- 测试准确率
- Top K 准确率
- 相似度 Top K 准确率
- 浮动距离中位数 Range K 准确率

```
plt.plot(accuracies['accuracy'], label='Accuracy')
plt.plot(accuracies['topk'], label='Top K')
plt.plot(accuracies['sim_topk'], label='Similar Top K')
plt.plot(accuracies['floating_median_acc_range_k'],
    label='Floating Median Range K Acc')
plt.plot(accuracies['floating_median_sim_acc_range_k'],
    label='Floating Median Sim Range K Acc')
plt.legend()
_ = plt.ylim()
```

各类准确率如图 3.18 所示。关于准确率我们后面再分析。

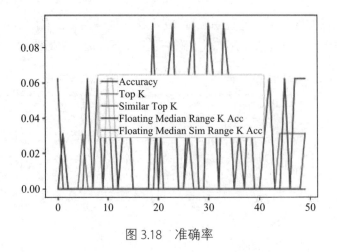

图 3.18 准确率

5. 显示预测结果和实际开奖结果

```
for batch_i, (x, y) in enumerate(test_batches):
    plt.plot(y, label='Targets')
    plt.plot(np.squeeze(probabilities.argmax(2)), label='Prediction')
    plt.legend()
    _ = plt.ylim()
```

结果如图 3.19 所示。

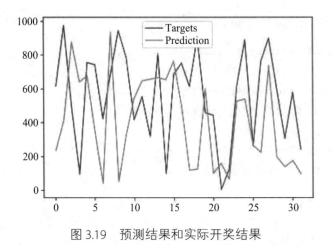

图 3.19 预测结果和实际开奖结果

从趋势上看还行,但实际结果是一个都没有猜中,有的简直错得离谱,南辕北辙。

6. 所有开奖号码的相似度分布

相似度分布如图 3.20 所示。

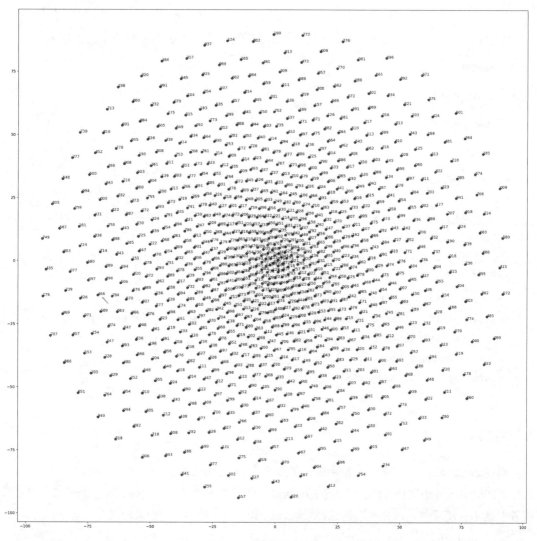

图 3.20 相似度分布

在训练不够充分的情况下，从所有号码相互之间的距离图可以看出，大部分数据点相互之间的距离基本上都差不多，只有一部分数据点之间的联系看起来比较紧密，对于这部分数据点我们还可以试试运气。

3.5.5 生成预测号码

1. 获取张量

使用 `get_tensor_by_name()` 函数从计算图中获取张量。取得的张量使用如下名字：

- "input:0"
- "initial_state:0"
- "final_state:0"
- "probs:0"

```
def get_tensors(loaded_graph):
    """
    从计算图loaded_graph中取得输入、初始状态、输出状态和softmax输出的概率向量
    :参数 loaded_graph: 计算图
    :返回: Tuple (InputTensor, InitialStateTensor, FinalStateTensor, ProbsTensor)
    """
    inputs = loaded_graph.get_tensor_by_name("input:0")
    initial_state = loaded_graph.get_tensor_by_name("initial_state:0")
    final_state = loaded_graph.get_tensor_by_name("final_state:0")
    probs = loaded_graph.get_tensor_by_name("probs:0")
    return inputs, initial_state, final_state, probs
```

2. 选择号码

实现 `pick_word()` 函数，从预测输出的概率向量 probabilities 或跟预测的号码具有最大相似度的向量 sim 中选择号码。算法是：根据概率向量中各个号码的概率，选择概率较大的号码，或者根据 Top K 个号码的概率随机选择号码，概率较大的号码会有较大的机会被选中。

pred_mode 表示选择号码的模式，有如下四种。

- sim：从相似度向量 Top K 中选号。
- median：从浮动距离中位数（相似度向量）Range K 中选号。
- topk：从概率向量 Top K 中选号。
- max：从概率向量中选择概率最大的号码。

```python
def pick_word(probabilities, sim, int_to_vocab, top_n = 5, pred_mode = 'sim'):
    """
    从向量中选出最有可能的号码
    :参数 probabilities: 预测输出的概率向量
    :参数 sim: 跟预测的号码相似度最大的向量
    :参数 int_to_vocab: 号码ID转号码字符串的字典
    :参数 top_n: Top数
    :参数 pred_mode: 选择号码的模式，默认使用跟预测的号码相似度最大的向量，因为如果
    网络训练得不好，则预测的概率向量不准确。如果网络训练得好，则可以使用topk或max模
    式
    :返回: 预测的号码
    """
    # 号码总数
    vocab_size = len(int_to_vocab)

    if pred_mode == 'sim':
        p = np.squeeze(sim)
        # 对概率向量中top_n以外的概率清零，只留下top_n个号码的概率
        p[np.argsort(p)[:-top_n]] = 0
        # 重新对概率向量归一化
        p = p / np.sum(p)
        # 根据top_n个号码的概率随机选择一个号码
        c = np.random.choice(vocab_size, 1, p=p)[0]
        return int_to_vocab[c]
    elif pred_mode == 'median':
        p = np.squeeze(sim)
        # 将以浮动距离中位数为中心、从Range K为半径的号码作为候选号码，清空候选号码
        # 以外的概率值
        p[np.argsort(p)[:floating_median_sim_idx - top_n]] = 0
        p[np.argsort(p)[floating_median_sim_idx + top_n:]] = 0
        # 重新归一化概率向量
        p = np.abs(p) / np.sum(np.abs(p))
        # 从候选号码中选择一个号码
        c = np.random.choice(vocab_size, 1, p=p)[0]
        return int_to_vocab[c]
    elif pred_mode == 'topk':
```

```
        p = np.squeeze(probabilities)
        # 对预测向量top_n以外的概率清零
        p[np.argsort(p)[:-top_n]] = 0
        # 重新归一化概率向量
        p = p / np.sum(p)
        # 从top_n个号码中随机选择一个号码
        c = np.random.choice(vocab_size, 1, p=p)[0]
        return int_to_vocab[c]
    elif pred_mode == 'max':
        # 只选择向量中概率最大的那个号码
        return int_to_vocab[probabilities.argmax()]
```

3. 生成彩票号码

现在开始预测彩票。

- `gen_length` 表示想生成多少期的号码。
- `prime_word` 表示前一期的号码。

算法是：根据训练好的网络参数，将前几期的号码作为初始输入送入网络中。初始号码可以多写一些往期的开奖号码，先不断送入网络中预热 LSTM 的状态。这部分循环预测的结果不用理会，只记录最后一次预测的结果即可。这样得到最后一次网络预测的概率向量，然后根据概率向量得到预测的号码。

假如想生成多期号码，则将刚刚预测得到的号码再次送入网络中，同样得到新一期的预测号码，如果还要生成下一期的号码，就再次将新预测的号码送入网络中，如此反复循环，直到多期号码生成完毕。

```
# 要生成17期的号码
gen_length = 17
# 将前几期的开奖号码作为初始号码
prime_word = ["623", "891", "262", "761", "900", "598", "306", "580", "243", "202"]

loaded_graph = tf.Graph()
with tf.Session(graph=loaded_graph) as sess:
    # 加载网络模型
    loader = tf.train.import_meta_graph(load_dir + '.meta')
    loader.restore(sess, load_dir)
```

```python
# 从模型中得到输入张量
input_text, initial_state, final_state, probs = get_tensors(loaded_graph)
# 从模型中得到归一化的嵌入矩阵，用于计算相似度
normalized_embedding = loaded_graph.get_tensor_by_name("truediv:0")
# 生成的号码序列
gen_sentences = []
# 先初始化RNN状态
prev_state = sess.run(initial_state, {input_text: np.array([[1]])})

# 将前几期的号码不断送入网络中，此时不用关心预测结果，只保留最后一次预测的结果
x = np.zeros((1, 1))
for word in prime_word:
    x[0,0] = vocab_to_int[word]
    probabilities, prev_state = sess.run(
        [probs, final_state],
        {input_text: x, initial_state: prev_state})

# 根据上面最后一次网络预测的概率向量，计算与向量中最大概率的号码的相似度向量
valid_embedding = tf.nn.embedding_lookup(normalized_embedding,
                                         probabilities.argmax())
valid_embedding = tf.reshape(valid_embedding, (1, embed_dim))

similarity = tf.matmul(valid_embedding, tf.transpose(normalized_embedding))
sim = similarity.eval()

# 传入生成号码函数，得到预测的号码。这里使用了topk模式，topk是5
pred_word = pick_word(probabilities, sim, int_to_vocab, 5, 'topk')
gen_sentences.append(pred_word)

# 循环生成多期号码
for n in range(gen_length):
    x[0,0] = pred_word

    # 得到网络预测的概率向量
    probabilities, prev_state = sess.run(
```

```
        [probs, final_state],
        {input_text: x, initial_state: prev_state})
    # 根据上面网络预测的概率向量，计算与向量中最大概率的号码的相似度向量
    valid_embedding = tf.nn.embedding_lookup(normalized_embedding,
                                             probabilities.argmax())
    valid_embedding = tf.reshape(valid_embedding, (1, embed_dim))
    similarity = tf.matmul(valid_embedding, tf.transpose(normalized_embedding))
    sim = similarity.eval()
    # 传入生成号码函数，得到预测的号码。这里使用了topk模式，topk是5
    pred_word = pick_word(probabilities, sim, int_to_vocab, 5, 'topk')
    # 将新生成的号码加入号码序列中，用于生成下一期号码
    gen_sentences.append(pred_word)

cp_script = ' '.join(gen_sentences)
cp_script = cp_script.replace('\n ', '\n')
cp_script = cp_script.replace('( ', '(')

print(cp_script)
```

得到的号码是：

202 463 972 359 345 106 221 321 792 408 830 035 127 267 790 589 476 904

4. 新的思路

从训练结果打印出的准确率，和往期开奖号码相互之间的距离图都可以看得出来，想进行彩票预测是不可行的。在"排列3"这种简单的、排列组合只有1000种（样本空间已经足够小了）的等概率事件上进行预测都如此困难，这也印证了数学的奇妙之处。都说彩票号码是等概率出现的，那么出现任何一种号码都是有可能的，没有规律可言。惊不惊喜？意不意外？

既然不能准确地预测，那么唯一能给我们提供思路的就是学习器学到的趋势，来看看下面的代码。

- int_sentences：里面保存着上面生成的若干期号码。
- val_data：最新几期的开奖号码，作为验证数据集。

```
int_sentences = [int(words) for words in gen_sentences]
int_sentences = int_sentences[1:]
```

```
val_data = [[103],[883],[939],[36],[435],[173],[572],[828],[509],[723],[145],[621],
 [535],[385], [98],[321],[427]]

plt.plot(int_sentences, label='History')
plt.plot(val_data, label='val_data')
plt.legend()
_ = plt.ylim()
```

趋势如图 3.21 所示,预测号码和实际号码之间的比较。

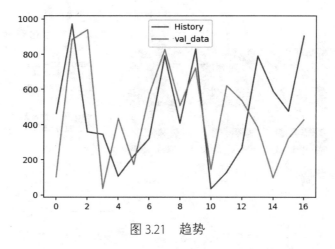

图 3.21 趋势

可以看出,虽然每期预测的号码都不对,但是下一期号码的大概范围以及若干期号码的变化趋势学习得还可以,剩下的就要靠运气了。

彩票预测的 TensorFlow 2.0 实现请参考本项目内的 cp_script_generation_tf2.ipynb 文件,稍有不同的是嵌入矩阵和 LSTM 的输出维度以及 epochs 训练次数。

3.6 文本生成

前面我们通过一个彩票预测案例讲解了 RNN 和 LSTM 的使用方法,但其实将其代码稍做修改就可以用于 RNN 的其他应用场景中,比如用于生成文本(各类文章、小说、古诗、歌词等)。当然,还有其他模型可以用于生成文本,比如 GAN 以及 OpenAI 发布的 GPT-2 等。不仅可以生成文本,甚至可以使用 RNN 来生成音乐或者程序代码。

现在以生成小说为例，我们可以从网上找到很多小说的文本，选择自己喜欢的任意风格的文本作为训练数据。根据上一节的代码，首先要提取出所有单词的个数，对于彩票预测是所有可能的号码，而对于文本生成就是所有出现的文字（不重复的）。同样要生成两个字典，即文字转 ID 的字典和 ID 转文字的字典。

需要注意的是 get_batches 函数，其中的参数 seq_length 可以不再输入 1 了，因为在进行彩票预测时，我们输入的是 batch×1 的一列号码，这里生成文本可以输入多列文本，这个数值是任意的，比如可以是 batch×5，这意味着每次输入给网络的就不再是一个字，而是 seq_length 个字了，如图 3.22 所示。

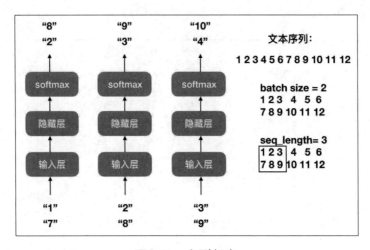

图 3.22　序列长度

网络模型由原来的输入一个号码预测出下一期号码，变成了输入一个句子中的每一个字，预测出句子中该字的下一个字。网络结构如图 3.23 所示。

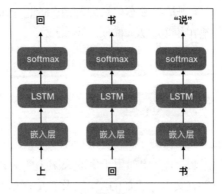

图 3.23　文本生成网络结构

网络没有任何变化，像这样将文本的每句话每一个字送入网络中进行训练，得到的是每一个字的嵌入向量，参照图 3.20 生成其相似度分布，可以看到训练的每一个字之间的距离远近。另外，在网络训练的代码中，相似度向量 sim 和距离中位数等代码都没有用了，因为这些是用来预测彩票的，生成文本只需要网络 softmax 输出的概率向量即可。

网络训练的代码如下：

```python
# 得到批量数据
batches = get_batches(int_text, batch_size, seq_length)

with tf.Session(graph=train_graph) as sess:
    sess.run(tf.global_variables_initializer())
    # 训练循环
    for epoch_i in range(num_epochs):
        # 初始化RNN状态
        state = sess.run(initial_state, {input_text: batches[0][0]})

        # 批量数据的循环
        for batch_i, (x, y) in enumerate(batches):
            feed = {
                input_text: x,
                targets: y,
                initial_state: state,
                lr: learning_rate}
            # 训练网络
            train_loss, state, _ = sess.run([cost, final_state, train_op], feed)

            # 打印训练数据
            if (epoch_i * len(batches) + batch_i) % show_every_n_batches == 0:
                print('Epoch {:>3} Batch {:>4}/{}   train_loss = {:.3f}'.format(
                    epoch_i,
                    batch_i,
                    len(batches),
                    train_loss))

    # 保存模型
    saver = tf.train.Saver()
```

```
saver.save(sess, save_dir)
print('Model Trained and Saved')
```

最后在生成文本的 pick_word 函数中，只使用 topk 一种模式即可，其他模式都可以删掉了。简化后的代码如下：

```
def pick_word(probabilities, int_to_vocab):
    p = np.squeeze(probabilities)
    top_k = 5
    p[np.argsort(p)[:-top_k]] = 0
    p = p / np.sum(p)
    idx = np.random.choice(len(p), 1, p=p)[0]
    # return int_to_vocab[probabilities.argmax()]
    return int_to_vocab[idx]
```

这里要注意一点，为什么生成文本时不能使用向量中概率最大的那个字呢？大家想一下。这里是在 top_k 中选择一个字，top_k 是 5，也可以修改这个数值，加大或者减小，最好别太小了。想清楚了吗？在生成文本时，训练后的网络已经学习了每一个字的特征以及字与字之间的关系，这样根据某个字预测出下一个字，最大可能的字永远是固定的，一旦使用了概率向量中概率最大的那个字，那么所生成的文本看起来根本不是小说，而是概率最大的字组合在一起的大段重复的文本。比如，"我"的下一个字最大可能是"爱"，而"爱"的下一个字最大可能是"你"，然后"你"的下一个字最大可能也是"爱"，这样所生成的文本很可能就是"我爱你爱你爱你爱你…"。所以，这里需要加入一点模糊概率，加入不确定性。比如上面的实现在最大概率的 5 个字中选择一个，这样所生成的文本每一个字都是五选一，文字重复的概率大大减小了，看起来也更像句子了。最后的文本生成跟彩票号码生成没有区别，感兴趣的读者可以尝试亲自实现。

3.7 生成古诗：基于 TensorFlow 2.0

本节我们基于 TensorFlow 2.0 来实现古诗生成，使用 3.6 节介绍过的代码，代码基本没有变化，换一个训练集即可。

3.7.1 数据预处理

笔者从网络上找到了一个古诗的数据集，已经放在 data 文件夹下了，名称为 poetry.txt。但是这个数据集需要先清理，里面有些影响训练的字符，比如 "_"、书名号、中括号，还有其他字符，所以需要预处理数据集。

```
import os
import re
# 用于正则表达式
pattern = '[a-zA-Z0-9''"#$%&\'()*+-./:;<=>@★...【】《》""''[\\]^_`{|}~]+'

def preprocess_poetry(outdir, datadir):
    # 处理后生成的新文档叫作new_poetry.txt
    with open(os.path.join(outdir, 'new_poetry.txt'), 'w') as out_f:
        with open(os.path.join(datadir, 'poetry.txt'), 'r') as f:
            # 读取每行文本
            for line in f:
                # 对文本行做处理
                content = line.strip().rstrip('\n').split(':')[1]
                content = content.replace(' ','')
                # 如果包含这些特殊字符, 这首诗就不要了
                if '}' in content or '_' in content or '(' in content or '（' in content or '《' in content or '[' in content:
                    continue
                # 如果一首诗太短, 就不要了
                if len(content) < 20:
                    continue
                # 使用正则表达式修改文本
                content=re.sub(pattern, '', content)
                out_f.write(content + '\n')
```

为了快速演示效果,这里只截取新文档前1000行左右的数据进行训练,这个文档的名称是 newtxt.txt。

同样先看数据情况。

不重复汉字的个数: 1020
诗词个数: 1027
行数: 1028
平均每行字数: 76.0

一共有 1020 个汉字，这样嵌入矩阵的行数就是 1020。

接下来要生成两个字典，即文字转 ID 的字典和 ID 转文字的字典。

```python
import numpy as np
from collections import Counter

def create_lookup_tables():
    # 文本中所有的字(不重复)
    vocab = sorted(set(text))
    # 文字转ID的字典
    vocab_to_int = {u:i for i, u in enumerate(vocab)}
    # 不重复汉字的个数，1020个
    
    int_to_vocab = np.array(vocab)
    # ID转文字的字典
    int_text = np.array([vocab_to_int[word] for word in text if word != '\n'])
    # 保存到本地
    pickle.dump((int_text, vocab_to_int, int_to_vocab), open('preprocess.p', 'wb'))
```

超参数如下：

```python
# 不重复汉字的个数，1020个
vocab_size = len(int_to_vocab)
# RNN的尺寸（LSTM输出的维度）
rnn_size = 1000
# 嵌入层的维度
embed_dim = 256
# 序列的长度，每个批次都传入[批次×seq_length]个数据，这里是[批次×15]
# 注意这里已经不是1了，在古诗生成里面这个数值可以大一些，比如100也是可以的
seq_length = 15
# 学习率
learning_rate = 0.01
# 表示经过多少批次以后打印训练信息
show_every_n_batches = 10


save_dir = './save'
```

3.7.2 构建网络

使用 Keras 构建网络模型很容易,第一层是嵌入层,指定嵌入矩阵维度(vocab_size × embed_dim),并且指定批量输入形状。第二层是 LSTM,指定 LSTM 的输出维度是 rnn_size,这里只用了一层 LSTM,也可以再加一层。回想 3.3 节介绍的 LSTM 输出门,它的输出是 h_t,t 是时间步,h_t 的输出维度就是我们传入的参数 rnn_size。如果在创建 LSTM 时将参数 return_sequences 设置为 false,则 LSTM 只返回最后一个时间步的输出,形状是 [batch_size, rnn_size];如果将 return_sequences 设置为 true,则 LSTM 返回的是每一个时间步的输出,形状是 [batch_size, time_step, rnn_size]。最后一层是全连接层,输出维度是文字个数。

```
import tensorflow as tf
import datetime
from tensorflow import keras
from tensorflow.python.ops import summary_ops_v2
import time


# 模型保存的路径
MODEL_DIR = "./poetry_models"


# 批量数据迭代器
train_batches = get_batches(int_text, batch_size, seq_length)
# 用于后面描画损失曲线
losses = {'train': []}


class poetry_network(object):
    # 默认批次大小是32
    def __init__(self, batch_size=32):

        self.batch_size = batch_size
        self.best_loss = 9999

        # 构建模型
        self.model = tf.keras.Sequential([
            tf.keras.layers.Embedding(vocab_size, embed_dim,
                                      batch_input_shape=[batch_size, None]),
            tf.keras.layers.LSTM(rnn_size,
```

```python
                                           return_sequences=True,
                                           stateful=True,
                                           recurrent_initializer='glorot_uniform'),
            tf.keras.layers.Dense(vocab_size)
        ])
        self.model.summary()

        # 使用Adam优化器
        self.optimizer = tf.keras.optimizers.Adam()
        # 分类交叉熵损失
        self.ComputeLoss = tf.keras.losses.SparseCategoricalCrossentropy(from_logits=True)

        # 创建保存模型的文件夹
        if tf.io.gfile.exists(MODEL_DIR):
            pass
        else:
            tf.io.gfile.makedirs(MODEL_DIR)

        train_dir = os.path.join(MODEL_DIR, 'summaries', 'train')

        # 用于记录日志
        self.train_summary_writer = summary_ops_v2.create_file_writer(train_dir, flush_millis=10000)

        checkpoint_dir = os.path.join(MODEL_DIR, 'checkpoints')
        self.checkpoint_prefix = os.path.join(checkpoint_dir, 'ckpt')
        self.checkpoint = tf.train.Checkpoint(model=self.model, optimizer=self.optimizer)

        # 加载已存在的模型参数
        self.checkpoint.restore(tf.train.latest_checkpoint(checkpoint_dir))

    @tf.function
    def train_step(self, x, y):
        # 单步训练函数，使用磁带记录计算损失的操作，后面使用磁带计算梯度
```

```python
        with tf.GradientTape() as tape:
            # 调用模型得到模型输出,参数training为True
            logits = self.model(x, training=True)
            # 计算分类交叉熵损失
            loss = self.ComputeLoss(y, logits)
        # 使用磁带计算模型参数的梯度
        grads = tape.gradient(loss, self.model.trainable_variables)
        self.optimizer.apply_gradients(zip(grads, self.model.trainable_variables))
        return loss, logits

    # 训练函数,默认epochs是1
    def training(self, epochs=1, log_freq=50):
        batchCnt = len(int_text) // (batch_size * seq_length)
        print("batchCnt : ", batchCnt)
        # 训练的循环
        for i in range(epochs):
            train_start = time.time()
            with self.train_summary_writer.as_default():
                start = time.time()
                # 定义平均指标用来计算平均损失
                avg_loss = tf.keras.metrics.Mean('loss', dtype=tf.float32)

                # 遍历训练数据
                for batch_i, (x, y) in enumerate(train_batches):
                    # 单步训练
                    loss, logits = self.train_step(x, y)
                    avg_loss(loss)
                    losses['train'].append(loss)

                    # 训练达到一定次数,打印日志
                    if tf.equal(self.optimizer.iterations % log_freq, 0):
                        summary_ops_v2.scalar('loss', avg_loss.result(),
                            step=self.optimizer.iterations)

                        rate = log_freq / (time.time() - start)
                        print('Step #{}\tLoss: {:0.6f} ({} steps/sec)'.format(
```

```
                    self.optimizer.iterations.numpy(), loss, rate))

                avg_loss.reset_states()

                start = time.time()

            # 保存模型参数
            self.checkpoint.save(self.checkpoint_prefix)
            print("save model\n")
```

3.7.3 开始训练

在预处理过的数据上训练神经网络。

```
net = poetry_network()
net.training(20)   # 训练20个epochs
```

模型结构和参数打印如下：

```
Layer (type)                 Output Shape              Param #
=================================================================
embedding_6 (Embedding)      (32, None, 256)           1901824
_____
unified_lstm_6 (UnifiedLSTM) (32, None, 1000)          5028000
_____
dense_6 (Dense)              (32, None, 7429)          7436429
=================================================================
Total params: 14,366,253
Trainable params: 14,366,253
Non-trainable params: 0
_____
```

3.7.4 生成古诗

现在加载已保存的模型，准备生成古诗。

```
restore_net=poetry_network(1)   # 这里传入1，表示设置batch_size为1
restore_net.model.build(tf.TensorShape([1, None]))
```

此时的模型结构和参数如下：

```
Layer (type)                 Output Shape              Param #
=================================================================
embedding_7 (Embedding)      (1, None, 256)            1901824
_____
unified_lstm_7 (UnifiedLSTM) (1, None, 1000)           5028000
_____
dense_7 (Dense)              (1, None, 7429)           7436429
=================================================================
Total params: 14,366,253
Trainable params: 14,366,253
Non-trainable params: 0
_____
```

注意此时网络结构的 batch_size 跟训练时是不一样的。

开始生成古诗了，先介绍一下参数。

- `prime_word`，给定开始的头几个字（起始字），类似于彩票预测中的前几期号码。
- `top_n`，从前 N 个候选汉字中随机选择，使用此参数的代码被注释掉了，这里使用了另外的方式来选择汉字。
- `rule`，默认是 7，表示七言绝句。
- `sentence_lines`，生成几句古诗，默认是 4 句（逗号和句号都算一句）。
- `hidden_head`，藏头诗的前几个字。

生成古诗的代码与彩票预测中生成号码的代码差不多，只是这里加了一些逗号、句号和藏头诗的处理。第一次生成先将前 N 个起始字传入网络中，只使用最后一个预测的字作为接续字，送入网络中继续生成，如此反复，最终得到整首诗。其间，判断生成的字数够不够一句，够了就加上标点符号。

根据模型输出的对数概率 logits 选择汉字使用了 tf.random.categorical 函数，在 TensorFlow 1.x 下要换成 tf.multinomial 函数。传入该函数的参数，一个是未归一化的对数概率 logits；另一个是抽取的样本数。将样本数设置为 1，只采样出一个汉字。

```
def gen_poetry(prime_word='白', top_n=5, rule=7, sentence_lines=4,
```

```python
hidden_head=None):
    # 计算出要生成的汉字个数：
    # 句子个数 * 每句字数(比如7个汉字+1个标点就是一句的字数) - 提前给出的起始字数
    gen_length = sentence_lines * (rule + 1) - len(prime_word)
    # 先把起始字放入保存生成的古诗的列表里
    gen_sentences = [prime_word] if hidden_head==None else [hidden_head[0]]
    temperature = 1.0

    # 将起始字转成ID
    dyn_input = [vocab_to_int[s] for s in prime_word]
    dyn_input = tf.expand_dims(dyn_input, 0)

    # 初始化RNN状态
    restore_net.model.reset_states()
    # index是当前字的位置
    index=len(prime_word) if hidden_head==None else 1
    # 生成古诗的循环
    for n in range(gen_length):
        index += 1
        # 调用函数得到预测的logits
        predictions = restore_net.model(np.array(dyn_input))
        # 移除批量的维度
        predictions = tf.squeeze(predictions, 0)

        # 满足if条件，说明应该是标点符号了
        if index!=0 and (index % (rule+1)) == 0:
            if ((index / (rule+1)) + 1) % 2 == 0:
                predicted_id=vocab_to_int[', ']
            else:
                predicted_id=vocab_to_int['。']
        else:
            # 如果是生成藏头诗，并且此时是句子的头一个字，则使用给定的藏头诗的字
            if hidden_head != None and (index-1)%(rule+1)==0 and (index-1)//(rule+1) < len(hidden_head):
                predicted_id=vocab_to_int[hidden_head[(index-1)//(rule+1)]]
            else:
```

```
        # 使用多项分布来预测模型返回的汉字,如果是标点符号则重新选择
        while True:
            predictions = predictions / temperature
            predicted_id = tf.random.categorical(predictions,
                num_samples=1)[-1, 0].numpy()
            # p = np.squeeze(predictions[-1].numpy())
            # p[np.argsort(p)[:-top_n]] = 0
            # p = p / np.sum(p)
            # c = np.random.choice(vocab_size, 1, p=p)[0]
            # predicted_id=c

            # 如果选择的是汉字,不是标点符号,则退出循环
            if(predicted_id != vocab_to_int[','] and predicted_id !=
            vocab_to_int['。'] ):
                break

        # 将新预测的汉字作为下一次预测的输入
        dyn_input = tf.expand_dims([predicted_id], 0)
        gen_sentences.append(int_to_vocab[predicted_id])

    poetry_script = ' '.join(gen_sentences)
    poetry_script = poetry_script.replace('\n ', '\n')
    poetry_script = poetry_script.replace('( ', '(')

    return poetry_script
```

下面来看看训练了 20 个 epochs 的古诗生成结果。

1. 给定开头

```
gen_poetry(prime_word='白日依山尽', top_n=10, rule=5, sentence_lines=4)
```

'白日依山尽 , 照岸终年矜 。 目尽飞猿笑 , 愁长断独声 。'

2. 七言绝句

```
gen_poetry(prime_word='月', top_n=10, rule=7, sentence_lines=4)
```

'月皎朝帐外山雪，上兵裁缝玉吹声。胡尘夜去凉汉路，金鲸泻佩如飞泉。'

3. 五言律诗

gen_poetry(prime_word='月', top_n=10, rule=5, sentence_lines=4)

'月来长河闭，苔疏连叶满。归来入汉使，居带天门台。'

4. 藏头诗

gen_poetry(prime_word='夏', top_n=10, rule=5, sentence_lines=4, hidden_head='春夏秋冬')

'春影掌庭傳，夏然別復多。秋然謫思老，冬水入澤春。'

在做数据预处理时，还可以在每首诗的首尾加上开始和结束标志，比如加上"["和"]"。这样网络可以学到一首诗开始和结束的特征，然后在生成古诗时就可以判断所生成的汉字是否是结束标志"]"，如果是就可以停止生成古诗的循环了。只不过这样生成的文本数量不可控，文本不一定规整。

3.8 自然语言处理

自然语言处理也是 RNN 大显身手的领域，能够使神经网络读懂人类的语言（比如机器翻译），甚至可以与人交流（比如聊天机器人）。本节将介绍"序列到序列"的基本概念，然后讲解基于注意力机制的序列处理模型 Transformer，最后了解基于双向 Transformer 的预训练语言模型 BERT。

3.8.1 序列到序列

序列到序列（Sequence to Sequence, Seq2Seq），它的思想是将一组序列转换为另一组序列，两个序列的长度可以不同，其结构如图 3.24 所示。

Seq2Seq 由编码器和解码器组成，编码器接收不定长的输入序列，学习序列的特征并将该序列映射到中间表征，叫作序列上下文（Context）。序列上下文是定长的，作为解码器的输入。解码器处理 Context 后，生成不定长的输出序列。

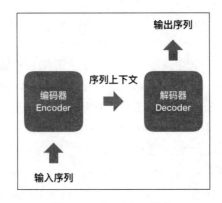

图 3.24　Seq2Seq 结构

序列，可以是序列化表示的任何数据，比如文字、图片或者语音等。以文字作为输入序列，就是将一句话、一段话输入给编码器，然后由解码器生成另一段文字。这样的场景最常见的就是机器翻译，输入的是中文，输出的是英文，可以将编码器输出的序列上下文理解成学习到的语义。再比如输入的文本是问题，解码器输出的可以是回答，这样就是一个问答（Q&A）模型或者客服系统（大家在网站上见到过客服机器人吧）。对于客服机器人来说，实际上就是用各种对话数据集来训练的，输入一句话，解码器生成回应的一句话，这就是一个聊天机器人（Chatbot）。还可以输入一篇文章，对应的输出是一段总结的文字或者文摘，这就是文档摘要生成器。如果输入的是图片，则可以输出一段描述该图片的文字，这就是图片描述生成器。以文字作为序列的应用场景特别广泛，人类之间的交流主要靠语音和文字传播，有极大的应用价值。

编码器和解码器是两个神经网络，你可以发挥想象力用各种方式来实现，最基本的实现方式是使用 RNN 实现编码器和解码器。对于基于时间序列的数据，使用 RNN 可以很好地学习到语义特征，正适用于 Seq2Seq 模型。

这样话，使用 RNN 来实现 Seq2Seq 模型，我们会感觉很熟悉，比如编码器和解码器都使用 LSTM 来实现，结构如图 3.25 所示（图片出自论文 Sequence to Sequence Learning with Neural Networks，论文地址：https://arxiv.org/abs/1409.3215）。

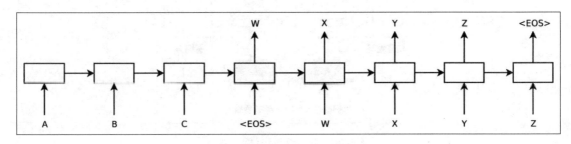

图 3.25　Seq2Seq—LSTM 结构

看起来跟文本生成网络结构很像是不是？之前的文本生成只有一个 LSTM 网络，而 Seq2Seq 模型分为编码器和解码器两个网络。图 3.25 中左侧输入 A、B、C 的就是编码器，这个跟我们之前实现的 LSTM 网络是一样的，先生成所有文本单词与 ID 之间的转换字典，然后将代表该文字的 ID 传入嵌入矩阵中，得到嵌入向量；再将嵌入向量传给 LSTM 网络，这样 LSTM 网络最后一个时间步输出的状态（Cell State 和 Hidden State）就是编码器的输出了，即语义（如图 3.25 中输入 C 之后的输出）。这个循环网络最后一个时间步输出的状态就是 tf.nn.dynamic_rnn 函数的第二个返回值。当语义传播到解码器之后，LSTM 网络经过时间序列的学习生成最后的输出 WXYZ，目标是编码器的输出要和学习目标 y 一致，仍然使用 sequence_loss 损失函数。解码器的构建方式跟编码器有些不同，调用的函数是 tf.contrib.seq2seq.dynamic_rnn_decoder。

3.8.2 Transformer

在实现 Seq2Seq 模型前，先来了解 Transformer。Transformer 是由 2017 年的论文 *Attention Is All You Need* 提出的序列处理模型，它的网络结构只基于注意力机制，而没有使用 RNN 或者 CNN。Transformer 模型结构如图 3.26 所示（图片出自论文 *Attention Is All You Need*，论文地址：https://arxiv.org/abs/1706.03762）。

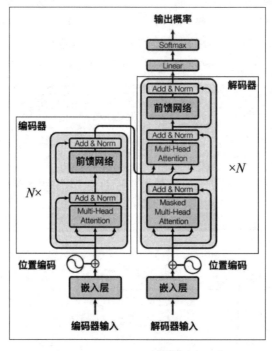

图 3.26　Transformer 模型结构

看起来比 Seq2Seq—LSTM 结构要复杂一些，Transformer 同样分为编码器和解码器。

1. 编码器

编码器模块由两部分组成，先是多头注意力模块，然后是前馈网络模块（全连接层）。这两个模块都使用了残差连接，这样一共 6 个（$N=6$）编码器模块堆叠在一起共同组成了 Transformer 的编码器。

编码器的输入比如是一串文字，先将输入序列转换成对应的 ID 输入给嵌入层，得到输入文本对应的嵌入向量。这里需要将嵌入向量与位置编码相加，得到编码器的第一个模块多头注意力（Multi-Head Attention）的输入。

通过残差连接，多头注意力模块的输出与输入相加（Add）之后做层归一化（Layer Normalization），得到前馈网络的输入。同样，前馈网络通过残差连接将输出与输入相加，然后做层归一化，这样就得到了编码器模块的输出。

2. 位置编码

现在对编码器的细节进行讲解，首先是位置编码。因为 Transformer 模型既没有使用 RNN 也没有使用 CNN，通过自注意力机制只能学习到序列内部的特征，而不能捕捉到序列的顺序。可以理解为，对于输入的序列，每一个字的意思模型都能看懂，但是却不能将其所理解的意思组成一句完整的话，因为模型没有办法学到顺序信息。RNN 的特点我们都知道，它就是为学习时序而生。CNN 通过多层卷积的感受野（Receptive Field），也能够感受到时序的特点。为了让模型能够利用序列的顺序信息，需要将单词在序列中的位置信息添加到输入中，这样注意力机制就可以识别不同位置的单词了。

关于位置编码，既可以通过算法学习出来，也可以使用固定的表达式计算出来。论文对两种方式都进行了实验，发现两者的结果差不多，所以最终选择了固定表达式计算，计算公式如下：

$$\text{PE}_{(pos,2i)} = \sin(pos/10000^{2i/d_{\text{model}}})$$

$$\text{PE}_{(pos,2i+1)} = \cos(pos/10000^{2i/d_{\text{model}}})$$

位置编码的维度跟嵌入向量的维度是一致的，都是 d_{model}，这样两者才能做加和，论文中 $d_{\text{model}} = 512$。公式中 pos 是单词在序列中的位置，i 是位置编码向量中的位置（下标），位置编码的表示如图 3.27 所示。

上面两个公式用来计算每一个单词的位置编码向量，其中 $2i$ 表示偶数位置，$2i+1$ 表示奇数位置。

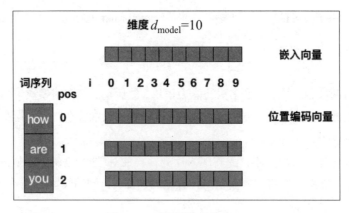

图 3.27　位置编码的表示

3. 注意力

注意力函数可以被描述为将查询（Query）和一组键值对（Key-Value）映射到输出，其中查询、键、值和输出都是向量。可以将 Query 看成是原序列，键值对是目标序列，注意力机制通过 Query 和 Key 的相似性得到 Query 和 Value 之间的关系。注意力结构如图 3.28 所示。

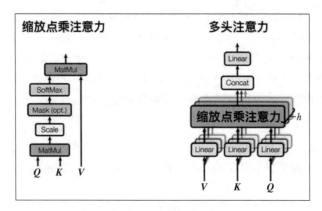

图 3.28　注意力结构

（1）多头注意力

Query、Key 和 Value 作为多头注意力模块的输入，首先进行线性变换，然后传给缩放点乘注意力模块。这里线性变换和缩放点乘注意力一共有 h 组，也就是说，把 Q、K 和 V 分成 h 份，分别计算 h 个不同的线性变换，然后执行 h 个不同的缩放点乘注意力。这里 h 就是头（head），多头就是多份（h 份），论文中 $h=8$。当然，你也可以将 Q、K 和 V 做一次全连接（线性变换），然后将结果分成 h 份，这样每份的维度是 $d_q = d_k = d_v = d_{model}/h = 512/8 = 64$。最后将多个点乘注意力的输出结果拼接起来，再做一次全连接，就得到了多头注意力的输出，输出维度经过拼接仍然是 $d_{\text{model}} = 64 \times 8 = 512$。

（2）缩放点乘注意力

终于到了真正计算注意力的地方了，缩放点乘注意力模块接收维度是 d_k 的 Query 和 Key，维度是 d_v 的 Value。首先计算 $\boldsymbol{Q}$ 和 $\boldsymbol{K}$ 的点乘，然后结果除以 $\sqrt{d_k}$，对点乘结果进行比例缩放。目的是为了减小点积的尺寸，防止结果太大。因为点积越大，则 softmax 的梯度越小，使得网络训练很慢、效果很差。经过缩放之后，计算 softmax，然后将结果与 $\boldsymbol{V}$ 做点乘，softmax 的结果相当于 $\boldsymbol{V}$ 的权重。

其中有一个可选的操作，是掩码（Mask）。这个操作是在解码器当中使用的，解码器的第一个模块就是掩码多头注意力（Masked Multi-Head Attention）。掩码的作用是防止在计算 $\boldsymbol{Q}$ 和 $\boldsymbol{K}$ 的相关性时看到未来单词的信息，只允许注意力机制看到有限的序列信息。比如当前时间步 t，那么只能看到时间步 t 以及之前时间步的信息。

缩放点乘注意力的公式如下：

$$\text{Attention}(\boldsymbol{Q}, \boldsymbol{K}, \boldsymbol{V}) = \text{softmax}(\frac{\boldsymbol{Q}\boldsymbol{K}^{\text{T}}}{\sqrt{d_k}})\boldsymbol{V}$$

回到编码器的处理流程上来，编码器的输入加上位置编码之后得到维度是 d_{model} 的输出，假设这个输出是 $\boldsymbol{x}$。接着要把 $\boldsymbol{x}$ 传给多头注意力模块，而多头注意力的输入是 $\boldsymbol{Q}$、$\boldsymbol{K}$ 和 $\boldsymbol{V}$，貌似个数有些对不上。此时传入的 $\boldsymbol{Q}$、$\boldsymbol{K}$ 和 $\boldsymbol{V}$ 都是 $\boldsymbol{x}$，用公式表示就是 $\text{Attention}(\boldsymbol{x}, \boldsymbol{x}, \boldsymbol{x})$，这就是自注意力机制。在处理点乘注意力时，对于 $\boldsymbol{Q}$ 和 $\boldsymbol{K}$ 的相关性计算，由于 $\boldsymbol{Q}$ 和 $\boldsymbol{K}$ 是相同的 $\boldsymbol{x}$，所以相当于计算输入序列内部的相关性，并根据这个相关性注意到自身 Value 的内部特征。

4. 解码器

清楚了编码器的过程之后，对解码器的理解就容易了。同样，首先是解码器的输入嵌入向量与位置编码向量做加和，得到解码器注意力模块的输入。解码器的输入是什么？打个比方，假如现在做的是机器翻译，编码器的输入是"你好吗"，那么解码器的输入就是"How are"，目标 y 是"are you"。

同样，解码器也是由 6 个（$N = 6$）解码器模块堆叠在一起共同组成的。其大体结构跟编码器一样，第一层的注意力模块和第三层的前馈网络模块与编码器相同，差别是第二层多了一个注意力模块。解码器的第一个模块是掩码自注意力模块，也就是说，在计算点乘注意力时，要掩盖未来时间步的信息。$\boldsymbol{Q}$、$\boldsymbol{K}$ 和 $\boldsymbol{V}$ 都是相同的值，都是解码器输入与位置编码的和。对于第二层的注意力模块，差别仅在于编码器的输出（称之为 memory）作为 $\boldsymbol{K}$ 和 $\boldsymbol{V}$，然后解码器第一层的掩码自注意力模块的输出（称之为 x）作为 Q，则第二层的注意力模块用公式表达就是 Attention(x, memory, memory)。

更多的细节读者可以看论文原文。

3.8.3 BERT

2018 年 Google 发表了基于双向 Transformer 的论文 *BERT: Pre-training of Deep Bidirectional Transformers for Language Understanding*，也就是 BERT（双向 Transformer 编码器表示，Bidirectional Encoder Representations from Transformers），论文地址为 https://arxiv.org/abs/1810.04805。它是预训练语言模型，该模型在 11 项自然语言处理任务中取得了最先进的结果。

BERT 在实现上只使用了 Transformer 模型的编码器部分，这意味着 BERT 模型主要用于理解文本语义。BERT 最大的贡献在于 Google 团队强大的计算力训练得到的预训练模型，我们可以直接使用该模型进行迁移学习，使用在自然语言处理任务上，就不用再做编码器的语义训练了。

Transformer 的编码器堆叠了 6 个编码器模块，而 BERT 训练了两个版本，一个是 $\text{BERT}_{\text{base}}$，另一个是 $\text{BERT}_{\text{large}}$。其中 $\text{BERT}_{\text{base}}$ 的配置是 12 个编码器模块，隐藏单元大小为 768（输出的维度），12 个 head；$\text{BERT}_{\text{large}}$ 的配置是 24 个编码器模块，隐藏单元大小为 1024（输出的维度），16 个 head。

与 Transformer 的一个重要区别是，BERT 没有解码器，因此注意力模块没有掩码部分。对 BERT 来说，能够看到序列前向和后向的每个单词。在处理当前时间步的单词时，可以充分利用前面时间步的单词和后面时间步的单词两个方向（双向）的信息。

1. 输入表示

BERT 的输入如图 3.29 所示（出自 *BERT: Pre-training of Deep Bidirectional Transformers for Language Understanding* 论文），由三个向量相加得到。第一个就是词嵌入向量，每个单词从嵌入矩阵中取得的嵌入向量。这里有几个特殊字符，每个句子的开头是固定的 [CLS] 符号，这个符号用于分类任务，表达整个输入序列的特征。将特殊符号 [SEP] 放在每句的结尾，用来分割两个句子，前一句会加上分段嵌入向量 A，后一句会加上分段嵌入向量 B。BERT 使用了经过学习的位置编码来表达每个单词的位置信息。

2. 预训练

BERT 有两个训练任务，一个是掩码语言模型（Masked LM），一个是下一句预测。

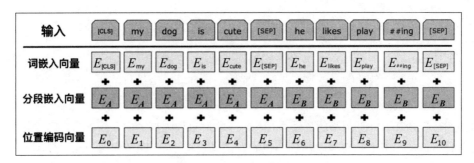

图 3.29　BERT 的输入

训练任务 1：掩码语言模型

为了训练网络双向表达信息的能力，BERT 采用的方法是随机屏蔽（Mask）一定比例的输入单词，然后仅预测那些被屏蔽的单词。随机屏蔽的过程是，首先随机选择整个序列中 15% 的单词，这些单词有 80% 的概率会被 [MASK] 这个特殊符号替代，10% 的概率会被随机选定的单词替代，10% 的概率保持不变，不会被替代。

比如：输入序列是"我的狗是毛毛"。

80% 概率：我的狗是 [MASK]。

10% 概率：我的狗是苹果。

10% 概率：我的狗是毛毛。

Transformer 编码器不知道要预测的是哪些单词，或者哪些单词已被随机单词替换掉，因此它被迫保持每个输入单词的上下文表示。此外，因为随机替换只发生在所有单词的 1.5%（即 15% 的 10%）上，这似乎不会损害模型的语言理解能力。由于每次只能预测 15% 的单词，模型收敛会比较慢。

训练任务 2：下一句预测

这是一个二分类的预测任务，目的是为了训练理解句子关系的模型。从训练集中选择句子 A 和 B 作为一个序列（A+B），B 有 50% 的概率是 A 的下一句，另 50% 的概率是随机的句子。如果 B 是 A 的下一句，标签就是 IsNext；否则，标签就是 NotNext，标签就是二分类任务的学习目标。

这些就是 BERT，是不是感觉很简洁？完全使用 Transformer 编码器，两个训练任务，判断 B 是 A 的下一句来训练理解句子间的关系，预测屏蔽的单词来训练理解句子内部单词之间的关系。

3. 微调

经过上面的介绍，我们了解了 BERT 的架构和训练。实际上，最重要的是 Google 为我们提供了预训练好的模型，包括中文模型，我们只要使用预训练模型进行微调，并应用于不同的自然语言处理任务就行了。

微调的过程是在 BERT 网络的最后再加一层，我们只训练这最后一层就行。对于句子级别的分类任务，可以使用 [CLS] 符号输出的向量 C、训练权重 W，即 $P = \text{softmax}(CW^T)$。如果你想使用 BERT 做 Seq2Seq，则需要再实现一个解码器。

BERT 的项目地址为 https://github.com/google-research/bert，现在引用官方文档中介绍的几个微调和预测方法。

（1）句子分类任务的微调和预测

对于句子级别的分类任务的微调，首先需要下载 GLUE 数据，下载数据的脚本地址请在官方 BERT 项目首页上查找。然后将下载的数据放在目录 $GLUE_DIR 中。接着下载 BERT-Base 预训练模型，解压缩后放在目录 $BERT_BASE_DIR 中。

预训练模型包含三个文件：

- vocab.txt：将 WordPiece 映射到单词 ID 的词汇文件。
- bert_config.json：指定模型的超参数的配置文件。
- bert_model.ckpt：预训练的权重的检查点文件。

示例代码在 Microsoft Research Paraphrase Corpus（MRPC）语料库上对 BERT-Base 进行微调，该语料库仅包含 3600 个样本，在大多数 GPU 上进行微调只需几分钟。

```
export BERT_BASE_DIR=/path/to/bert/uncased_L-12_H-768_A-12
export GLUE_DIR=/path/to/glue

python run_classifier.py \
  --task_name=MRPC \
  --do_train=true \
  --do_eval=true \
  --data_dir=$GLUE_DIR/MRPC \
  --vocab_file=$BERT_BASE_DIR/vocab.txt \
  --bert_config_file=$BERT_BASE_DIR/bert_config.json \
  --init_checkpoint=$BERT_BASE_DIR/bert_model.ckpt \
  --max_seq_length=128 \
  --train_batch_size=32 \
```

```
  --learning_rate=2e-5 \
  --num_train_epochs=3.0 \
  --output_dir=/tmp/mrpc_output/
```

参数就不解释了,输出结果如下:

```
***** Eval results *****
  eval_accuracy = 0.845588
  eval_loss = 0.505248
  global_step = 343
  loss = 0.505248
```

注意:你可能会看到"Running train on CPU"的提示信息,这只表示它没有运行在云 TPU 上。

训练完分类器后,可以使用 --do_predict=true 命令进行推理。在输入文件夹中需要有一个名为 test.tsv 的文件。输出将在输出文件夹中创建一个名为 test_results.tsv 的文件,每行包含每个样本的输出,列是分类概率。

命令如下:

```
export BERT_BASE_DIR=/path/to/bert/uncased_L-12_H-768_A-12
export GLUE_DIR=/path/to/glue
export TRAINED_CLASSIFIER=/path/to/fine/tuned/classifier

python run_classifier.py \
  --task_name=MRPC \
  --do_predict=true \
  --data_dir=$GLUE_DIR/MRPC \
  --vocab_file=$BERT_BASE_DIR/vocab.txt \
  --bert_config_file=$BERT_BASE_DIR/bert_config.json \
  --init_checkpoint=$TRAINED_CLASSIFIER \
  --max_seq_length=128 \
  --output_dir=/tmp/mrpc_output/
```

(2)斯坦福问答数据集 1.1 的微调和预测

首先需要下载数据集,下载地址请在 BERT 项目首页上查找,下载后放到目录 $SQUAD_DIR 中。训练命令如下:

```
python run_squad.py \
  --vocab_file=$BERT_BASE_DIR/vocab.txt \
```

```
  --bert_config_file=$BERT_BASE_DIR/bert_config.json \
  --init_checkpoint=$BERT_BASE_DIR/bert_model.ckpt \
  --do_train=True \
  --train_file=$SQUAD_DIR/train-v1.1.json \
  --do_predict=True \
  --predict_file=$SQUAD_DIR/dev-v1.1.json \
  --train_batch_size=12 \
  --learning_rate=3e-5 \
  --num_train_epochs=2.0 \
  --max_seq_length=384 \
  --doc_stride=128 \
  --output_dir=/tmp/squad_base/
```

开发集预测结果将被保存在输出目录（output_dir）下名为 predictions.json 的文件中。

执行如下命令：

```
python $SQUAD_DIR/evaluate-v1.1.py $SQUAD_DIR/dev-v1.1.json ./squad/predictions.json
```

会看到这样的输出：

```
{"f1": 88.41249612335034, "exact_match": 81.2488174077578}
```

（3）使用 BERT 提取语义特征向量

除微调模型以外，我们还可以提取出 BERT 模型学习到的语义特征向量，方法同提取嵌入向量。

使用 extract_features.py 脚本提取特征向量：

```
echo 'Who was Jim Henson ? ||| Jim Henson was a puppeteer' > /tmp/input.txt

python extract_features.py \
  --input_file=/tmp/input.txt \
  --output_file=/tmp/output.jsonl \
  --vocab_file=$BERT_BASE_DIR/vocab.txt \
  --bert_config_file=$BERT_BASE_DIR/bert_config.json \
  --init_checkpoint=$BERT_BASE_DIR/bert_model.ckpt \
  --layers=-1,-2,-3,-4 \
  --max_seq_length=128 \
  --batch_size=8
```

句子 A 和句子 B 由 "|||" 分隔，对于单句子的输入，每行一个句子，不使用分隔符。

这将创建一个 JSON 文件（每行输入占一行），其中包含由 --layers 指定的每个 Transformer 层的 BERT 激活值（−1 是 Transformer 的最后一个隐藏层，依此类推）。注意，此脚本将生成非常大的输出文件。

（4）使用 BERT 进行预训练

官方提供了在任意文本语料上做"掩码语言模型"和"预测下一句"训练的代码。首先要生成预训练数据，输入是纯文本文件，每行一个句子，由空行分隔，输出的预训练数据是 TFRecord 文件。

你可以使用现成的 NLP 工具包（如 spaCy）执行句子分段。create_pretraining_data.py 脚本将连接这些段，直到它们达到最大序列长度，以最大限度地减少填充的计算浪费。

此脚本将整个输入文件的所有样本存储在内存中，因此对于大型数据文件，则应该对输入文件进行分片并多次调用脚本（可以将文件 glob 传递给 run_pretraining.py，例如，tf_examples.tf_record*）。

max_predictions_per_seq 是每个序列的掩码语言模型预测的最大数量，应该将其设置为 max_seq_length × masked_lm_prob。

生成预训练数据的命令如下：

```
python create_pretraining_data.py \
  --input_file=./sample_text.txt \
  --output_file=/tmp/tf_examples.tfrecord \
  --vocab_file=$BERT_BASE_DIR/vocab.txt \
  --do_lower_case=True \
  --max_seq_length=128 \
  --max_predictions_per_seq=20 \
  --masked_lm_prob=0.15 \
  --random_seed=12345 \
  --dupe_factor=5
```

以下介绍如何进行预训练。如果想从头开始进行预训练，请不要包含参数 init_checkpoint。模型配置（包括词汇大小）在 bert_config_file 中指定。此演示代码训练步骤很少（只有 20 步），想提高训练步数可以修改参数 num_train_steps。传递给 run_pretraining.py 的参数 max_seq_length 和 max_predictions_per_seq 必须与 create_pretraining_data.py 相同。预训练命令如下：

```
python run_pretraining.py \
```

```
--input_file=/tmp/tf_examples.tfrecord \
--output_dir=/tmp/pretraining_output \
--do_train=True \
--do_eval=True \
--bert_config_file=$BERT_BASE_DIR/bert_config.json \
--init_checkpoint=$BERT_BASE_DIR/bert_model.ckpt \
--train_batch_size=32 \
--max_seq_length=128 \
--max_predictions_per_seq=20 \
--num_train_steps=20 \
--num_warmup_steps=10 \
--learning_rate=2e-5
```

输出如下：

```
***** Eval results *****
  global_step = 20
  loss = 0.0979674
  masked_lm_accuracy = 0.985479
  masked_lm_loss = 0.0979328
  next_sentence_accuracy = 1.0
  next_sentence_loss = 3.45724e-05
```

3.9 本章小结

本章通过彩票预测和古诗生成两个案例介绍了 RNN 和 LSTM 的使用，希望读者能够学会 RNN 和 LSTM 的使用方法，或者说体会到其使用场景的特点。当然，RNN 的应用场景还有很多，只要具有时间序列的特征就可以，比如用于声音的处理，推荐给大家一个开源项目 DeepBach，用来生成巴赫的音乐。论文地址：https://arxiv.org/pdf/1612.01010.pdf。

本章关于自然语言处理（NLP）的部分，讲解了 Seq2Seq、Transformer 和 BERT 的原理，最后通过引用 BERT 官方的应用介绍学习到了如何使用 BERT 进行微调、预测和预训练。自然语言处理的应用场景很丰富，也很有研究价值，如果你对自然语言处理感兴趣，则可以研究 Rasa 这个关于聊天机器人的开源库。

希望本章的项目能够给你带来继续探索 RNN 和 NLP 的乐趣。

4 个性化推荐系统

4.1 概述

推荐系统在日常的网络应用中无处不在，比如网上购物、新闻 App、社交网络、音乐网站、电影网站等，有人的地方就有推荐。根据个人的喜好、相同喜好人群的习惯等信息进行个性化的内容推荐。比如打开新闻类的 App，因为有了个性化的内容，每个人看到的新闻首页都是不一样的。

这当然是很有用的，在信息爆炸的今天，获取信息的途径和方式多种多样，人们花费时间最多的不再是去哪里获取信息，而是要在众多的信息中寻找自己感兴趣的，这就是信息超载问题。为了解决这个问题，便有了推荐系统。

协同过滤是推荐系统中应用较广泛的技术，该方法搜集用户的历史记录、个人喜好等信息，计算与其他用户的相似度，利用相似用户的评价来预测目标用户对特定项目的喜好程度。其优点是会给用户推荐未浏览过的项目；缺点是对于新用户来说，没有任何与商品的交互记录和个人喜好等信息，存在冷启动问题，导致模型无法找到相似的用户或商品。

为了解决冷启动的问题，通常的做法是对于刚注册的用户，要求其先选择自己感兴趣的话题、群组、商品、性格、喜欢的音乐类型等信息。

本章先对 MovieLens 数据集的结构进行分析，并讲解数据的预处理，以便在训练神经网络时使用。然后讲解推荐系统的网络模型设计思路，以及文本卷积神经网络的结构。以上理论介绍完之后，我们就开始动手实践了，通过讲解代码带领大家一步步实现网络训练和各种推荐功能。本章使用文本卷积神经网络，并结合 MovieLens（https://grouplens.org/datasets/movielens/）数据集完成电影推荐的任务。

本章的项目地址是：https://github.com/chengstone/movie_recommender，TensorFlow 1.x 的代码文件是 `movie_recommender.ipynb`，TensorFlow 2.0 的代码文件是 `movie_recommender_tf2.ipynb`。

本章的模型参考自百度 PaddlePaddle 的文章《个性化推荐》，地址是：http://paddlepaddle.org/documentation/docs/zh/1.3/beginners_guide/basics/recommender_system/index.html。

4.2 MovieLens 1M 数据集分析

4.2.1 下载数据集

首先下载数据集，地址是：http://files.grouplens.org/datasets/movielens/ml-1m.zip，并解压缩到项目当前文件夹下。

本项目使用的 MovieLens 1M 数据集，包含 6000 个用户在近 4000 部电影上的 1 亿条评论。

数据集分为三个文件：users.dat（用户数据）、movies.dat（电影数据）和 ratings.dat（评分数据）。

4.2.2 用户数据

用户数据有用户 ID、性别、年龄、职业 ID 和邮政编码字段。

数据中的格式为：UserID::Gender::Age::Occupation::Zip-code。

- 性别（Gender）："M" 表示男性，"F" 表示女性。
- 年龄（Age）从以下范围中选择。
    - 1："18 岁以下"
    - 18："18~24 岁"
    - 25："25~34 岁"
    - 35："35~44 岁"
    - 45："45~49 岁"
    - 50："50~55 岁"
    - 56："56 岁以上"
- 职业 ID 有以下选择。

- 0:"其他"或未指定
- 1:"学术/教育家"
- 2:"艺术家"
- 3:"文书/管理员"
- 4:"大学生/研究生"
- 5:"客户服务"
- 6:"医生/医疗保健"
- 7:"行政/管理"
- 8:"农民"
- 9:"家庭主妇"
- 10:"K-12 学生"（在美国，公立学校提供的基础教育统称为 K-12 教育）
- 11:"律师"
- 12:"程序员"
- 13:"退休"
- 14:"销售/营销"
- 15:"科学家"
- 16:"个体经营"
- 17:"技术人员/工程师"
- 18:"商人/工匠"
- 19:"失业人员"
- 20:"作家"

运行下面代码查看部分数据，如表 4.1 所示。

```
users_title = ['用户ID','性别','年龄','职业ID','邮政编码']
users = pd.read_table('./ml-1m/users.dat', sep='::', header=None, names=users_title
, engine = 'python')
users.head()
```

表 4.1　部分用户数据

序号	用户 ID	性别	年龄	职业 ID	邮政编码
0	1	F	1	10	48067
1	2	M	56	16	70072
2	3	M	25	15	55117
3	4	M	45	7	02460
4	5	M	25	20	55455

可以看出用户 ID、性别、年龄和职业 ID 都是类别字段，而邮政编码字段我们不使用。

4.2.3　电影数据

电影数据有电影 ID、标题和电影风格字段。

数据中的格式为：MovieID::Title::Genres。

- 标题与 IMDB 提供的标题相同（包括发行年份）。
- 电影风格有以下选择：

    动作、冒险、动画、儿童、喜剧、犯罪、纪录片、戏剧、奇幻、黑色电影（Film-Noir）、恐怖、音乐剧、神秘、浪漫、科幻、惊悚、战争、西部片。

    查看部分数据，如表 4.2 所示。

```
movies_title = ['电影ID', '标题', '电影风格']
movies = pd.read_table('./ml-1m/movies.dat', sep='::', header=None, names=
movies_title, engine = 'python')
movies.head()
```

表 4.2　部分电影数据

序号	电影 ID	标题	电影风格
0	1	Toy Story (1995)	Animation|Children's|Comedy
1	2	Jumanji (1995)	Adventure|Children's|Fantasy
2	3	Grumpier Old Men (1995)	Comedy|Romance
3	4	Waiting to Exhale (1995)	Comedy|Drama
4	5	Father of the Bride Part II (1995)	Comedy

电影 ID 是类别字段，标题字段是文本，电影风格也是类别字段。

4.2.4 评分数据

评分数据有用户 ID、电影 ID、评分和时间戳字段。

数据中的格式为：UserID::MovieID::Rating::Timestamp。

- 用户 ID 在 1 到 6040 之间。
- 电影 ID 在 1 到 3952 之间。
- 评分是以星级为单位的，共 5 颗星。
- 时间戳。
- 每个用户至少有 20 条影评。

查看部分数据，如表 4.3 所示。

```
ratings_title = ['用户ID','电影ID', '评分', '时间戳']
ratings = pd.read_table('./ml-1m/ratings.dat', sep='::', header=None, names=
ratings_title, engine = 'python')
ratings.head()
```

表 4.3 部分评分数据

序号	用户 ID	电影 ID	评分	时间戳
0	1	1193	5	978300760
1	1	661	3	978302109
2	1	914	3	978301968
3	1	3408	4	978300275
4	1	2355	5	978824291

评分字段 Rating 就是神经网络要学习的目标（Target），时间戳字段我们不使用。

4.3 数据预处理

下面是预处理要做的事情。

- 用户 ID、职业 ID 和电影 ID 不用变，可以直接传给嵌入层使用。

- 性别字段：需要将"F"和"M"转换成0和1。
- 年龄字段：年龄虽然是类别字段，但是并不连续，需要转换成7个数字0~6。
- 电影风格字段：电影风格是类别字段，要转换成数字。首先将电影风格中的所有类别转换成字符串到数字的字典，然后将每部电影的电影风格转换成数字列表，因为有些电影是多种风格的组合。
- 标题字段：其处理方式跟电影风格字段一样，首先创建文本到数字的字典，然后将标题中的描述转换成数字列表。另外，标题中的年份我们不使用，也需要去掉。
- 电影风格和标题字段需要将长度统一，这样在神经网络中方便处理。空白部分用特殊字符 <PAD> 对应的数字填充。

4.3.1 代码实现

代码实现如下：

```python
def load_data():
    # 读取用户数据集
    users_title = ['UserID', 'Gender', 'Age', 'JobID', 'Zip-code']  # 设置表格标题
    users = pd.read_table('./ml-1m/users.dat', sep='::', header=None,
names=users_title, engine = 'python')  # 读取数据到表格中
    # 过滤标题，去掉邮政编码字段
    users = users.filter(regex='UserID|Gender|Age|JobID')
    users_orig = users.values  # 得到用户数据
    # 改变用户数据中的性别和年龄
    gender_map = {'F':0, 'M':1}
    users['Gender'] = users['Gender'].map(gender_map)  # 将性别F和M映射成0和1

    # 计算年龄到序号的映射关系
    age_map = {val:ii for ii,val in enumerate(set(users['Age']))}
    users['Age'] = users['Age'].map(age_map) # 将分散的年龄字段映射成连续的数字

    # 读取电影数据集
    movies_title = ['MovieID', 'Title', 'Genres']
    movies = pd.read_table('./ml-1m/movies.dat', sep='::', header=None,
names=movies_title, engine = 'python')
    movies_orig = movies.values
```

```python
# 将标题中的年份去掉
pattern = re.compile(r'^(.*)\(((\d+)\)$')

title_map = {val:pattern.match(val).group(1) for ii,val in enumerate(set(movies
    ['Title']))} # 得到标题字段中的所有单词与序号的字典
# 将每个标题中的单词序列转换成数字序列
movies['Title'] = movies['Title'].map(title_map)


# 电影风格转数字字典，循环得到电影风格所有单词的集合
genres_set = set()
for val in movies['Genres'].str.split('|'):
    genres_set.update(val)
# 将表示空白含义的<PAD>追加到集合中
genres_set.add('<PAD>')
# 生成电影风格单词与序号的字典
genres2int = {val:ii for ii, val in enumerate(genres_set)}

# 将数据中电影风格转换成等长的数字列表，长度是18，空白由<PAD>填充
# 最右侧循环是取得每条电影数据的电影风格字段值
# 中间循环是将每个电影风格字段值拆开，因为一部电影有多种风格
# 最后就是生成电影风格的数字列表和序号的映射字典
genres_map = {val:[genres2int[row] for row in val.split('|')] for ii,val in
    enumerate(set(movies['Genres']))}
# 循环这个字典，对长度不够的字段填充<PAD>
for key in genres_map:
    for cnt in range(max(genres2int.values()) - len(genres_map[key])):
        genres_map[key].insert(len(genres_map[key]) + cnt,genres2int['<PAD>'])

movies['Genres'] = movies['Genres'].map(genres_map)


# 电影标题转数字字典
title_set = set()
# 同样的做法，循环取得所有标题的单词集合
for val in movies['Title'].str.split():
    title_set.update(val)
# 将表示空白含义的<PAD>追加到集合中
```

```python
    title_set.add('<PAD>')
    # 生成电影标题与序号的字典
    title2int = {val:ii for ii, val in enumerate(title_set)}

    # 将电影标题转换成等长的数字列表,长度是15,做法跟电影风格的处理相同
    title_count = 15
    title_map = {val:[title2int[row] for row in val.split()] for ii,val in
         enumerate(set(movies['Title']))}

    for key in title_map:
        for cnt in range(title_count - len(title_map[key])):
            title_map[key].insert(len(title_map[key]) + cnt,title2int['<PAD>'])

    movies['Title'] = movies['Title'].map(title_map)

    # 读取评分数据集
    ratings_title = ['UserID','MovieID', 'ratings', 'timestamps']
    ratings = pd.read_table('./ml-1m/ratings.dat', sep='::', header=None,
names=ratings_title, engine = 'python')
    ratings = ratings.filter(regex='UserID|MovieID|ratings')

    # 合并三个表
    data = pd.merge(pd.merge(ratings, users), movies)

    # 将数据分成X和y两个表
    target_fields = ['ratings']
    features_pd, targets_pd = data.drop(target_fields, axis=1), data[target_fields]

    features = features_pd.values
    targets_values = targets_pd.values

    return title_count, title_set, genres2int, features, targets_values, ratings, users, movies, data, movies_orig, users_orig
```

4.3.2 加载数据并保存到本地

各个变量的含义如下。

- title_count：Title 字段的长度（15）。
- title_set：Title 文本的集合。
- genres2int：电影风格转数字字典。
- features：输入 X。
- targets_values：学习目标 y。
- ratings：评分数据集的 Pandas 对象。
- users：用户数据集的 Pandas 对象。
- movies：电影数据集的 Pandas 对象。
- data：三个数据集组合在一起的 Pandas 对象。
- movies_orig：没有做数据处理的原始电影数据。
- users_orig：没有做数据处理的原始用户数据。

```
# 加载数据
title_count, title_set, genres2int, features, targets_values, ratings, users,
movies, data, movies_orig, users_orig = load_data()
# 保存到本地
pickle.dump((title_count, title_set, genres2int, features, targets_values, ratings,
users, movies, data, movies_orig, users_orig), open('preprocess.p', 'wb'))
```

预处理后的数据如表 4.4 和表 4.5 所示。

表 4.4　预处理后的部分用户数据

No.	UserID	Gender	Age	JobID
0	1	0	0	10
1	2	1	5	16
2	3	1	6	15
3	4	1	2	7
4	5	1	6	20

表 4.5 预处理后的部分电影数据

No.	MovieID	Title	Genres
0	1	[3001, 5100, 275, 275, 275, 275, 275, 275, 275, ⋯]	[3, 6, 2, 17, 17, 17, 17, 17, 17, 17, 17, 17, ⋯]
1	2	[2280, 275, 275, 275, 275, 275, 275, 275, 275, ⋯]	[16, 6, 14, 17, 17, 17, 17, 17, 17, 17, 17, ⋯]
2	3	[4339, 3338, 348, 275, 275, 275, 275, 275, 275, ⋯]	[2, 13, 17, 17, 17, 17, 17, 17, 17, 17, 17, ⋯]
3	4	[4507, 4093, 596, 275, 275, 275, 275, 275, 275, ⋯]	[2, 4, 17, 17, 17, 17, 17, 17, 17, 17, 17, ⋯]
4	5	[2123, 4479, 2698, 2221, 4495, 3997, 275, 275, ⋯]	[2, 17, 17, 17, 17, 17, 17, 17, 17, 17, 17, ⋯]

```
users.head()

movies.head()

movies.values[0]

array([1,
    list([3001, 5100, 275, 275, 275, 275, 275, 275, 275, 275, 275, 275,
    275, 275]),
    list([3, 6, 2, 17, 17, 17, 17, 17, 17, 17, 17, 17, 17, 17, 17, 17])
    ], dtype=object)
```

4.3.3 从本地读取数据

从本地读取数据，代码如下：

```
title_count,title_set,genres2int,features,targets_values,ratings,users,movies,
data, movies_orig, users_orig = pickle.load(open('preprocess.p', mode='rb'))
```

4.4 神经网络模型设计

神经网络模型结构如图 4.1 所示。

4 个性化推荐系统

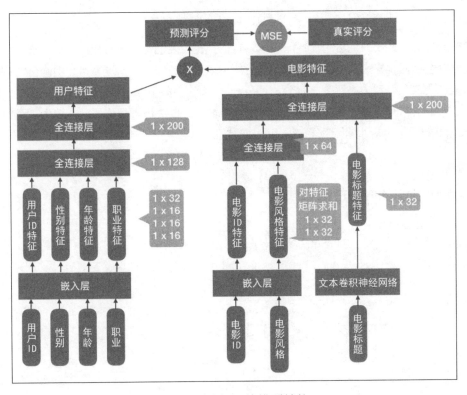

图 4.1 神经网络模型结构

通过研究数据集中的字段类型，我们发现有一些是类别字段，通常的处理是将这些字段转换成 one-hot 编码，但是像用户 ID、电影 ID 这样的字段就会变得非常稀疏，输入的维度急剧膨胀，这是我们不愿意见到的。

所以，在预处理数据时将这些字段转换成了数字，我们将这个数字当作嵌入矩阵的索引，在网络的第一层使用了嵌入层，维度是（N, 32）和（N, 16）。

对电影风格的处理要多一步，有时一部电影有多种电影风格，这样从嵌入矩阵索引出的是一个（n, 32）的矩阵，我们要将这个矩阵求和，变成（1, 32）的向量。

对电影标题的处理比较特殊，没有使用循环神经网络，而是使用了文本卷积神经网络，下文会进行说明。

从嵌入层索引出特征以后，将各特征传入全连接层，将输出再次传入全连接层，最终分别得到（1, 200）的用户特征和电影特征两个特征向量。

这里读者可能要问：对于性别、年龄字段是否一定要用嵌入层，以及全连接层和嵌入层的维度是否一定要设置成 16、32、128 和 200？这一点请不必纠结，在讨论如何选择维度和层数等超

129

参数时，有条件的话可以多试试各种不同的参数组合。

我们的目的就是要训练出用户特征和电影特征，在实现推荐功能时使用。得到这两个特征以后，就可以选择任意方式来拟合评分了。笔者使用了两种方式，一是图 4.2 中画出的将两个特征做向量点乘求和得到预测评分，将预测评分与真实评分做回归，采用 MSE 优化损失，因为本质上这是一个回归问题；另一种是将两个特征作为输入，再次传入全连接层，输出一个值作为预测评分，将输出值回归到真实评分，采用 MSE 优化损失。

4.5　文本卷积神经网络

文本卷积神经网络结构如图 4.2 所示，图片引用自 Kim Yoon 的论文：*Convolutional Neural Networks for Sentence Classification*（论文地址：https://arxiv.org/abs/1408.5882）。

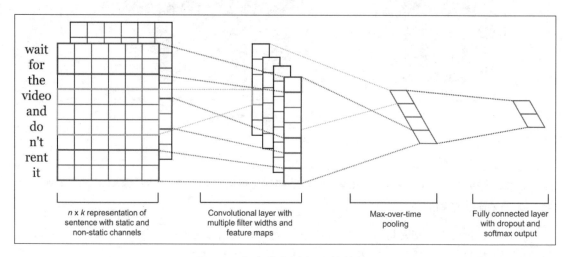

图 4.2　文本卷积神经网络结构

网络的第一层是词嵌入层，由每一个单词的嵌入向量组成的嵌入矩阵。

第二层使用多个不同尺寸（窗口大小）的卷积核在嵌入矩阵上做卷积，窗口大小指的是每次卷积覆盖几个单词。这里跟对图像做卷积不太一样，图像卷积通常用 2×2、3×3、5×5 之类的尺寸，而文本卷积要覆盖整个单词的嵌入向量，所以尺寸是（单词数，向量维度），比如每次滑动 3 个、4 个或者 5 个单词。

第三层是最大池化层，得到一个长向量，最后使用 Dropout 做正则化，最终得到了电影 Title 的特征。

4.6 实现电影推荐：基于 TensorFlow 1.x

网络结构和思想都介绍完了，现在我们来实现它。

4.6.1 构建计算图

构建计算图，代码实现如下：

```
import tensorflow as tf
import os
import pickle


def save_params(params):
    """
    保存参数到本地文件中
    """
    pickle.dump(params, open('params.p', 'wb'))


def load_params():
    """
    从本地文件中加载参数
    """
    return pickle.load(open('params.p', mode='rb'))

# 嵌入向量的维度
embed_dim = 32
# 用户ID个数
uid_max = max(features.take(0,1)) + 1 # 6040
# 性别个数
gender_max = max(features.take(2,1)) + 1 # 1 + 1 = 2
# 年龄类别个数
age_max = max(features.take(3,1)) + 1 # 6 + 1 = 7
# 职业个数
job_max = max(features.take(4,1)) + 1# 20 + 1 = 21


# 电影ID个数
```

```python
movie_id_max = max(features.take(1,1)) + 1 # 3952
# 电影风格个数
movie_categories_max = max(genres2int.values()) + 1 # 18 + 1 = 19
# 电影标题单词个数
movie_title_max = len(title_set) # 5216

# 对电影风格嵌入向量做加和操作的标志
combiner = "sum"

# 电影标题长度
sentences_size = title_count # = 15
# 文本卷积滑动窗口,分别滑动2, 3, 4, 5个单词
window_sizes = {2, 3, 4, 5}
# 文本卷积核数量
filter_num = 8

# 电影ID转下标的字典,在数据集中电影ID跟下标不一致,比如第5行的数据电影ID不一定是5
movieid2idx = {val[0]:i for i, val in enumerate(movies.values)}
```

超参数如下:

```python
# 训练次数
num_epochs = 5
# 批量大小
batch_size = 256
# 丢弃率
dropout_keep = 0.5
# 学习率
learning_rate = 0.0001
# 设置每处理N个批量数据显示一次统计信息
show_every_n_batches = 20
# 训练好的模型文件存放位置
save_dir = './save'
```

定义输入占位符:

```python
def get_inputs():
    # 用户ID
```

```python
uid = tf.placeholder(tf.int32, [None, 1], name="uid")
# 性别
user_gender = tf.placeholder(tf.int32, [None, 1], name="user_gender")
# 年龄
user_age = tf.placeholder(tf.int32, [None, 1], name="user_age")
# 职业
user_job = tf.placeholder(tf.int32, [None, 1], name="user_job")

# 电影ID
movie_id = tf.placeholder(tf.int32, [None, 1], name="movie_id")
# 电影风格
movie_categories=tf.placeholder(tf.int32, [None, 18], name="movie_categories")
# 电影标题
movie_titles = tf.placeholder(tf.int32, [None, 15], name="movie_titles")

# 学习目标
targets = tf.placeholder(tf.int32, [None, 1], name="targets")
# 学习率
LearningRate = tf.placeholder(tf.float32, name = "LearningRate")
# 丢弃率
dropout_keep_prob = tf.placeholder(tf.float32, name = "dropout_keep_prob")
return uid, user_gender, user_age, user_job, movie_id, movie_categories,
movie_titles, targets, LearningRate, dropout_keep_prob
```

定义用户的嵌入矩阵:

```python
def get_user_embedding(uid, user_gender, user_age, user_job):
    with tf.name_scope("user_embedding"):
        # embed_dim是嵌入向量的维度,我们设置为32。uid_max是用户ID的总个数
        uid_embed_matrix=tf.Variable(tf.random_uniform([uid_max,embed_dim], -1, 1),
                                     name = "uid_embed_matrix")
        # 定义用户ID的嵌入矩阵
        uid_embed_layer = tf.nn.embedding_lookup(uid_embed_matrix, uid,
                                                 name = "uid_embed_layer")

        # 对性别的处理跟用户ID处理是一样的,(2, 16)的矩阵
        gender_embed_matrix = tf.Variable(
```

```
            tf.random_uniform([gender_max, embed_dim // 2], -1, 1),
                name= "gender_embed_matrix")
        gender_embed_layer = tf.nn.embedding_lookup(gender_embed_matrix,
user_gender, name = "gender_embed_layer")

        # 对年龄也做同样的处理，（7，16）的矩阵
        age_embed_matrix = tf.Variable(
            tf.random_uniform([age_max, embed_dim // 2], -1, 1), name="
            age_embed_matrix")
        age_embed_layer = tf.nn.embedding_lookup(age_embed_matrix, user_age,
                                        name="age_embed_layer")

        # 对职业的处理也相同，（21，16）的矩阵
        job_embed_matrix = tf.Variable(tf.random_uniform([job_max, embed_dim // 2],
-1, 1), name = "job_embed_matrix")
        job_embed_layer = tf.nn.embedding_lookup(job_embed_matrix, user_job,
                                        name = "job_embed_layer")
    return uid_embed_layer, gender_embed_layer, age_embed_layer, job_embed_layer
```

将用户的嵌入矩阵一起全连接生成用户的特征：

```
def get_user_feature_layer(uid_embed_layer, gender_embed_layer, age_embed_layer,
job_embed_layer):
    with tf.name_scope("user_fc"):
        # 第一层全连接层
        # 将用户ID、性别、年龄和职业的嵌入向量传给全连接层，输出维度都是32
        uid_fc_layer = tf.layers.dense(uid_embed_layer, embed_dim,
                                name = "uid_fc_layer",
                                activation=tf.nn.relu)
        gender_fc_layer = tf.layers.dense(gender_embed_layer, embed_dim,
                                    name = "gender_fc_layer",
                                    activation=tf.nn.relu)
        age_fc_layer = tf.layers.dense(age_embed_layer, embed_dim,
                                name ="age_fc_layer",
                                activation=tf.nn.relu)
        job_fc_layer = tf.layers.dense(job_embed_layer, embed_dim,
                                name = "job_fc_layer",
```

```
                                    activation=tf.nn.relu)

    # 第二层全连接层
    # 将第一层全连接层的输出连接在一起，再次传给全连接层，输出维度是200
    user_combine_layer = tf.concat([uid_fc_layer, gender_fc_layer,
                                   age_fc_layer, job_fc_layer], 2)  #(?, 1,
                                   128)
    user_combine_layer=tf.contrib.layers.fully_connected(user_combine_layer,
200, tf.tanh)  #(?, 1, 200)

    user_combine_layer_flat = tf.reshape(user_combine_layer, [-1, 200])
    return user_combine_layer, user_combine_layer_flat
```

定义电影ID的嵌入矩阵：

```
def get_movie_id_embed_layer(movie_id):
    with tf.name_scope("movie_embedding"):
        # 对电影ID的处理跟用户ID处理是一样的，（3952, 32）的矩阵
        movie_id_embed_matrix = tf.Variable(
            tf.random_uniform([movie_id_max, embed_dim], -1, 1),
            name = "movie_id_embed_matrix")
        movie_id_embed_layer = tf.nn.embedding_lookup(movie_id_embed_matrix,
movie_id, name = "movie_id_embed_layer")
    return movie_id_embed_layer
```

对电影风格的多个嵌入向量做加和：

```
def get_movie_categories_layers(movie_categories):
    with tf.name_scope("movie_categories_layers"):
        # 首先定义电影风格的嵌入矩阵，（19, 32）
        movie_categories_embed_matrix = tf.Variable(
            tf.random_uniform([movie_categories_max, embed_dim], -1, 1),
            name = "movie_categories_embed_matrix")
        movie_categories_embed_layer = tf.nn.embedding_lookup(
            movie_categories_embed_matrix, movie_categories,
            name = "movie_categories_embed_layer")
        # 因为每部电影的风格有可能是多种组合，所以从嵌入矩阵输出的是多个向量
        # 这里对多个嵌入向量求和，当然你也可以尝试求平均值
```

```python
        if combiner == "sum":
            movie_categories_embed_layer = tf.reduce_sum(
                                             movie_categories_embed_layer,
                                             axis=1, keepdims=True)

    return movie_categories_embed_layer
```

电影标题的文本卷积神经网络实现：

```python
def get_movie_cnn_layer(movie_titles):
    # 从嵌入矩阵中得到电影标题对应的各个单词的嵌入向量
    with tf.name_scope("movie_embedding"):
        # 电影标题的嵌入矩阵维度是（5216, 32）
        movie_title_embed_matrix = tf.Variable(
            tf.random_uniform([movie_title_max, embed_dim], -1, 1),
            name = "movie_title_embed_matrix")
        movie_title_embed_layer = tf.nn.embedding_lookup(movie_title_embed_matrix,
                                                        movie_titles,
                                                        name = "
                                                        movie_title_embed_layer")
        movie_title_embed_layer_expand=tf.expand_dims(movie_title_embed_layer, -1)

    # 对文本嵌入层使用不同尺寸的卷积核做卷积和最大池化
    pool_layer_lst = []
    for window_size in window_sizes:
        with tf.name_scope("movie_txt_conv_maxpool_{}".format(window_size)):
            # 前文说过，因为文本卷积要覆盖整个单词的嵌入向量，所以尺寸是（窗口大
            # 小，向量维度32）
            # 文本卷积核数量filter_num是8
            filter_weights = tf.Variable(
                tf.truncated_normal([window_size, embed_dim, 1, filter_num],
                                    stddev=0.1), name = "filter_weights")
            filter_bias = tf.Variable(tf.constant(0.1, shape=[filter_num]),
                                      name="filter_bias")
            # 卷积
            conv_layer = tf.nn.conv2d(movie_title_embed_layer_expand,
                                      filter_weights,
```

```python
                            [1,1,1,1], padding="VALID", name="conv_layer")
            relu_layer = tf.nn.relu(tf.nn.bias_add(conv_layer,filter_bias),
                            name ="relu_layer")
            # 最大池化
            maxpool_layer = tf.nn.max_pool(relu_layer,
                            [1,sentences_size - window_size + 1 ,1,1],
                            [1,1,1,1], padding="VALID", name="maxpool_layer")
            pool_layer_lst.append(maxpool_layer)

    # Dropout层
    with tf.name_scope("pool_dropout"):
        pool_layer = tf.concat(pool_layer_lst, 3, name ="pool_layer")
        # 每一个窗口都有filter_num（卷积核数量）个结果，所以最终是（窗口个数*卷积核
        # 数量）个结果
        max_num = len(window_sizes) * filter_num
        pool_layer_flat = tf.reshape(pool_layer , [-1, 1, max_num],
                            name = "pool_layer_flat")

        dropout_layer = tf.nn.dropout(pool_layer_flat, dropout_keep_prob,
                            name = "dropout_layer")
    return pool_layer_flat, dropout_layer
```

将电影的各个层一起做全连接：

```python
def get_movie_feature_layer(movie_id_embed_layer, movie_categories_embed_layer,
dropout_layer):
    with tf.name_scope("movie_fc"):
        # 第一层全连接层
        # 将电影ID和电影风格的嵌入向量传入全连接层，输出维度都是32
        movie_id_fc_layer = tf.layers.dense(movie_id_embed_layer, embed_dim,
                                    name = "movie_id_fc_layer",
                                    activation=tf.nn.relu)
        movie_categories_fc_layer = tf.layers.dense(movie_categories_embed_layer,
                            embed_dim,
                            name = "movie_categories_fc_layer",
                            activation=tf.nn.relu)
```

```python
    # 第二层全连接层
    # 将第一层全连接层的输出和文本卷积的输出一起传入全连接层，输出维度是200
    movie_combine_layer = tf.concat([movie_id_fc_layer,
                                     movie_categories_fc_layer,
                                     dropout_layer], 2)  #(?, 1, 96)
    movie_combine_layer=tf.contrib.layers.fully_connected(movie_combine_layer,
200, tf.tanh)  #(?, 1, 200)

    movie_combine_layer_flat = tf.reshape(movie_combine_layer, [-1, 200])
    return movie_combine_layer, movie_combine_layer_flat
```

构建计算图，代码实现如下：

```python
tf.reset_default_graph()
train_graph = tf.Graph()
with train_graph.as_default():
    # 获取输入占位符
    uid, user_gender, user_age, user_job, movie_id, movie_categories, movie_titles,
targets, lr, dropout_keep_prob = get_inputs()
    # 获取用户的4个嵌入向量
    uid_embed_layer, gender_embed_layer, age_embed_layer,
job_embed_layer = get_user_embedding(uid, user_gender, user_age, user_job)
    # 得到用户特征
    user_combine_layer, user_combine_layer_flat = get_user_feature_layer(
uid_embed_layer, gender_embed_layer, age_embed_layer, job_embed_layer)
    # 获取电影ID的嵌入向量
    movie_id_embed_layer = get_movie_id_embed_layer(movie_id)
    # 获取电影风格的嵌入向量
    movie_categories_embed_layer = get_movie_categories_layers(movie_categories)
    # 获取电影标题的特征向量
    pool_layer_flat, dropout_layer = get_movie_cnn_layer(movie_titles)
    # 得到电影特征
    movie_combine_layer, movie_combine_layer_flat = get_movie_feature_layer(
movie_id_embed_layer, movie_categories_embed_layer, dropout_layer)
    # 计算出评分，要注意两个不同的方案，inference的名字（name值）是不一样的
    # 后面做推荐时要根据name获取张量
    with tf.name_scope("inference"):
```

```
#         将用户特征和电影特征作为输入，经过全连接，输出一个值的方案
#         inference_layer = tf.concat([user_combine_layer_flat,
# movie_combine_layer_flat], 1)
#         inference = tf.layers.dense(inference_layer, 1,
#             kernel_initializer=tf.truncated_normal_initializer(stddev=0.01),
#             kernel_regularizer=tf.nn.l2_loss, name="inference")
        # 简单地将用户特征和电影特征做矩阵点乘求和得到一个预测评分
        inference = tf.reduce_sum(user_combine_layer_flat * movie_combine_layer_flat,axis=1)
        inference = tf.expand_dims(inference, axis=1)

    with tf.name_scope("loss"):
        # MSE损失，将计算值回归到评分
        cost = tf.losses.mean_squared_error(targets, inference )
        loss = tf.reduce_mean(cost)
    # 优化损失
    global_step = tf.Variable(0, name="global_step", trainable=False)
    optimizer = tf.train.AdamOptimizer(lr)
    gradients = optimizer.compute_gradients(loss)
    train_op = optimizer.apply_gradients(gradients, global_step=global_step)
```

迭代批量数据：

```
def get_batches(Xs, ys, batch_size):
    for start in range(0, len(Xs), batch_size):
        end = min(start + batch_size, len(Xs))
        yield Xs[start:end], ys[start:end]
```

4.6.2 训练网络

训练网络，代码实现如下：

```
%matplotlib inline
%config InlineBackend.figure_format = 'retina'
import matplotlib.pyplot as plt
import time
import datetime
```

```python
losses = {'train':[], 'test':[]}

with tf.Session(graph=train_graph) as sess:

    # 收集数据给TensorBoard用
    grad_summaries = []
    for g, v in gradients:
        if g is not None:
            grad_hist_summary = tf.summary.histogram("{}/grad/hist".format(
                v.name.replace(':', '_')), g)
            sparsity_summary = tf.summary.scalar("{}/grad/sparsity".format(
                v.name.replace(':', '_')), tf.nn.zero_fraction(g))
            grad_summaries.append(grad_hist_summary)
            grad_summaries.append(sparsity_summary)
    grad_summaries_merged = tf.summary.merge(grad_summaries)

    # 摘要的输出目录
    timestamp = str(int(time.time()))
    out_dir = os.path.abspath(os.path.join(os.path.curdir, "runs", timestamp))
    print("Writing to {}\n".format(out_dir))

    # 损失的摘要
    loss_summary = tf.summary.scalar("loss", loss)

    # 训练摘要
    train_summary_op = tf.summary.merge([loss_summary, grad_summaries_merged])
    train_summary_dir = os.path.join(out_dir, "summaries", "train")
    train_summary_writer = tf.summary.FileWriter(train_summary_dir, sess.graph)

    # 推理摘要
    inference_summary_op = tf.summary.merge([loss_summary])
    inference_summary_dir = os.path.join(out_dir, "summaries", "inference")
    inference_summary_writer = tf.summary.FileWriter(inference_summary_dir, sess.graph)
```

```python
sess.run(tf.global_variables_initializer())
saver = tf.train.Saver()
for epoch_i in range(num_epochs):

    # 将数据集分成训练集和测试集,随机种子不固定
    train_X,test_X, train_y, test_y = train_test_split(features,
                                                       targets_values,
                                                       test_size = 0.2,
                                                       random_state = 0)

    train_batches = get_batches(train_X, train_y, batch_size)
    test_batches = get_batches(test_X, test_y, batch_size)

    # 训练的迭代,保存训练损失
    for batch_i in range(len(train_X) // batch_size):
        # 取得批量数据
        x, y = next(train_batches)

        # 准备传入给计算图的数据
        categories = np.zeros([batch_size, 18])
        for i in range(batch_size):
            categories[i] = x.take(6,1)[i]

        titles = np.zeros([batch_size, sentences_size])
        for i in range(batch_size):
            titles[i] = x.take(5,1)[i]

        feed = {
            uid: np.reshape(x.take(0,1), [batch_size, 1]),
            user_gender: np.reshape(x.take(2,1), [batch_size, 1]),
            user_age: np.reshape(x.take(3,1), [batch_size, 1]),
            user_job: np.reshape(x.take(4,1), [batch_size, 1]),
            movie_id: np.reshape(x.take(1,1), [batch_size, 1]),
            movie_categories: categories,
            movie_titles: titles,
            targets: np.reshape(y, [batch_size, 1]),
```

```python
                    dropout_keep_prob: dropout_keep,
                    lr: learning_rate}
            # 开始训练
            step, train_loss, summaries, _ = sess.run([global_step, loss,
train_summary_op, train_op], feed)
            # 保存损失
            losses['train'].append(train_loss)
            # 保存摘要给TensorBoard用
            train_summary_writer.add_summary(summaries, step)

            # 每处理show_every_n_batches个批量数据显示一次统计信息
            if (epoch_i * (len(train_X) // batch_size) + batch_i) %
show_every_n_batches == 0:
                time_str = datetime.datetime.now().isoformat()
                print('{}: Epoch {:>3} Batch {:>4}/{}   train_loss = {:.3f}'.format(
                    time_str,
                    epoch_i,
                    batch_i,
                    (len(train_X) // batch_size),
                    train_loss))

        # 使用测试数据的迭代,处理同上
        for batch_i  in range(len(test_X) // batch_size):
            x, y = next(test_batches)

            categories = np.zeros([batch_size, 18])
            for i in range(batch_size):
                categories[i] = x.take(6,1)[i]

            titles = np.zeros([batch_size, sentences_size])
            for i in range(batch_size):
                titles[i] = x.take(5,1)[i]

            feed = {
                uid: np.reshape(x.take(0,1), [batch_size, 1]),
```

```
                    user_gender: np.reshape(x.take(2,1), [batch_size, 1]),
                    user_age: np.reshape(x.take(3,1), [batch_size, 1]),
                    user_job: np.reshape(x.take(4,1), [batch_size, 1]),
                    movie_id: np.reshape(x.take(1,1), [batch_size, 1]),
                    movie_categories: categories,
                    movie_titles: titles,
                    targets: np.reshape(y, [batch_size, 1]),
                    dropout_keep_prob: 1,
                    lr: learning_rate}

            step, test_loss, summaries = sess.run([global_step, loss,
inference_summary_op], feed)

            # 保存测试损失
            losses['test'].append(test_loss)
            inference_summary_writer.add_summary(summaries, step)

            time_str = datetime.datetime.now().isoformat()
            if (epoch_i * (len(test_X) // batch_size) + batch_i) %
show_every_n_batches == 0:
                print('{}: Epoch {:>3} Batch {:>4}/{}   test_loss = {:.3f}'.format(
                    time_str,
                    epoch_i,
                    batch_i,
                    (len(test_X) // batch_size),
                    test_loss))

    # 保存模型
    saver.save(sess, save_dir)
    print('Model Trained and Saved')
```

在 TensorBoard 中查看可视化结果，如图 4.3 所示。

```
tensorboard --logdir /PATH_TO_CODE/runs/1513402825/summaries/
```

保存参数 save_dir，在生成预测评分时使用。

```
save_params((save_dir))
```

```
load_dir = load_params()
```

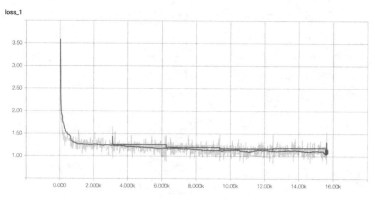

图 4.3　可视化结果

4.6.3　实现个性化推荐

网络训练好之后,就可以进行预测了。

使用 get_tensor_by_name() 函数从 `loaded_graph` 中获取张量,后面的推荐功能要用到。

```
def get_tensors(loaded_graph):

    uid = loaded_graph.get_tensor_by_name("uid:0")
    user_gender = loaded_graph.get_tensor_by_name("user_gender:0")
    user_age = loaded_graph.get_tensor_by_name("user_age:0")
    user_job = loaded_graph.get_tensor_by_name("user_job:0")
    movie_id = loaded_graph.get_tensor_by_name("movie_id:0")
    movie_categories = loaded_graph.get_tensor_by_name("movie_categories:0")
    movie_titles = loaded_graph.get_tensor_by_name("movie_titles:0")
    targets = loaded_graph.get_tensor_by_name("targets:0")
    dropout_keep_prob = loaded_graph.get_tensor_by_name("dropout_keep_prob:0")
    lr = loaded_graph.get_tensor_by_name("LearningRate:0")
    # 计算预测评分的两种不同方案使用不同的name获取张量inference
#     inference = loaded_graph.get_tensor_by_name("inference/inference/BiasAdd:0")
    inference = loaded_graph.get_tensor_by_name("inference/ExpandDims:0")
    movie_combine_layer_flat=loaded_graph.get_tensor_by_name("movie_fc/Reshape:0")
```

```
        user_combine_layer_flat = loaded_graph.get_tensor_by_name("user_fc/Reshape:0")
    return uid, user_gender, user_age, user_job, movie_id, movie_categories,
            movie_titles, targets, lr, dropout_keep_prob, inference,
            movie_combine_layer_flat, user_combine_layer_flat
```

1. 指定用户和电影进行评分

这部分就是对网络做正向传播，计算得到预测评分。

```
def rating_movie(user_id_val, movie_id_val):
    loaded_graph = tf.Graph()
    with tf.Session(graph=loaded_graph) as sess:
        # 加载模型
        loader = tf.train.import_meta_graph(load_dir + '.meta')
        loader.restore(sess, load_dir)

        # 从加载的模型中获取张量
        uid, user_gender, user_age, user_job, movie_id, movie_categories,
        movie_titles, targets, lr, dropout_keep_prob, inference,_, __ = get_tensors
(loaded_graph)

        categories = np.zeros([1, 18])
        categories[0] = movies.values[movieid2idx[movie_id_val]][2]

        titles = np.zeros([1, sentences_size])
        titles[0] = movies.values[movieid2idx[movie_id_val]][1]

        feed = {
            uid: np.reshape(users.values[user_id_val-1][0], [1, 1]),
            user_gender: np.reshape(users.values[user_id_val-1][1], [1, 1]),
            user_age: np.reshape(users.values[user_id_val-1][2], [1, 1]),
            user_job: np.reshape(users.values[user_id_val-1][3], [1, 1]),
            movie_id: np.reshape(movies.values[movieid2idx[movie_id_val]][0], [1,
                1]),
            movie_categories: categories,
            movie_titles: titles,
            dropout_keep_prob: 1}
```

```
        # 预测
        inference_val = sess.run([inference], feed)

        return (inference_val)
```

现在来试一下，给定用户 ID 234、电影 ID 1401，得到评分 4.27963877。

```
rating_movie(234, 1401)

[array([[ 4.27963877]], dtype=float32)]
```

2. 生成电影特征矩阵和用户特征矩阵

（1）生成电影特征矩阵

用训练好的模型将每一部电影的特征向量提取出来，然后将电影特征矩阵保存到本地。

```
loaded_graph = tf.Graph()
movie_matrics = []
with tf.Session(graph=loaded_graph) as sess:
    # 加载模型
    loader = tf.train.import_meta_graph(load_dir + '.meta')
    loader.restore(sess, load_dir)

    # 从加载的模型中获取张量
    uid, user_gender, user_age, user_job, movie_id, movie_categories, movie_titles,
     targets, lr, dropout_keep_prob, _, movie_combine_layer_flat, __ = get_tensors(
loaded_graph)

    # 循环遍历每一部电影
    for item in movies.values:
        categories = np.zeros([1, 18])
        categories[0] = item.take(2)

        titles = np.zeros([1, sentences_size])
        titles[0] = item.take(1)

        feed = {
```

```python
            movie_id: np.reshape(item.take(0), [1, 1]),
            movie_categories: categories,
            movie_titles: titles,
            dropout_keep_prob: 1}
        # 得到电影特征
        movie_combine_layer_flat_val = sess.run([movie_combine_layer_flat], feed)
        movie_matrics.append(movie_combine_layer_flat_val)
# 保存电影特征矩阵到本地
pickle.dump((np.array(movie_matrics).reshape(-1, 200)), open('movie_matrics.p', 'wb'))
movie_matrics = pickle.load(open('movie_matrics.p', mode='rb'))


movie_matrics = pickle.load(open('movie_matrics.p', mode='rb'))
```

（2）生成用户特征矩阵

将训练好的用户特征提取出来，然后将用户特征矩阵保存到本地。

```python
loaded_graph = tf.Graph()
users_matrics = []
with tf.Session(graph=loaded_graph) as sess:
    # 加载模型
    loader = tf.train.import_meta_graph(load_dir + '.meta')
    loader.restore(sess, load_dir)

    # 从加载的模型中获取张量
    uid, user_gender, user_age, user_job, movie_id, movie_categories, movie_titles,
     targets, lr, dropout_keep_prob, _, __,user_combine_layer_flat = get_tensors(
loaded_graph)

    for item in users.values:

        feed = {
            uid: np.reshape(item.take(0), [1, 1]),
            user_gender: np.reshape(item.take(1), [1, 1]),
            user_age: np.reshape(item.take(2), [1, 1]),
            user_job: np.reshape(item.take(3), [1, 1]),
            dropout_keep_prob: 1}
```

```
        # 得到用户特征
        user_combine_layer_flat_val = sess.run([user_combine_layer_flat], feed)
        users_matrics.append(user_combine_layer_flat_val)
# 保存用户特征矩阵到本地
pickle.dump((np.array(users_matrics).reshape(-1, 200)), open('users_matrics.p', 'wb'))
users_matrics = pickle.load(open('users_matrics.p', mode='rb'))

users_matrics = pickle.load(open('users_matrics.p', mode='rb'))
```

3. 开始推荐电影

使用所生成的用户特征矩阵和电影特征矩阵做电影推荐。

（1）推荐相同风格的电影

推荐相同风格的电影，思路是计算当前看的电影特征向量与整个电影特征矩阵的余弦相似度，取相似度最大的 top_k 个，这里加入了随机选择，保证每次的推荐稍稍有些不同。

```
def recommend_same_type_movie(movie_id_val, top_k = 20):

    loaded_graph = tf.Graph()
    with tf.Session(graph=loaded_graph) as sess:
        # 加载模型
        loader = tf.train.import_meta_graph(load_dir + '.meta')
        loader.restore(sess, load_dir)

        norm_movie_matrics = tf.sqrt(tf.reduce_sum(tf.square(movie_matrics), 1,
                            keepdims=True))
        normalized_movie_matrics = movie_matrics / norm_movie_matrics

        # 推荐相同风格的电影
        # 先根据电影ID从电影特征矩阵中索引指定电影的特征
        probs_embeddings = (movie_matrics[movieid2idx[movie_id_val]]).reshape([1,
                            200])
        # 计算跟所有电影的相似度
```

```python
        probs_similarity = tf.matmul(probs_embeddings, tf.transpose(
normalized_movie_matrics))
        sim = (probs_similarity.eval())
        # results = (-sim[0]).argsort()[0:top_k]
        # print(results)

        print("您看的电影是: {}".format(movies_orig[movieid2idx[movie_id_val]]))
        print("以下是给您的推荐: ")
        p = np.squeeze(sim)
        # 先排序，然后将top_k以外的相似度清零
        p[np.argsort(p)[:-top_k]] = 0
        p = p / np.sum(p)
        results = set()
        # 从top_k中随机选5个，3883是电影数据的个数
        while len(results) != 5:
            c = np.random.choice(3883, 1, p=p)[0]
            results.add(c)
        for val in (results):
            print(val)
            print(movies_orig[val])

        return results

recommend_same_type_movie(1401, 20)

您看的电影是: [1401 'Ghosts of Mississippi (1996)' 'Drama']
以下是给您的推荐:
3385
[3454 'Whatever It Takes (2000)' 'Comedy|Romance']
707
[716 'Switchblade Sisters (1975)' 'Crime']
2351
[2420 'Karate Kid, The (1984)' 'Drama']
2189
[2258 'Master Ninja I (1984)' 'Action']
2191
```

```
[2260 'Wisdom (1986)' 'Action|Crime']
Out[60]:
{707, 2189, 2191, 2351, 3385}
```

(2)推荐您喜欢的电影

推荐您喜欢的电影,思路是使用用户特征向量与电影特征矩阵计算所有电影的评分,取评分最高的top_k个,同样加入了随机选择。

```
def recommend_your_favorite_movie(user_id_val, top_k = 10):

    loaded_graph = tf.Graph()
    with tf.Session(graph=loaded_graph) as sess:
        # 加载模型
        loader = tf.train.import_meta_graph(load_dir + '.meta')
        loader.restore(sess, load_dir)

        # 推荐您喜欢的电影
        # 先根据用户ID取得用户特征
        probs_embeddings = (users_matrics[user_id_val-1]).reshape([1, 200])
        # 计算用户对所有电影的评分
        probs_similarity = tf.matmul(probs_embeddings, tf.transpose(movie_matrics))
        sim = (probs_similarity.eval())
        # print(sim.shape)
        # results = (-sim[0]).argsort()[0:top_k]
        # print(results)

        # sim_norm = probs_norm_similarity.eval()
        # print((-sim_norm[0]).argsort()[0:top_k])

        print("以下是给您的推荐:")
        p = np.squeeze(sim)
        # 留下评分最高的top_k个
        p[np.argsort(p)[:-top_k]] = 0
        p = p / np.sum(p)
        results = set()
        while len(results) != 5:
            c = np.random.choice(3883, 1, p=p)[0]
```

```
            results.add(c)
        for val in (results):
            print(val)
            print(movies_orig[val])
            
        return results

recommend_your_favorite_movie(234, 10)
```

以下是给您的推荐：
1642
[1688 'Anastasia (1997)' "Animation|Children's|Musical"]
994
[1007 'Apple Dumpling Gang, The (1975)' "Children's|Comedy|Western"]
667
[673 'Space Jam (1996)' "Adventure|Animation|Children's|Comedy|Fantasy"]
1812
[1881 'Quest for Camelot (1998)' "Adventure|Animation|Children's|Fantasy"]
1898
[1967 'Labyrinth (1986)' "Adventure|Children's|Fantasy"]
Out[62]:
{667, 994, 1642, 1812, 1898}

（3）看过这部电影的人还看了（喜欢）哪些电影

看过这部电影的人还看了（喜欢）哪些电影，思路如下：

- 为了得到"看过这部电影的人"，首先要找出喜欢这部电影的 top_k 个人，得到这几个人的用户特征向量。如何找呢？实际上就是计算所有人对这部电影的评分，评分最高的 top_k 个人就是我们要找的。
- 然后计算这几个人对所有电影的评分。
- 选择每个人评分最高的电影进行推荐。
- 同样加入了随机选择。

```
import random

def recommend_other_favorite_movie(movie_id_val, top_k = 20):
    loaded_graph = tf.Graph()
```

```python
with tf.Session(graph=loaded_graph) as sess:
    # 加载模型
    loader = tf.train.import_meta_graph(load_dir + '.meta')
    loader.restore(sess, load_dir)
    # 根据电影ID得到这部电影的特征
    probs_movie_embeddings = (movie_matrics[movieid2idx[movie_id_val]]).reshape
        ([1, 200])
    # 计算所有人对这部电影的评分
    probs_user_favorite_similarity = tf.matmul(probs_movie_embeddings,
                                              tf.transpose(users_matrics))
    # 得到top_k个人的用户ID
    favorite_user_id = np.argsort(probs_user_favorite_similarity.eval())[0][-
        top_k:]
    # print(normalized_users_matrics.eval().shape)
    # print(probs_user_favorite_similarity.eval()[0][favorite_user_id])
    # print(favorite_user_id.shape)

    print("您看的电影是：{}".format(movies_orig[movieid2idx[movie_id_val]]))

    print("喜欢看这部电影的人是：{}".format(users_orig[favorite_user_id-1]))
    # 得到这几个人的用户特征
    probs_users_embeddings = (users_matrics[favorite_user_id-1]).reshape([-1,
        200])
    # 计算这几个人对所有电影的评分
    probs_similarity = tf.matmul(probs_users_embeddings, tf.transpose(
        movie_matrics))
    sim = (probs_similarity.eval())
    # results = (-sim[0]).argsort()[0:top_k]
    # print(results)

    # print(sim.shape)
    # print(np.argmax(sim, 1))

    # 每个人最喜欢看的电影
    p = np.argmax(sim, 1)
    print("喜欢看这部电影的人还喜欢看：")
```

```python
        # 从top_k中随机选5部电影进行推荐
        results = set()
        while len(results) != 5:
            c = p[random.randrange(top_k)]
            results.add(c)
        for val in (results):
            print(val)
            print(movies_orig[val])

        return results

recommend_other_favorite_movie(1401, 20)
```

您看的电影是：[1401 'Ghosts of Mississippi (1996)' 'Drama']
喜欢看这部电影的人是：[[5782 'F' 35 0]
 [5767 'M' 25 2]
 [3936 'F' 35 12]
 [3595 'M' 25 0]
 [1696 'M' 35 7]
 [2728 'M' 35 12]
 [763 'M' 18 10]
 [4404 'M' 25 1]
 [3901 'M' 18 14]
 [371 'M' 18 4]
 [1855 'M' 18 4]
 [2338 'M' 45 17]
 [450 'M' 45 1]
 [1130 'M' 18 7]
 [3035 'F' 25 7]
 [100 'M' 35 17]
 [567 'M' 35 20]
 [5861 'F' 50 1]
 [4800 'M' 18 4]
 [3281 'M' 25 17]]
喜欢看这部电影的人还喜欢看：

```
1779
[1848 'Borrowers, The (1997)' "Adventure|Children's|Comedy|Fantasy"]
1244
[1264 'Diva (1981)' 'Action|Drama|Mystery|Romance|Thriller']
1812
[1881 'Quest for Camelot (1998)' "Adventure|Animation|Children's|Fantasy"]
1742
[1805 'Wild Things (1998)' 'Crime|Drama|Mystery|Thriller']
2535
[2604 'Let it Come Down: The Life of Paul Bowles (1998)' 'Documentary']
Out[64]:
{1244, 1742, 1779, 1812, 2535}
```

4.7 实现电影推荐：基于 TensorFlow 2.0

4.7.1 构建模型

现在使用 TensorFlow 2.0 实现电影推荐。本节的模型有些特别，因为有多个输入（7个），但是不要紧，构建步骤跟之前是一样的，本节使用函数式方式构建模型，我们来看代码实现。

首先定义输入占位符，跟上一节没有区别，只是换成了 Keras 的函数而已。

```
def get_inputs():
    # 用户ID
    uid = tf.keras.layers.Input(shape=(1,), dtype='int32', name='uid')
    # 性别
    user_gender = tf.keras.layers.Input(shape=(1,), dtype='int32', name='user_gender')
    # 年龄
    user_age = tf.keras.layers.Input(shape=(1,), dtype='int32', name='user_age')
    # 职业
    user_job = tf.keras.layers.Input(shape=(1,), dtype='int32', name='user_job')

    # 电影ID
    movie_id = tf.keras.layers.Input(shape=(1,), dtype='int32', name='movie_id')
    # 电影风格
```

```
    movie_categories = tf.keras.layers.Input(shape=(18,), dtype='int32',
                                    name='movie_categories')
    # 电影标题
    movie_titles = tf.keras.layers.Input(shape=(15,), dtype='int32',
name='movie_titles')
    return uid, user_gender, user_age, user_job, movie_id, movie_categories,
movie_titles
```

定义用户的嵌入矩阵，使用的是第 3 章介绍的 Embedding 函数，其中参数 input_length 要跟输入的张量形状一致。

```
def get_user_embedding(uid, user_gender, user_age, user_job):
    # 定义用户ID的嵌入矩阵，uid_max是用户ID的总个数，每个向量的维度是32
    uid_embed_layer = tf.keras.layers.Embedding(uid_max, embed_dim, input_length=1,
                                    name='uid_embed_layer')(uid)
    # 对性别的处理跟用户ID处理是一样的，（2，16）的矩阵
    gender_embed_layer = tf.keras.layers.Embedding(gender_max, embed_dim // 2,
                                    input_length=1,
                                    name='gender_embed_layer')(
                                    user_gender)
    # 对年龄也做同样的处理，（7，16）的矩阵
    age_embed_layer = tf.keras.layers.Embedding(age_max, embed_dim // 2,
                                    input_length=1,
                                    name='age_embed_layer')(user_age)
    # 对职业的处理也相同，（21，16）的矩阵
    job_embed_layer = tf.keras.layers.Embedding(job_max, embed_dim // 2,
                                    input_length=1,
                                    name='job_embed_layer')(user_job)
    return uid_embed_layer, gender_embed_layer, age_embed_layer, job_embed_layer
```

将用户的嵌入矩阵一起全连接生成用户的特征：

```
def get_user_feature_layer(uid_embed_layer, gender_embed_layer, age_embed_layer,
job_embed_layer):
    # 第一层全连接层
    # 将用户ID、性别、年龄和职业的嵌入向量传给全连接层，输出维度都是32
    uid_fc_layer = tf.keras.layers.Dense(embed_dim, name="uid_fc_layer",
                                    activation='relu')(uid_embed_layer)
```

```python
gender_fc_layer = tf.keras.layers.Dense(embed_dim, name="gender_fc_layer",
                                        activation='relu')(gender_embed_layer)
age_fc_layer = tf.keras.layers.Dense(embed_dim, name="age_fc_layer",
                                     activation='relu')(age_embed_layer)
job_fc_layer = tf.keras.layers.Dense(embed_dim, name="job_fc_layer",
                                     activation='relu')(job_embed_layer)

# 第二层全连接层
# 将第一层全连接层的输出连接在一起,再次传给全连接层,输出维度是200
user_combine_layer = tf.keras.layers.concatenate([uid_fc_layer,
                                                  gender_fc_layer,
                                                  age_fc_layer,
                                                  job_fc_layer], 2)  #(?, 1,
                                                  128)
user_combine_layer = tf.keras.layers.Dense(
    200, activation='tanh')(user_combine_layer)  #(?, 1, 200)

user_combine_layer_flat = tf.keras.layers.Reshape(
    [200], name="user_combine_layer_flat")(user_combine_layer)
return user_combine_layer, user_combine_layer_flat
```

定义电影 ID 的嵌入矩阵:

```python
def get_movie_id_embed_layer(movie_id):
    # 对电影ID的处理跟用户ID处理是一样的,(3952, 32)的矩阵
    movie_id_embed_layer = tf.keras.layers.Embedding(movie_id_max,
                                                     embed_dim,
                                                     input_length=1,
                                                     name='movie_id_embed_layer')(
                                                     movie_id)
    return movie_id_embed_layer
```

合并电影风格的多个嵌入向量:

```python
def get_movie_categories_layers(movie_categories):
    # 首先定义电影风格的嵌入矩阵,(19, 32)
    movie_categories_embed_layer =
        tf.keras.layers.Embedding(movie_categories_max, embed_dim,
```

```
            input_length=18, name='movie_categories_embed_layer')(movie_categories)
    # 因为每部电影的风格有可能是多种风格的组合，所以从嵌入矩阵输出的是多个向量
    # 这里跟上一节的实现有些不同，没有对多个嵌入向量求和，而是将多个向量链接在一起
    # 如果要实现求和的话，可以用下面语句替代
    # 需要注意，在构建模型的每一层时一定要用keras.layers中的函数
    # 所以要使用Lambda函数将tf.reduce_sum包装起来
    # movie_categories_embed_layer = tf.keras.layers.Lambda(lambda layer:
    #       tf.reduce_sum(layer, axis=1, keepdims=True))(
    # movie_categories_embed_layer)
    movie_categories_embed_layer =
        tf.keras.layers.Reshape([1, 18 * embed_dim])(movie_categories_embed_layer)

    return movie_categories_embed_layer
```

电影标题的文本卷积神经网络实现：

```
def get_movie_cnn_layer(movie_titles):
    # 从嵌入矩阵中得到电影标题对应的各个单词的嵌入向量，电影标题的嵌入矩阵维度是
    # （5216，32）
    movie_title_embed_layer =
        tf.keras.layers.Embedding(movie_title_max, embed_dim, input_length=15,
                                  name='movie_title_embed_layer')(movie_titles)
    sp = movie_title_embed_layer.shape
    movie_title_embed_layer_expand =
        tf.keras.layers.Reshape([sp[1], sp[2], 1])(movie_title_embed_layer)
    # 对文本嵌入层使用不同尺寸的卷积核做卷积和最大池化
    pool_layer_lst = []
    for window_size in window_sizes:
        # 前文说过，因为文本卷积要覆盖整个单词的嵌入向量，所以尺寸是（窗口大小，向
        # 量维度32）
        # 文本卷积核数量filter_num是8
        conv_layer =
            tf.keras.layers.Conv2D(filter_num, (window_size, embed_dim),
                                   1, activation='relu')(
                                   movie_title_embed_layer_expand)
        maxpool_layer = tf.keras.layers.MaxPooling2D(
            pool_size=(sentences_size - window_size + 1 ,1), strides=1)(conv_layer)
```

```
        pool_layer_lst.append(maxpool_layer)

    pool_layer = tf.keras.layers.concatenate(pool_layer_lst, 3, name ="pool_layer")
    # 每一个窗口都有filter_num（卷积核数量）个结果，所以最终是（窗口个数*卷积核数
    # 量）个结果
    max_num = len(window_sizes) * filter_num
    pool_layer_flat = tf.keras.layers.Reshape([1, max_num],
                                       name = "pool_layer_flat")(pool_layer)

    dropout_layer = tf.keras.layers.Dropout(dropout_keep,
                            name = "dropout_layer")(pool_layer_flat
                            )
    return pool_layer_flat, dropout_layer
```

将电影的各个层一起做全连接：

```
def get_movie_feature_layer(movie_id_embed_layer, movie_categories_embed_layer,
dropout_layer):
    # 第一层全连接层
    # 将电影ID和电影风格的嵌入向量传入全连接层，输出维度都是32
    movie_id_fc_layer =
        tf.keras.layers.Dense(embed_dim, name="movie_id_fc_layer",
                              activation='relu')(movie_id_embed_layer)
    movie_categories_fc_layer =
        tf.keras.layers.Dense(embed_dim, name="movie_categories_fc_layer",
                              activation='relu')(movie_categories_embed_layer)

    # 第二层全连接层
    # 将第一层全连接层的输出和文本卷积的输出一起传入全连接层，输出维度是200
    movie_combine_layer =
        tf.keras.layers.concatenate([movie_id_fc_layer,
                              movie_categories_fc_layer, dropout_layer], 2)
    movie_combine_layer = tf.keras.layers.Dense(200, activation='tanh')(
        movie_combine_layer)

    movie_combine_layer_flat =
        tf.keras.layers.Reshape([200], name="movie_combine_layer_flat")(
```

```
            movie_combine_layer)
    return movie_combine_layer, movie_combine_layer_flat
```

定义电影推荐类,包含模型构建和训练函数:

```
import tensorflow as tf
import datetime
from tensorflow import keras
from tensorflow.python.ops import summary_ops_v2
import time

MODEL_DIR = "./models"

class mv_network(object):
    def __init__(self, batch_size=256):
        self.batch_size = batch_size
        self.best_loss = 9999
        self.losses = {'train': [], 'test': []}

        # 获取输入占位符
        uid, user_gender, user_age, user_job, movie_id, movie_categories,
movie_titles = get_inputs()
        # 获取用户的4个嵌入向量
        uid_embed_layer, gender_embed_layer, age_embed_layer, job_embed_layer =
            get_user_embedding(uid, user_gender, user_age, user_job)
        # 得到用户特征
        user_combine_layer, user_combine_layer_flat =
            get_user_feature_layer(uid_embed_layer, gender_embed_layer,
                             age_embed_layer, job_embed_layer)
        # 获取电影ID的嵌入向量
        movie_id_embed_layer = get_movie_id_embed_layer(movie_id)
        # 获取电影风格的嵌入向量
        movie_categories_embed_layer=get_movie_categories_layers(movie_categories)
        # 获取电影标题的特征向量
        pool_layer_flat, dropout_layer = get_movie_cnn_layer(movie_titles)
        # 得到电影特征
        movie_combine_layer, movie_combine_layer_flat = get_movie_feature_layer(
```

```python
                    movie_id_embed_layer, movie_categories_embed_layer, dropout_layer)
# 计算出评分
# 将用户特征和电影特征作为输入，经过全连接，输出一个值的方案
inference_layer = tf.keras.layers.concatenate(
    [user_combine_layer_flat, movie_combine_layer_flat], 1)  # (?, 400)
# 可以再加一层全连接层，像下面这样，试试效果
# inference_layer = tf.keras.layers.Dense(64, kernel_regularizer=tf.nn.
# l2_loss, activation='relu')(inference_layer)
inference = tf.keras.layers.Dense(1, name="inference")(inference_layer)
# 当然，你也可以使用将用户特征和电影特征做矩阵乘法得到一个预测评分的方案
# inference = tf.keras.layers.Lambda(
#     lambda layer: tf.reduce_sum(layer[0] * layer[1], axis=1),
#     name="inference")((user_combine_layer_flat,movie_combine_layer_flat))
# inference = tf.keras.layers.Lambda(
#     lambda layer: tf.expand_dims(layer, axis=1))(inference)

# 构建模型，一共有七个输入、一个输出
self.model = tf.keras.Model(
    inputs=[uid, user_gender, user_age, user_job,
            movie_id, movie_categories, movie_titles],
    outputs=[inference])

self.model.summary()

# Adam优化器
self.optimizer = tf.keras.optimizers.Adam(learning_rate)
# MSE损失
self.ComputeLoss = tf.keras.losses.MeanSquaredError()
# MAE指标
self.ComputeMetrics = tf.keras.metrics.MeanAbsoluteError()

# 创建保存模型的文件夹
if tf.io.gfile.exists(MODEL_DIR):
    pass
else:
    tf.io.gfile.makedirs(MODEL_DIR)
```

```python
        train_dir = os.path.join(MODEL_DIR, 'summaries', 'train')
        test_dir = os.path.join(MODEL_DIR, 'summaries', 'eval')

#         self.train_summary_writer =
#             summary_ops_v2.create_file_writer(train_dir, flush_millis=10000)
#         self.test_summary_writer =
#             summary_ops_v2.create_file_writer(test_dir, flush_millis=10000,
# name='test')

        checkpoint_dir = os.path.join(MODEL_DIR, 'checkpoints')
        self.checkpoint_prefix = os.path.join(checkpoint_dir, 'ckpt')
        self.checkpoint = tf.train.Checkpoint(model=self.model,
                                              optimizer=self.optimizer)

        # 加载保存的模型
        self.checkpoint.restore(tf.train.latest_checkpoint(checkpoint_dir))

    def compute_loss(self, labels, logits):
        return tf.reduce_mean(tf.keras.losses.mse(labels, logits))

    def compute_metrics(self, labels, logits):
        return tf.keras.metrics.mae(labels, logits)
# 单步训练函数
@tf.function
def train_step(self, x, y):
    # 使用tf.GradientTape()的磁带记录计算损失的操作
    with tf.GradientTape() as tape:
        # 调用模型得到预测评分
        logits = self.model([x[0],
                             x[1],
                             x[2],
                             x[3],
                             x[4],
                             x[5],
                             x[6]], training=True)
```

```python
            # 计算损失
            loss = self.ComputeLoss(y, logits)
            # loss = self.compute_loss(labels, logits)
            # 计算MAE指标
            self.ComputeMetrics(y, logits)
            # metrics = self.compute_metrics(labels, logits)
        # 使用磁带计算模型参数的梯度
        grads = tape.gradient(loss, self.model.trainable_variables)
        self.optimizer.apply_gradients(zip(grads, self.model.trainable_variables))
        return loss, logits

    # 训练函数
    def training(self, features, targets_values, epochs=5, log_freq=50):

        for epoch_i in range(epochs):
            # 将数据集分成训练集和测试集，随机种子不固定
            train_X, test_X, train_y, test_y = train_test_split(features,
                                                                targets_values,
                                                                test_size=0.2,
                                                                random_state=0)

            train_batches = get_batches(train_X, train_y, self.batch_size)
            batch_num = (len(train_X) // self.batch_size)

            train_start = time.time()
            #   with self.train_summary_writer.as_default():
            if True:
                start = time.time()
                # 定义平均指标用来计算平均损失
                avg_loss = tf.keras.metrics.Mean('loss', dtype=tf.float32)
                # avg_mae = tf.keras.metrics.Mean('mae', dtype=tf.float32)

                # 遍历训练数据
                for batch_i in range(batch_num):
                    x, y = next(train_batches)
                    # 准备输入数据
```

```python
        categories = np.zeros([self.batch_size, 18])
        for i in range(self.batch_size):
            categories[i] = x.take(6, 1)[i]

        titles = np.zeros([self.batch_size, sentences_size])
        for i in range(self.batch_size):
            titles[i] = x.take(5, 1)[i]

        # 单步训练
        loss, logits = self.train_step(
       [np.reshape(x.take(0, 1),[self.batch_size,1]).astype(np.float32),
        np.reshape(x.take(2, 1),[self.batch_size,1]).astype(np.float32),
        np.reshape(x.take(3, 1),[self.batch_size,1]).astype(np.float32),
        np.reshape(x.take(4, 1),[self.batch_size,1]).astype(np.float32),
        np.reshape(x.take(1, 1),[self.batch_size,1]).astype(np.float32),
        categories.astype(np.float32), titles.astype(np.float32)],
        np.reshape(y, [self.batch_size, 1]).astype(np.float32))
    avg_loss(loss)
#     avg_mae(metrics)
        self.losses['train'].append(loss)

        # 训练达到一定次数，打印日志
        if tf.equal(self.optimizer.iterations % log_freq, 0):
    #     summary_ops_v2.scalar('loss', avg_loss.result(),
    #         step=self.optimizer.iterations)
    #     summary_ops_v2.scalar('mae',self.ComputeMetrics.result(),
    #         step=self.optimizer.iterations)
    #     summary_ops_v2.scalar('mae', avg_mae.result(),
    #         step=self.optimizer.iterations)

            rate = log_freq / (time.time() - start)
            print('Step #{}\tEpoch {:>3} Batch {:>4}/{}   Loss: {:0.6f}
             mae: {:0.6f} ({} steps/sec)'.format(
                self.optimizer.iterations.numpy(),
                epoch_i,
                batch_i,
```

```python
                            batch_num,
                            loss, (self.ComputeMetrics.result()), rate))

                        avg_loss.reset_states()
                        self.ComputeMetrics.reset_states()
                        # avg_mae.reset_states()
                        start = time.time()

            train_end = time.time()
            print(
                '\nTrain time for epoch #{} ({} total steps): {}'.format(
                    epoch_i + 1, self.optimizer.iterations.numpy(),
                    train_end - train_start))
            # 使用测试集评估网络
            # with self.test_summary_writer.as_default():
            self.testing((test_X, test_y), self.optimizer.iterations)

        self.export_path = os.path.join(MODEL_DIR, 'export')
        tf.saved_model.save(self.model, self.export_path)
# 测试函数
def testing(self, test_dataset, step_num):
    test_X, test_y = test_dataset
    test_batches = get_batches(test_X, test_y, self.batch_size)

    # 定义平均指标
    avg_loss = tf.keras.metrics.Mean('loss', dtype=tf.float32)
    # avg_mae = tf.keras.metrics.Mean('mae', dtype=tf.float32)

    # 遍历测试集
    batch_num = (len(test_X) // self.batch_size)
    for batch_i in range(batch_num):
        x, y = next(test_batches)
        # 准备输入数据
        categories = np.zeros([self.batch_size, 18])
        for i in range(self.batch_size):
            categories[i] = x.take(6, 1)[i]
```

```python
            titles = np.zeros([self.batch_size, sentences_size])
            for i in range(self.batch_size):
                titles[i] = x.take(5, 1)[i]
            # 调用模型得到预测评分
            logits = self.model([
                np.reshape(x.take(0, 1), [self.batch_size, 1]).astype(np.float32),
                np.reshape(x.take(2, 1), [self.batch_size, 1]).astype(np.float32),
                np.reshape(x.take(3, 1), [self.batch_size, 1]).astype(np.float32),
                np.reshape(x.take(4, 1), [self.batch_size, 1]).astype(np.float32),
                np.reshape(x.take(1, 1), [self.batch_size, 1]).astype(np.float32),
                categories.astype(np.float32),
                titles.astype(np.float32)], training=False)
            # 计算损失
            test_loss = self.ComputeLoss(np.reshape(y, [self.batch_size, 1]).
                astype(np.float32), logits)
            avg_loss(test_loss)
            # 保存测试损失
            self.losses['test'].append(test_loss)
            self.ComputeMetrics(np.reshape(y, [self.batch_size, 1]).astype(np.
                float32), logits)
            # avg_loss(self.compute_loss(labels, logits))
            # avg_mae(self.compute_metrics(labels, logits))

print('Model test set loss: {:0.6f} mae: {:0.6f}'.format(avg_loss.result(),
 self.ComputeMetrics.result()))
# print('Model test set loss: {:0.6f} mae: {:0.6f}'.format(avg_loss.result
# (), avg_mae.result()))
# summary_ops_v2.scalar('loss', avg_loss.result(), step=step_num)
# summary_ops_v2.scalar('mae', self.ComputeMetrics.result(), step=step_num)
# summary_ops_v2.scalar('mae', avg_mae.result(), step=step_num)

# 如果当前损失是最小的,则更新best_loss,保存模型参数
if avg_loss.result() < self.best_loss:
    self.best_loss = avg_loss.result()
    print("best loss = {}".format(self.best_loss))
```

```
            self.checkpoint.save(self.checkpoint_prefix)

    # 定义预测函数
    def forward(self, xs):
        # 调用模型得到预测评分
        predictions = self.model(xs)

        return predictions
```

整个 TensorFlow 2.0 的实现流程跟前几章的实现是一样的，模型的构建同样跟上一节内容相似。

4.7.2 训练网络

训练网络，代码如下：

```
mv_net=mv_network()
mv_net.training(features, targets_values, epochs=5)
```

将用户特征和电影特征作为输入，经过全连接，输出一个值的训练结果：

```
Model test set loss: 0.835090 mae: 0.720690
```

将用户特征和电影特征做矩阵乘法，得到一个预测评分的训练结果：

```
Model test set loss: 0.797666 mae: 0.705161
```

4.7.3 实现个性化推荐

1. 指定用户和电影进行评分

这部分就是对网络做正向传播，计算得到预测评分。

```
def rating_movie(mv_net, user_id_val, movie_id_val):
    # 准备输入数据
    categories = np.zeros([1, 18])
    categories[0] = movies.values[movieid2idx[movie_id_val]][2]
```

```
        titles = np.zeros([1, sentences_size])
        titles[0] = movies.values[movieid2idx[movie_id_val]][1]
        # 调用模型得到预测评分
        inference_val = mv_net.model([
                np.reshape(users.values[user_id_val-1][0], [1, 1]),
                np.reshape(users.values[user_id_val-1][1], [1, 1]),
                np.reshape(users.values[user_id_val-1][2], [1, 1]),
                np.reshape(users.values[user_id_val-1][3], [1, 1]),
                np.reshape(movies.values[movieid2idx[movie_id_val]][0], [1, 1]),
                categories,
                titles])

        return (inference_val.numpy())
```

试一下，给定用户 ID 234、电影 ID 1401，得到评分 3.7995248。

```
rating_movie(mv_net, 234, 1401)

array([[3.7995248]], dtype=float32)
```

2. 生成电影特征矩阵和用户特征矩阵

（1）生成电影特征矩阵

用训练好的模型将每一部电影的特征向量提取出来，然后将电影特征矩阵保存到本地。

```
# 模型一共有七个输入，其中最后三个输入是电影特征的输入
# input[4]是movie_id, input[5]是movie_categories, input[6]是movie_titles)
# 经过多层网络处理之后，输出的电影特征张量名是movie_combine_layer_flat
# 根据张量名得到电影特征作为输出，这样传入三个输入和一个输出就得到生成电影特征矩阵
# 的模型了
movie_layer_model = keras.models.Model(
    inputs=[mv_net.model.input[4], mv_net.model.input[5], mv_net.model.input[6]],
    outputs=mv_net.model.get_layer("movie_combine_layer_flat").output)
movie_matrics = []
# 遍历所有的电影
for item in movies.values:
    categories = np.zeros([1, 18])
    categories[0] = item.take(2)
```

```
    titles = np.zeros([1, sentences_size])
    titles[0] = item.take(1)
    # 调用模型得到电影特征
    movie_combine_layer_flat_val = movie_layer_model([np.reshape(item.take(0), [1,
        1]), categories, titles])
    movie_matrics.append(movie_combine_layer_flat_val)
# 保存到本地
pickle.dump((np.array(movie_matrics).reshape(-1, 200)), open('movie_matrics.p', 'wb
    '))
movie_matrics = pickle.load(open('movie_matrics.p', mode='rb'))
```

（2）生成用户特征矩阵

将训练好的用户特征提取出来，然后将用户特征矩阵保存到本地。

```
# 跟生成电影特征矩阵的处理相似，模型的前四个输入是用户特征的输入
# 分别是uid, user_gender, user_age, user_job
# 用户特征的输出是user_combine_layer_flat，传入四个输入和一个输出得到生成用户特征
# 的模型
user_layer_model = keras.models.Model(inputs=[mv_net.model.input[0],
    mv_net.model.input[1], mv_net.model.input[2], mv_net.model.input[3]],
    outputs=mv_net.model.get_layer("user_combine_layer_flat").output)
users_matrics = []
# 遍历用户数据
for item in users.values:
    # 调用模型得到用户特征
    user_combine_layer_flat_val = user_layer_model([
                                        np.reshape(item.take(0), [1, 1]),
                                        np.reshape(item.take(1), [1, 1]),
                                        np.reshape(item.take(2), [1, 1]),
                                        np.reshape(item.take(3), [1, 1])])
    users_matrics.append(user_combine_layer_flat_val)
# 保存到本地
pickle.dump((np.array(users_matrics).reshape(-1, 200)), open('users_matrics.p',
    'wb'))
users_matrics = pickle.load(open('users_matrics.p', mode='rb'))
```

当用户特征矩阵和电影特征矩阵提取出来之后，后面的推荐电影部分就没有区别了。这里的重点就是用训练好的模型，利用网络的输入和不同层的输出构建子模型。

4.8 本章小结

本章介绍了常用的推荐功能的实现，将网络模型作为回归问题进行训练，得到训练好的用户特征矩阵和电影特征矩阵进行推荐。现在 MovieLens 还提供了新的数据集，数据的特征发生了变化，感兴趣的读者可以尝试构建一个使用最新数据集训练的推荐模型。

5 广告点击率预估：Kaggle 实战

5.1 概述

点击率预估用来判断一条广告被用户点击的概率，对每次广告的点击做出预测，把用户最有可能点击的广告找出来，是广告技术最重要的算法之一。本章通过 Kaggle 上的 CTR 挑战案例，讲解场感知分解机（FFM）、因子分解机（FM）和梯度提升决策树（GBDT）的概念及使用，讨论如何搭建一个用于广告点击率预估的模型和算法思路，并去实践预测广告点击率。

本章的项目地址是：https://github.com/chengstone/kaggle_criteo_ctr_challenge-。

5.2 下载数据集

我们使用 Kaggle 上的 Display Advertising Challenge（https://www.kaggle.com/c/criteo-display-ad-challenge/）挑战的 criteo 数据集（如图 5.1 所示）。

图 5.1 Kaggle CTR 挑战（选自 Kaggle Display Advertising Challenge 首页）

下载数据集，请在终端输入下面命令（脚本文件路径：./data/download.sh）：

```
wget --no-check-certificate https://s3-eu-west-1.amazonaws.com/kaggle-display-advertising-challenge-dataset/dac.tar.gz

tar zxf dac.tar.gz

rm -f dac.tar.gz

mkdir raw

mv ./*.txt raw/
```

解压缩以后，train.txt 文件大小为 11.7GB，test.txt 文件大小为 1.35GB。

数据量太大了，我们只使用前 100 万条数据，输入以下命令：

```
head -n 1000000 test.txt > test_sub100w.txt

head -n 1000000 train.txt > train_sub100w.txt
```

然后将文件名重新命名为 train_sub100w.txt 和 test_sub100w.txt，文件位置不变。

5.3 数据字段的含义

文件中每行数据都是由标签（Label）字段、13 个整型数据（I1~I13）和 26 个分类数据（C1~C26）组成的。

Label：1 表示广告被点击，0 表示广告没有被点击。

I1~I13：总共 13 列数值特征（主要是计数特征，相当于 Dense Input）。

C1~C26：总共 26 列分类特征（相当于 Sparse Input）。为了匿名化，这些特征的值被散列成 32 位的数据。

数据看起来是这样的：

```
1    0    127  1    3    1683    19   26   17   475  0    9    0    3    05db9164    8947f767
     11c9d79e    52a787c8    4cf72387    fbad5c96    18671b18    0b153874
a73ee510    ceb10289    77212bd7    79507c6b    7203f04e    07d13a8f    2c14c412
     49013ffe    8efede7f    bd17c3da    f6a3e43b    a458ea53    35cd95c9
ad3062eb    c7dc6720    3fdb382b    010f6491    49d68486
```

文件中每行数据并不都是完整的，有些字段的值会缺少。

5.4 点击率预估的实现思路

我们先来了解要使用的几个组件。

5.4.1 梯度提升决策树

梯度提升决策树（Gradient Boosting Decision Tree, GBDT）是一种迭代的决策树算法，每次迭代都会在损失函数的梯度方向建立一棵新决策树，这意味着迭代多少次就会生成多少棵决策树。每棵决策树都可以被看成是一个弱学习器，训练的目标是减少当前弱学习器的损失，然后下一棵树会重点关注之前学习器做错的样本，这样经过多次迭代之后，每棵树都会对之前犯的错进行修正，准确度会持续提高，而整体的损失会持续下降。最后，通过对所有树的结论做加权累加的方式得到最终的学习器。

GBDT 可以用来发现多种有区分性的特征和特征组合，完全省去了人工做特征选择的步骤。更多内容请参考 Facebook 的论文 *Practical Lessons from Predicting Clicks on Ads at Facebook*（论文地址：http://quinonero.net/Publications/predicting-clicks-facebook.pdf）。论文中描述的模型结构如图 5.2 所示，输入特征 x 通过决策树进行转换，每棵树的输出作为分类特征输入给线性分类器。梯度提升决策树被证明具有非常强大的特征变换能力。

传入训练数据后，GBDT 会训练出若干棵树，我们要使用的是 GBDT 中每棵树输出的叶子节点（可以理解为有效的特征组合），将这些叶子节点作为分类特征输入给 FM。下一节我们来介绍 FM。

5.4.2 因子分解机

因子分解机（FM，Factorization Machine）用来解决数据量大并且特征稀疏下的特征组合问题，比如在跟购物相关的特征中，女性可能更关注化妆品或者首饰之类的物品，男性可能更关注体育用品或者电子产品等商品，这说明特征组合训练是有意义的。而商品特征可能存在几百上千种分类，通常我们将分类特征转换成 one-hot 编码的形式，这样一个特征就要变成几百维的特征，再加上其他的分类特征，这导致输入的特征空间急剧膨胀，所以数据的稀疏性是实际问题中不可避免的挑战。

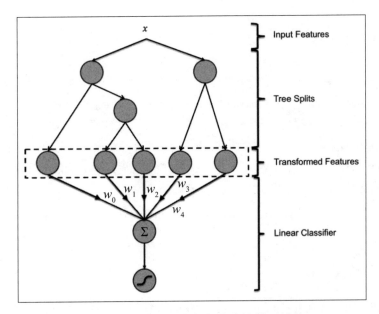

图 5.2 Facebook 论文中描述的模型结构

我们先来看看公式（只考虑二阶多项式的情况）：n 代表样本的特征数量，x_i 是第 i 个特征的值，w_0、w_i、w_{ij} 是模型参数。

$$y(x) = w_0 + \sum_{i=1}^{n} w_i x_i + \sum_{i=1}^{n} \sum_{j=i+1}^{n} w_{ij} x_i x_j \qquad (5\text{-}1)$$

从公式可以看出，这是在线性模型的基础上添加了特征组合 $x_i x_j$，当然，只有在特征 x_i 和 x_j 都不为 0 时才有意义。然而，在实际的应用场景中，训练组合特征的参数是很困难的。因为输入数据存在稀疏性，这导致 x_i 和 x_j 大部分情况都是 0，而组合特征的参数 w_{ij}，只有在特征不为 0 时才能训练出有意义的值。

为了解决二次项参数训练的问题，引入了矩阵分解的概念。在第 4 章中我们讨论的是电影推荐系统，构造了用户特征矩阵和电影特征矩阵，如果将用户特征矩阵和电影特征矩阵相乘就会得到所有用户对所有电影的评分矩阵。

如果将上面的过程反过来看，实际上对于评分矩阵，可以分解成用户矩阵和电影矩阵，而评分矩阵中每一个数据点就相当于上面讨论的组合特征的参数 w_{ij}。

对于参数矩阵 $\boldsymbol{W}$，我们采用矩阵分解的方法，将每一个参数 w_{ij} 分解成两个向量（称之为隐向量）的点积。这样矩阵就可以分解为 $\boldsymbol{W} = \boldsymbol{V}\boldsymbol{V}^\mathrm{T}$，而对于每个参数 $w_{ij} = <\boldsymbol{v}_i, \boldsymbol{v}_j>$，$\boldsymbol{v}_i$ 是 x_i 的隐向量，$\boldsymbol{v}_j$ 是 x_j 的隐向量，这样 FM 的二阶公式就变成：

$$y(x) = w_0 + \sum_{i=1}^{n} w_i x_i + \sum_{i=1}^{n} \sum_{j=i+1}^{n} <\boldsymbol{v}_i, \boldsymbol{v}_j> x_i x_j \qquad （5-2）$$

假设有两个组合特征 $x_h x_i$ 和 $x_i x_j$，在多项式公式（5-1）中，它们的参数 w_{hi} 和 w_{ij} 是相互独立的，彼此没有联系。将参数分解成隐向量后，$x_h x_i$ 的参数 w_{hi} 变成了 $<\boldsymbol{v}_h, \boldsymbol{v}_i>$，$x_i x_j$ 的参数 w_{ij} 变成了 $<\boldsymbol{v}_i, \boldsymbol{v}_j>$，参数中都含有 $\boldsymbol{v}_i$。这意味着，参数分解后使得参数之间不再是相互独立的，所有包含 x_i 的非零组合特征的数据都可以用来训练隐向量 $\boldsymbol{v}_i$。这在一定程度上避免了数据稀疏性带来的影响。用隐向量来表示特征，这样通过训练隐向量 $\boldsymbol{v}$ 能够更好地学习到特征间的相互关系，这就是 FM 模型的思想。

关于 FM 的论文，请参考：https://www.ismll.uni-hildesheim.de/pub/pdfs/Rendle2010FM.pdf。

5.4.3 场感知分解机

现在介绍场感知分解机（Field-aware Factorization Machine，FFM）。FFM 在 FM 的基础上增加了一个 Field 的概念，其特点是相同性质的特征同属于一个 Field。比如商品类型字段是一个分类特征，它可以用不同的值来表示不同类型的商品。这些值都是用来表示商品所属的类型的，所以它们属于同一个 Field，或者说同一个类别的分类特征都可以被放到同一个 Field 中。

这可以看成是一对多的关系，打个比方，比如职业字段，这是一个分类特征，经过 one-hot 编码以后，变成了 N 个特征。这 N 个特征其实都属于职业，所以职业就是一个 Field。总的来说，同一个分类特征，经过 one-hot 编码生成的特征都可以归于同一个 Field。如图 5.3 所示，职业字段经过 one-hot 编码之后，变成了三个特征，这三个特征就都属于同一个 Field。

	同一个Field		
	特征1	特征2	特征3
职业1	1	0	0
职业2	0	1	0
职业3	0	0	1

图 5.3　FFM Field 图示

我们要通过特征组合来训练隐向量，这样每一维特征 x_i 都会与其他特征的每一种 Field f_j 学习一个隐向量 $\boldsymbol{v}_{i,f_j}$。这样的话，隐向量不仅与特征有关，还与 Field 有关。模型的公式如下：

$$y(x) = w_0 + \sum_{i=1}^{n} w_i x_i + \sum_{i=1}^{n} \sum_{j=i+1}^{n} <\boldsymbol{v}_{i,f_j}, \boldsymbol{v}_{j,f_i}> x_i x_j$$

关于 FFM 的论文，请参考：https://www.csie.ntu.edu.tw/~cjlin/papers/ffm.pdf。

5.4.4 网络模型

前面我们介绍了 GBDT、FM 和 FFM 三个组件，使用这些组件的目的就在于代替人工的方式进行特征选择，并解决特征稀疏性的问题。我们要设计的解决方案是，将神经网络和自动特征选择结合在一起，就是将数据输入给 GBDT、FM 和 FFM 做特征选择，然后将特征选择的结果与 DNN（深度神经网络）的输出一起作为输入，做 Logistics 回归，通过 Sigmoid 激活函数计算得到该条广告被点击的概率。

网络模型结构如图 5.4 所示。整个网络分为两大部分，一部分是传统的 DNN，将数据直接传给 DNN 即可；另一部分又分为两个小网络，一个是单独使用 FFM 的网络，另一个是 GBDT 和 FM 的组合网络。GBDT 和 FM 组合在一起的意图是先通过 GBDT 做特征选择，然后通过 FM 的隐向量更好地学习特征间的关系，并且也能够解决数据量大和数据稀疏性的问题。我们将 GBDT 输出的叶子节点作为训练数据的输入来训练 FM 模型。这样对于网络，则需要分别训练 GBDT、FM 和 FFM。

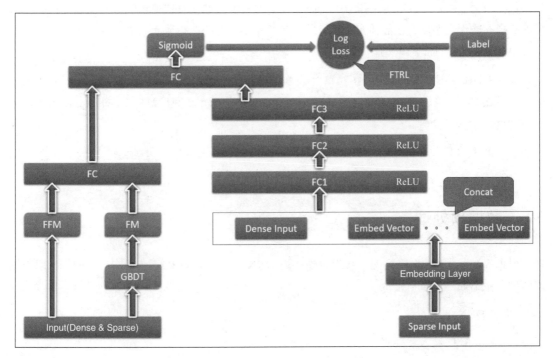

图 5.4 网络模型结构

我们来看 DNN 的部分。将输入数据分成两部分，一部分是数值特征（Dense Input）对应 I1~I13；另一部分是分类特征（Sparse Input）对应 C1~C26。我们仍然不使用 one-hot 编码，将分类特征传入嵌入层，得到多个嵌入向量，再将这些嵌入向量和数值特征连接在一起，传入全连接层，一共三层全连接层，使用 ReLU 激活函数。然后将第三层全连接层的输出和 FFM、FM 的全连接层的输出连接在一起，传入最后一层全连接层。

神经网络的学习目标 Label 表示广告是否被点击了，只有 1（被点击）和 0（没有被点击）两种状态。所以网络的最后一层要做 Logistics 回归（Logistics Regression, LR），在最后一层全连接层使用 Sigmoid 激活函数，得到广告被点击的概率。

网络最终是做 LR，使用 LogLoss 作为损失函数，使用 FTRL（Follow-The-Regularized-Leader）优化损失。

简单地说，LR 在批量处理超大规模数据集和在线数据流时无法有效处理这些数据，处理速度慢，不适合做在线学习。在批量算法中每次迭代都要对所有训练数据进行计算，比如全局梯度，但是在处理超大规模数据集上，计算全局梯度的代价就太高了，而且也没法使用在线数据流做在线学习。FTRL 算法的优点是具有非常好的稀疏性和收敛性。这个稀疏性很重要，也就是说，当训练好的特征参数不具有稀疏性时，每个特征都有自己的参数，包括重要的和不重要的特征。这些参数都不为 0，都要参与到计算当中，而在线学习要求模型拥有快速地响应和更新的能力。如果能够提高或保证精度，还能让参数稀疏化，进行更少的计算，得到更快的响应结果，这就是我们想要的，也是 FTRL 的思想。

关于 FTRL 算法，Google 公司 2013 年发表的论文 *Ad click prediction: a view from the trenches* 中做了描述（论文地址：https://www.researchgate.net/publication/262412214_Ad_click_prediction_a_view_from_the_trenches）。

关于 FM、FFM 和 GBDT 三个组件的训练，我们会在下文进行说明。

5.5 数据预处理

根据 5.4 节设计的网络模型，我们首先要为 DNN、FFM 和 GBDT 准备输入数据，所以数据预处理要做下面三件事情：

- 生成 DNN 的输入数据。
- 生成 FFM 的输入数据。
- 生成 GBDT 的输入数据。

对于数值特征，我们将 I1~I13 转换成 0~1 之间的小数。对于分类特征，我们将某类别使用次数少于 cutoff 次（超参数之一）的特征忽略掉，留下使用次数多的特征作为某类别字段的特征，然后将这些特征以各自字段为组进行编号。

比如有 C1 和 C2 两个类别字段，C1 下面有特征 a（大于 cutoff 次）、b（少于 cutoff 次）、c（大于 cutoff 次），C2 下面有特征 x 和 y（均大于 cutoff 次），这样留下来的特征就是 C1：a、c 和 C2：x、y。然后以各自字段为组进行编号，对于 C1 字段，a 和 c 的特征 id 对应 0 和 1；对于 C2 字段，x 和 y 的特征 id 也对应 0 和 1。

对于分类特征的输入数据处理，FFM 和 GBDT 各不相同，我们分别来讲。

5.5.1 GBDT 的输入数据处理

GBDT 的输入数据处理要简单一些，将 C1~C26 每个字段各自的特征 id（相当于序号，在每个类别内独立编号）作为输入即可。GBDT 的输入数据格式是：Label I1~I13 C1~C26，所以实际输入数据可能是这样的：0 小数 1 小数 2 ~ 小数 13 1（C1 特征 id）0（C2 特征 id）~ C26 特征 id，其中 C1 特征 id 是 1，按照之前假设的例子，说明此处 C1 字段的特征是 c，而 C2 字段的特征是 x。

下面是生成的一段真实的数据：

```
0    0.05     0.004983     0.05     0    0.021594     0.008    0.15    0.04    0.362
0.166667     0.2  0.04    2    3    0    0    1    1    0    3    1    0    0    0    3
        0    0    1    4    1    3    0    0    2    0    1    0
```

5.5.2 FFM 的输入数据处理

FFM 的输入数据处理要复杂一些，详情可以参考官方 GitHub 上的说明（https://github.com/guestwalk/libffm）。下面引用 GitHub 上的部分描述解释 FFM 的输入数据。

理解字段（`field`）和特征（`feature`）的概念很重要，假设有如下原始数据：

```
点击      广告客户      发布商
=====   ==========   =========
0       Nike         CNN
1       ESPN         BBC
```

这样就有：

- 2个字段(fields):广告客户(Advertiser)和发布商(Publisher)。
- 4个特征(features):广告客户—Nike,广告客户—ESPN,发布商—CNN,发布商—BBC。

我们需要建立两个字典:一个字段字典和一个特征字典,如下所示。

```
DictField[Advertiser] -> 0
DictField[Publisher]  -> 1

DictFeature[Advertiser-Nike] -> 0
DictFeature[Publisher-CNN]   -> 1
DictFeature[Advertiser-ESPN] -> 2
DictFeature[Publisher-BBC]   -> 3
```

然后,生成FFM格式的数据:

```
0 0:0:1 1:1:1
1 0:2:1 1:3:1
```

格式是:fields:features:values。

注意:因为这些都是分类特征,所以这里的值都是1。

对于fields应该很好理解,features的划分跟GBDT有些不一样,在GBDT的处理中是在每个类别内独立编号的,C1有features $0 \sim n$,C2有features $0 \sim n$,而FFM是将所有的features统一起来编号的。在例子中,C1是Advertiser,有两个特征,C2是Publisher,有两个特征,统一起来编号就是0~3。而在GBDT的处理中要独立编号,看起来像这样:

```
DictFeature[Advertiser-Nike] -> 0
DictFeature[Advertiser-ESPN] -> 1
DictFeature[Publisher-CNN]   -> 0
DictFeature[Publisher-BBC]   -> 1
```

现在假设有第三条数据,我们看看如何生成FFM的输入数据。

点击	广告客户	发布商
0	Nike	CNN
1	ESPN	BBC
0	Lining	CNN

按照规则,应该像下面这样:

```
DictFeature[Advertiser-Nike]       -> 0
DictFeature[Publisher-CNN]         -> 1
DictFeature[Advertiser-ESPN]       -> 2
DictFeature[Publisher-BBC]         -> 3
DictFeature[Advertiser-Lining]     -> 4
```

在 FFM 的输入数据处理中,跟上面略微有些区别,每个类别编号以后,下一个类别继续编号,所以最终的 features 编号是这样的:

```
DictFeature[Advertiser-Nike]       -> 0
DictFeature[Advertiser-ESPN]       -> 1
DictFeature[Advertiser-Lining]     -> 2
DictFeature[Publisher-CNN]         -> 3
DictFeature[Publisher-BBC]         -> 4
```

数据是从 I1 开始编号的,I1~I13,所以 C1 的 features 编号要从加 13 开始。

这是生成的一段真实的 FFM 输入数据:

```
0    0:0:0.05      1:1:0.004983    2:2:0.05      3:3:0     4:4:0.021594    5:5:0.008
6:6:0.15       7:7:0.04      8:8:0.362     9:9:0.166667     10:10:0.2     11:11:0  12:12:0.04
    13:15:1  14:29:1  15:64:1  16:76:1  17:92:1  18:101:1     19:107:1     20:122:1
21:131:1      22:133:1      23:143:1     24:166:1     25:179:1     26:209:1     27:216:1
    28:243:1      29:260:1     30:273:1     31:310:1     32:317:1     33:318:1
34:333:1      35:340:1     36:348:1     37:368:1     38:381:1
```

5.5.3 DNN 的输入数据处理

DNN 的输入数据处理就没有那么复杂了,仍然是 I1~I13 的小数和 C1~C26 的统一编号,就像 FFM 一样,只是 C1~C26 的 features 编号不需要从加 13 开始了,最后是 Label。真实数据就像这样:

```
0.05,0.004983,0.05,0,0.021594,0.008,0.15,0.04,0.362,0.166667,0.2,0,0.04,2,16,51,
63,79,88,94,109,118,120,130,153,166,196,203,230,247,260,297,304,305,320,
327,335,355,368,0
```

5.5.4 数据预处理的实现

要说明的就这么多了，我们来看看代码吧。因为要同时生成训练数据、验证数据和测试数据，所以要运行一段时间。

```python
import os
import sys
import click
import random
import collections

import numpy as np
import lightgbm as lgb
import json
import lightgbm as lgb
import pandas as pd
from sklearn.metrics import mean_squared_error

import pickle


def save_params(params):
    """
    保存参数到文件中
    """
    pickle.dump(params, open('params.p', 'wb'))


def load_params():
    """
    从文件中加载参数
    """
    return pickle.load(open('params.p', mode='rb'))


def save_params_with_name(params, name):
    """
    保存参数到文件中
    """
    pickle.dump(params, open('{}.p'.format(name), 'wb'))
```

```python
def load_params_with_name(name):
    """
    从文件中加载参数
    """
    return pickle.load(open('{}.p'.format(name), mode='rb'))
```

以下代码出自百度 deep_fm 的 preprocess.py（https://github.com/PaddlePaddle/models/blob/develop/legacy/deep_fm/preprocess.py），只是添加了一些代码，我们就不重复"造轮子"了。

```python
# 13个数值特征和26个分类特征在数据文件每行数据中的下标
# 之前在介绍数据集时说过，每行数据是：Label I1~I13 C1~C26
continous_features = range(1, 14)
categorial_features = range(14, 40)

# 对数值特征做裁剪。每个数值特征的裁剪点来自每个特征的95%分位数
# 95%分位数的含义是，在一组数据所构成的样本集合中，小于某个值的
# 样本集合占整个样本集合的95%。也就是说，把裁剪点作为某个数值特征的最大值
continous_clip = [20, 600, 100, 50, 64000, 500, 100, 50, 500, 10, 10, 10, 50]

# 数值特征生成器
class ContinuousFeatureGenerator:
    """
    用min-max归一化将数值特征标准化为[0, 1]
    """

    def __init__(self, num_feature):
        # 用来保存每一个特征的最大值和最小值
        self.num_feature = num_feature
        self.min = [sys.maxsize] * num_feature
        self.max = [-sys.maxsize] * num_feature

    def build(self, datafile, continous_features):
        # 读取数据文件，计算得到每一个数值特征的最大值和最小值
        with open(datafile, 'r') as f:
            for line in f:
                features = line.rstrip('\n').split('\t')
```

```python
            for i in range(0, self.num_feature):
                # 得到每一个特征字段
                val = features[continous_features[i]]
                # 因为数据中有些字段是空的,所以要进行判断
                if val != '':
                    val = int(val)
                    # 截断数据,大于裁剪点的用裁剪点的值代替
                    if val > continous_clip[i]:
                        val = continous_clip[i]
                    self.min[i] = min(self.min[i], val)
                    self.max[i] = max(self.max[i], val)

    def gen(self, idx, val):
        # 生成新数据,对原数据做归一化
        if val == '':
            return 0.0
        val = float(val)
        return (val - self.min[idx]) / (self.max[idx] - self.min[idx])


# 分类特征生成器
class CategoryDictGenerator:
    """
    为每个分类特征生成字典,key是原值,value是编号
    比如:分类特征C1,有4个值{"aa", "bb", "cc", "dd"}
    生成字典后就是:
    dict[C1]["aa"] = 1
    dict[C1]["bb"] = 2
    dict[C1]["cc"] = 3
    dict[C1]["dd"] = 4
    多加了一个未知项"<unk>",值是0
    dict[C1]["<unk>"] = 0
    """

    def __init__(self, num_feature):
        # 初始化每个分类的字典
        self.dicts = []
```

```python
        self.num_feature = num_feature
        for i in range(0, num_feature):
            self.dicts.append(collections.defaultdict(int))

    def build(self, datafile, categorial_features, cutoff=0):
        with open(datafile, 'r') as f:
            for line in f:
                features = line.rstrip('\n').split('\t')
                for i in range(0, self.num_feature):
                    if features[categorial_features[i]] != '':
                        # 先对每个特征的每个分类特征进行计数
                        self.dicts[i][features[categorial_features[i]]] += 1
        for i in range(0, self.num_feature):
            # 对分类特征进行裁剪,当分类特征在数据中出现次数超过cutoff次时就留下
            # 小于cutoff次则说明该分类不重要,可以抛弃
            self.dicts[i] = filter(lambda x: x[1] >= cutoff,
                                   self.dicts[i].items())

            self.dicts[i] = sorted(self.dicts[i], key=lambda x: (-x[1], x[0]))
            # vocabs就是每个特征字段裁剪后的所有分类的值
            vocabs, _ = list(zip(*self.dicts[i]))
            # 生成字典,key是vocabs,value就是vocabs值的序号,从1开始
            self.dicts[i] = dict(zip(vocabs, range(1, len(vocabs) + 1)))
            # 追加未知项,因为数据中会出现裁剪掉的分类值,这些都会被映射成0
            self.dicts[i]['<unk>'] = 0

    def gen(self, idx, key):
        # 生成分类数据,传入原数据key,返回字典中对应的value(序号)
        if key not in self.dicts[idx]:
            res = self.dicts[idx]['<unk>']
        else:
            res = self.dicts[idx][key]
        return res

    def dicts_sizes(self):
        # 返回每个字典(分类特征C1~C26)的长度
```

```python
        return list(map(len, self.dicts))

# 数据预处理主函数
def preprocess(datadir, outdir):
    """
    对下载的数据集进行预处理，用来生成FFM和GBDT的输入数据。如果忘记了数据格式，则可
    以翻阅5.5.1节和5.5.2节中的描述
    """
    # 请注意这里两个变量名很相似，dists和dicts，不要弄混
    # 准备好数值特征生成器
    dists = ContinuousFeatureGenerator(len(continous_features))
    dists.build(os.path.join(datadir, 'train.txt'), continous_features)

    # 准备好分类特征生成器，裁剪点cutoff设置为200
    dicts = CategoryDictGenerator(len(categorial_features))
    dicts.build(
        os.path.join(datadir, 'train.txt'), categorial_features, cutoff=200)

    # 这部分代码记录每个分类特征的起始序号，因为对于FFM是要所有分类特征统一编号的
    # 比如有三个分类特征C1、C2、C3，每个特征的字典长度是2、3、4
    # 表明C1只有2个值，C2有3个值，C3有4个值
    # 那么最终categorial_feature_offset保存的值就是[0, 2, 5, 9]
    # C1的2个值编号就是0和1，C2的3个值的编号就从2开始，是2、3、4，C3的4个值的编号是
    # 从5开始的
    dict_sizes = dicts.dicts_sizes()
    categorial_feature_offset = [0]
    for i in range(1, len(categorial_features)):
        offset = categorial_feature_offset[i - 1] + dict_sizes[i - 1]
        categorial_feature_offset.append(offset)

    random.seed(0)

    # 90%的数据用于训练，10%的数据用于验证
    # 我们要生成用于DNN、FFM和GBDT的训练集和验证集，lgb就是GBDT
    train_ffm = open(os.path.join(outdir, 'train_ffm.txt'), 'w')
    valid_ffm = open(os.path.join(outdir, 'valid_ffm.txt'), 'w')
```

```python
train_lgb = open(os.path.join(outdir, 'train_lgb.txt'), 'w')
valid_lgb = open(os.path.join(outdir, 'valid_lgb.txt'), 'w')

# 生成思想就是从原数据一行一行地读进来，然后通过之前定义的特征生成器将每
# 个字段转换成新的值，最后根据要生成的DNN、FFM和GBDT的数据格式生成数据集
# train和valid就是用于DNN的数据
with open(os.path.join(outdir, 'train.txt'), 'w') as out_train:
    with open(os.path.join(outdir, 'valid.txt'), 'w') as out_valid:
        with open(os.path.join(datadir, 'train.txt'), 'r') as f:
            for line in f:
                features = line.rstrip('\n').split('\t')
                # continous_feats用于lgb
                continous_feats = []
                # continous_vals用于DNN和FFM
                continous_vals = []
                # 转换每行数据中的数值特征得到归一化后的值
                for i in range(0, len(continous_features)):

                    val = dists.gen(i, features[continous_features[i]])
                    continous_vals.append(
                        "{0:.6f}".format(val).rstrip('0').rstrip('.'))
                    continous_feats.append(
                        "{0:.6f}".format(val).rstrip('0').rstrip('.'))


                # categorial_vals用于DNN和FFM
                categorial_vals = []
                # categorial_lgb_vals用于lgb
                categorial_lgb_vals = []
                # 处理每行的分类特征数据
                for i in range(0, len(categorial_features)):
                    # 因为DNN和FFM中的分类特征数据要统一编号，所以要加上
                    # categorial feature_offset[i]的值
                    val = dicts.gen(i, features[categorial_features[i]]) + \
                        categorial_feature_offset[i]
                    categorial_vals.append(str(val))
                    # GDBT不用统一编号，直接使用生成器返回的值即可
```

```python
            val_lgb = dicts.gen(i, features[categorial_features[i]])
            categorial_lgb_vals.append(str(val_lgb))

    continous_vals = ','.join(continous_vals)
    categorial_vals = ','.join(categorial_vals)
    label = features[0]
    # 随机数，有90%的机会进入训练集
    if random.randint(0, 9999) % 10 != 0:
        # 按照之前讨论的DNN、FFM和GBDT的数据格式分别写入文件中
        out_train.write(','.join(
            [continous_vals, categorial_vals, label]) + '\n')
        train_ffm.write('\t'.join(label) + '\t')
        train_ffm.write('\t'.join(
            ['{}:{}:{}'.format(ii, ii, val) for ii,val in enumerate
            (continous_vals.split(','))]) + '\t')
        train_ffm.write('\t'.join(
            ['{}:{}:1'.format(ii + 13, str(np.int32(val) + 13)) for
             ii, val in enumerate(categorial_vals.split(','))]) +
            '\n')

        train_lgb.write('\t'.join(label) + '\t')
        train_lgb.write('\t'.join(continous_feats) + '\t')
        train_lgb.write('\t'.join(categorial_lgb_vals) + '\n')

    else:     # 有10%的机会进入验证集
        out_valid.write(','.join(
            [continous_vals, categorial_vals, label]) + '\n')
        valid_ffm.write('\t'.join(label) + '\t')
        valid_ffm.write('\t'.join(
            ['{}:{}:{}'.format(ii, ii, val) for ii,val in enumerate
            (continous_vals.split(','))]) + '\t')
        valid_ffm.write('\t'.join(
            ['{}:{}:1'.format(ii + 13, str(np.int32(val) + 13)) for
             ii, val in enumerate(categorial_vals.split(','))]) +
            '\n')
```

```python
                valid_lgb.write('\t'.join(label) + '\t')
                valid_lgb.write('\t'.join(continous_feats) + '\t')
                valid_lgb.write('\t'.join(categorial_lgb_vals) + '\n')

train_ffm.close()
valid_ffm.close()

train_lgb.close()
valid_lgb.close()

# 测试集的处理跟训练集是一样的
test_ffm = open(os.path.join(outdir, 'test_ffm.txt'), 'w')
test_lgb = open(os.path.join(outdir, 'test_lgb.txt'), 'w')

with open(os.path.join(outdir, 'test.txt'), 'w') as out:
    with open(os.path.join(datadir, 'test.txt'), 'r') as f:
        for line in f:
            features = line.rstrip('\n').split('\t')

            continous_feats = []
            continous_vals = []
            for i in range(0, len(continous_features)):
                val = dists.gen(i, features[continous_features[i] - 1])
                continous_vals.append(
                    "{0:.6f}".format(val).rstrip('0').rstrip('.'))
                continous_feats.append(
                    "{0:.6f}".format(val).rstrip('0').rstrip('.'))

            categorial_vals = []
            categorial_lgb_vals = []
            for i in range(0, len(categorial_features)):
                val = dicts.gen(i, features[categorial_features[i] -
                        1]) + categorial_feature_offset[i]
                categorial_vals.append(str(val))

                val_lgb = dicts.gen(i, features[categorial_features[i] - 1])
```

```
                categorial_lgb_vals.append(str(val_lgb))

            continous_vals = ','.join(continous_vals)
            categorial_vals = ','.join(categorial_vals)

            out.write(','.join([continous_vals, categorial_vals]) + '\n')

            test_ffm.write('\t'.join(['{}:{}:{}'.format(ii, ii, val) for ii,val
             in enumerate(continous_vals.split(','))]) + '\t')
            test_ffm.write('\t'.join(
                ['{}:{}:1'.format(ii + 13, str(np.int32(val) + 13)) for ii, val
                 in enumerate(categorial_vals.split(','))]) + '\n')

            test_lgb.write('\t'.join(continous_feats) + '\t')
            test_lgb.write('\t'.join(categorial_lgb_vals) + '\n')

    test_ffm.close()
    test_lgb.close()
    return dict_sizes

# 调用预处理函数，生成训练数据
dict_sizes = preprocess('./data/raw','./data')
save_params_with_name((dict_sizes), 'dict_sizes')
dict_sizes = load_params_with_name('dict_sizes')
```

5.6 训练 FFM

数据准备好了，开始训练 FFM 模型，这里我们使用 LibFFM 进行训练。

首先要安装 LibFFM，可以从 GitHub 地址 https://github.com/guestwalk/libffm 下载源码。因为笔者对 LibFFM 做了一些修改，所以请从笔者的代码库进行下载，地址是：https://github.com/chengstone/kaggle_criteo_ctr_challenge-/tree/master/libffm/libffm。

其安装也很简单，打开命令行，进入源码目录，执行 make 命令即可。这样 LibFFM 就安装好了，在训练这一步我们使用 ffm-train 这个文件。

ffm-train 的命令行格式是：

```
ffm-train [options] training_set_file [model_file]
```

options 参数说明如下。

- `-l <lambda>`：设置正则化参数（默认值为 0.00002）。
- `-k <factor>`：设置潜在因子（默认值为 4）。
- `-t <iteration>`：设置迭代次数（默认值为 15）。
- `-r <eta>`：设置学习率（默认值为 0.2）。
- `-s <nr_threads>`：设置线程数（默认值为 1）。
- `-p <path>`：设置验证集的路径。
- `--quiet`：静默模式（没有输出）。
- `--no-norm`：禁用 instance-wise 正则化。
- `--auto-stop`：达到最好的验证损失时停止训练（必须与-p 一起使用）。

我们使用的学习率（Learning Rate）是 0.1，迭代 32 次，验证集文件是 valid_ffm.txt，训练集文件是 train_ffm.txt，训练好后保存的模型文件是 model_ffm。

命令行如下：

```
./libffm/libffm/ffm-train --auto-stop -r 0.1 -t 32 -p ./data/valid_ffm.txt ./data/train_ffm.txt model_ffm
```

Python 代码如下：

```python
import subprocess, sys, os, time

NR_THREAD = 1

cmd = './libffm/libffm/ffm-train --auto-stop -r 0.1 -t 32 -s {nr_thread} -p ./data/valid_ffm.txt ./data/train_ffm.txt model_ffm'.format(nr_thread=NR_THREAD)
os.popen(cmd).readlines()
```

训练过程打印的结果如下：

```
['First check if the text file has already been converted to binary format (1.3 seconds)\n',
 'Binary file found. Skip converting text to binary\n',
 'First check if the text file has already been converted to binary format (0.2 seconds)\n',
```

```
'Binary file found. Skip converting text to binary\n',
'iter     tr_logloss    va_logloss        tr_time\n',
'   1       0.49339       0.48196          12.8\n',
'   2       0.47621       0.47651          25.9\n',
'   3       0.47149       0.47433          39.0\n',
'   4       0.46858       0.47277          51.2\n',
'   5       0.46630       0.47168          63.0\n',
'   6       0.46447       0.47092          74.7\n',
'   7       0.46269       0.47038          86.4\n',
'   8       0.46113       0.47000          98.0\n',
'   9       0.45960       0.46960         109.6\n',
'  10       0.45811       0.46940         121.2\n',
'  11       0.45660       0.46913         132.5\n',
'  12       0.45509       0.46899         144.3\n',
'  13       0.45366       0.46903\n',
'Auto-stop. Use model at 12th iteration.\n']
```

FFM 模型训练好了，我们把训练、验证和测试数据输入给 FFM，通过前向传播得到 FFM 层的输出，LibFFM 做预测得到输出是通过 ffm-predict 完成的。命令行格式如下：

`usage: ffm-predict test_file model_file output_file`

其中 test_file 是数据文件；model_file 是训练好的模型文件；output_file 是保存 FFM 输出结果的文件。

有一点要注意，原版 LibFFM 的输出是 Sigmoid 激活函数值。但是笔者不想要 Sigmoid 激活函数值，而是想要激活函数前一层的 logits 值，作为传输给下一层网络的输入，所以修改了 LibFFM 的代码，将 Sigmoid 激活函数前一层的值作为输出并保存到文件名为 *.out.logit 的文件中（tr_ffm.out.logit、va_ffm.out.logit 和 te_ffm.out.logit）。

还有一点，原版 LibFFM 做预测时传入的数据文件要求有 Label 字段，就是学习目标 Y。而我们还要将没有 Y 的测试数据传给 ffm-predict，所以相应的代码也需要修改。

修改代码 ffm-predict.cpp，我们要在 LibFFM 原命令行的基础上添加一个新的参数，用来表示要传入的数据是否是没有 Y 的测试数据。首先在 Option 结构体中添加一个成员变量 withoutY_flag，用来保存命令行的参数，如果传入的是测试数据，这个参数值就应该是字符串"true"。

```
struct Option {
    string test_path, model_path, output_path, withoutY_flag;
};
```

然后修改 parse_option 函数，接收我们定义的新参数：

```
Option parse_option(int argc, char **argv) {
    vector<string> args;
    for(int i = 0; i < argc; i++)
        args.push_back(string(argv[i]));

    if(argc == 1)
        throw invalid_argument(predict_help());

    Option option;

    if(argc != 4 && argc != 5)
        throw invalid_argument("cannot parse argument");

    option.test_path = string(args[1]);
    option.model_path = string(args[2]);
    option.output_path = string(args[3]);
    // 添加开始
    if(argc == 5){  // 保存传入的参数
        option.withoutY_flag = string(args[4]);
    } else {
        option.withoutY_flag = "";
    }
    // 添加结束

    return option;
}
```

接下来修改 main 函数，判断传入的新参数是否是"true"，如果是则转去执行不需要 Y 的 predict 函数，否则执行原来的函数。

```
int main(int argc, char **argv) {
    Option option;
    try {
        option = parse_option(argc, argv);
    } catch(invalid_argument const &e) {
        cout << e.what() << endl;
```

```cpp
        return 1;
    }
    // 修改开始
    if(argc == 5 && option.withoutY_flag.compare("true") == 0){
        // 新添加的不使用Y的predict函数，其他参数都一样
        predict_withoutY(option.test_path, option.model_path, option.output_path);
    } else {
        // 原函数
        predict(option.test_path, option.model_path, option.output_path);
    }
    // 修改结束
    return 0;
}
```

我们先说说对 LibFFM 中 predict 函数的修改。因为要保存 Sigmoid 激活函数前一层的输出值，所以要对 predict 函数进行修改。修改的地方并不多，只添加了 4 句代码。

```cpp
void predict(string test_path, string model_path, string output_path) {
    int const kMaxLineSize = 1000000;

    FILE *f_in = fopen(test_path.c_str(), "r");
    ofstream f_out(output_path);
    // 添加 1:定义输出文件流，用于保存输出
    ofstream f_out_t(output_path + ".logit");
    char line[kMaxLineSize];
    // 加载模型
    ffm_model model = ffm_load_model(model_path);

    ffm_double loss = 0;
    vector<ffm_node> x;
    ffm_int i = 0;

    // 循环处理每行数据
    for(; fgets(line, kMaxLineSize, f_in) != nullptr; i++) {
        x.clear();
        // 最先读出的是Y
        char *y_char = strtok(line, " \t");
```

```cpp
        ffm_float y = (atoi(y_char)>0)? 1.0f : -1.0f;

        // 循环读取其他FFM字段保存到x中，数据格式是：fields:features:values
        while(true) {
            char *field_char = strtok(nullptr,":");
            char *idx_char = strtok(nullptr,":");
            char *value_char = strtok(nullptr," \t");
            if(field_char == nullptr || *field_char == '\n')
                break;

            ffm_node N;
            N.f = atoi(field_char);
            N.j = atoi(idx_char);
            N.v = atof(value_char);

            x.push_back(N);
        }

        // 传入x做预测，得到y
        ffm_float y_bar = ffm_predict(x.data(), x.data()+x.size(), model);
        // 添加 2:调用自定义的函数ffm_get_wTx，返回Sigmoid激活函数前一层的输出值
        // ffm_get_wTx的代码就是复制的ffm_predict函数，把最后一句Sigmoid函数去掉
        ffm_float ret_t = ffm_get_wTx(x.data(), x.data()+x.size(), model);
        loss -= y==1? log(y_bar) : log(1-y_bar);
        // 添加 3:将返回值写入文件中
        f_out_t << ret_t << "\n";
        f_out << y_bar << "\n";
    }

    loss /= i;

    cout << "logloss = " << fixed << setprecision(5) << loss << endl;
    // 添加 4:关闭文件句柄
    fclose(f_in);
}
```

ffm_get_wTx 函数被定义在 ffm.cpp 文件中，它跟 ffm_predict 相比变化只有一处，就是最后

一句 return 返回的不同。

```cpp
ffm_float ffm_get_wTx(ffm_node *begin, ffm_node *end, ffm_model &model){
    ffm_float r = 1;
    // 计算正则项
    if(model.normalization) {
        r = 0;
        for(ffm_node *N = begin; N != end; N++)
            r += N->v*N->v;
        r = 1/r;
    }
    // 通过模型的参数矩阵与特征x计算ffm
    ffm_float t = wTx(begin, end, r, model);

    // 在ffm_predict函数中返回的是1/(1+exp(-t))，而此处只返回t即可
    return t;
}
```

不要忘记在 ffm.h 中加入函数的声明：

```cpp
ffm_float ffm_get_wTx(ffm_node *begin, ffm_node *end, ffm_model &model);
```

最后来看看函数 predict_withoutY，这个函数只做预测，数据中不含有 Y。

```cpp
void predict_withoutY(string test_path, string model_path, string output_path) {
    int const kMaxLineSize = 1000000;

    FILE *f_in = fopen(test_path.c_str(), "r");
    ofstream f_out(output_path);
    ofstream f_out_t(output_path + ".logit");
    char line[kMaxLineSize];

    ffm_model model = ffm_load_model(model_path);
    // 删除1:删除loss变量，因为数据中没有Y，不计算loss
    // ffm_double loss = 0;
    vector<ffm_node> x;
    ffm_int i = 0;

    for(; fgets(line, kMaxLineSize, f_in) != nullptr; i++) {
```

```
x.clear();
// 删除2：读取每行数据时，开头不再是Y了
// char *y_char = strtok(line, " \t");
// ffm_float y = (atoi(y_char)>0)? 1.0f : -1.0f;

// 添加开始
// 其实就是复制的下面循环中的代码，读取的第一个数据不再是Y了，所以要按照FFM
// 的格式读取
char *field_char = strtok(line,":");
char *idx_char = strtok(nullptr,":");
char *value_char = strtok(nullptr," \t");
if(field_char == nullptr || *field_char == '\n')
    continue;

ffm_node N;
N.f = atoi(field_char);
N.j = atoi(idx_char);
N.v = atof(value_char);

x.push_back(N);
// 添加结束

while(true) {
    char *field_char = strtok(nullptr,":");
    char *idx_char = strtok(nullptr,":");
    char *value_char = strtok(nullptr," \t");
    if(field_char == nullptr || *field_char == '\n')
        break;

    ffm_node N;
    N.f = atoi(field_char);
    N.j = atoi(idx_char);
    N.v = atof(value_char);

    x.push_back(N);
}
```

```cpp
        ffm_float y_bar = ffm_predict(x.data(), x.data()+x.size(), model);
        ffm_float ret_t = ffm_get_wTx(x.data(), x.data()+x.size(), model);
        // 删除3:跟loss相关的代码一律删除
        // loss -= y==1? log(y_bar) : log(1-y_bar);

        f_out_t << ret_t << "\n";
        f_out << y_bar << "\n";
    }
    // 删除4:删除loss计算
    // loss /= i;

    // 删除5:也不用打印loss了
    //cout << "logloss = " << fixed << setprecision(5) << loss << endl;
    cout << "done!" << endl;

    fclose(f_in);
}
```

以上就是对 LibFFM 的修改。因为本节重点不是讨论 FFM 的算法，在此就不展开介绍了，感兴趣的读者可以继续深入，对照 FFM 的公式看看函数 wTx 是如何实现的。

当模型训练好之后，通过 ffm-predict 生成 FFM 的输出。Python 代码如下：

```python
# 传入训练集数据，生成训练集的输出文件tr_ffm.out.logit
cmd = './libffm/libffm/ffm-predict ./data/train_ffm.txt model_ffm tr_ffm.out'.
    format(nr_thread=NR_THREAD)
os.popen(cmd).readlines()

# 传入验证集数据，得到输出文件va_ffm.out.logit
cmd = './libffm/libffm/ffm-predict ./data/valid_ffm.txt model_ffm va_ffm.out'.
    format(nr_thread=NR_THREAD)
os.popen(cmd).readlines()

# 传入测试集数据，得到输出文件te_ffm.out.logit
cmd = './libffm/libffm/ffm-predict ./data/test_ffm.txt model_ffm te_ffm.out true'.
    format(nr_thread=NR_THREAD)
os.popen(cmd).readlines()
```

5.7 训练 GBDT

关于 GBDT 的实现有多种选择，可以使用 XGBoost，也可以使用 Scikit-learn，还可以使用 LightGBM。

本节我们使用 LightGBM 来训练 GBDT 模型，使用如下命令进行安装：

`pip install lightgbm`

当然，你也可以通过官方发布的 whl 包安装，访问如下地址：

https://github.com/Microsoft/LightGBM/releases

因为决策树较容易过拟合，所以我们设置树的棵数为 32，叶子节点数为 30，深度就不设置了，学习率设为 0.05。当然，你也可以尝试其他的超参数组合。

传给 LightGBM 的参数说明：

```
# 任务是训练
task = train
# 训练方式
boosting_type = gbdt
# 目标:二分类
objective = binary
# 多个度量指标
metric = {'l2', 'auc', 'logloss'}
# 树的棵数
num_trees = 32
# 学习率
learning_rate = 0.05
# 一棵树上的叶子节点数
num_leaves = 30

# 特征子采样，设置为 0.9，在每棵树训练之前随机选择 90% 的特征
feature_fraction = 0.9

# 数据样本子采样，在不进行重采样的情况下随机选择部分数据
bagging_fraction = 0.8

# bagging 的频率，0 表示禁用 bagging；5 表示每 5 次迭代执行一次bagging
```

```python
bagging_freq = 5

import lightgbm as lgb
import pandas as pd

def lgb_pred(tr_path, va_path, _sep = '\t', iter_num = 32):
    # 加载训练集和验证集数据
    print('Load data...')
    df_train = pd.read_csv(tr_path, header=None, sep=_sep)
    df_test = pd.read_csv(va_path, header=None, sep=_sep)

    y_train = df_train[0].values
    y_test = df_test[0].values
    X_train = df_train.drop(0, axis=1).values
    X_test = df_test.drop(0, axis=1).values

    # 为lightgbm创建数据集
    lgb_train = lgb.Dataset(X_train, y_train)
    lgb_eval = lgb.Dataset(X_test, y_test, reference=lgb_train)

    # 设置传给lgb的超参数
    params = {
        'task': 'train',
        'boosting_type': 'gbdt',
        'objective': 'binary',
        'metric': {'l2', 'auc', 'logloss'},
        'num_leaves': 30,
        'num_trees': 32,
        'learning_rate': 0.05,
        'feature_fraction': 0.9,
        'bagging_fraction': 0.8,
        'bagging_freq': 5,
        'verbose': 0
    }

    print('Start training...')
    # 开始训练
```

```python
        gbm = lgb.train(params,
                        lgb_train,
                        num_boost_round=iter_num,
                        valid_sets=lgb_eval,
                        feature_name=["I1","I2","I3","I4","I5","I6","I7","I8","I9","I10
                        ","I11","I12","I13","C1","C2","C3","C4","C5","C6","C7","C8","C9
                        ","C10","C11","C12","C13","C14","C15","C16","C17","C18","C19","
                        C20","C21","C22","C23","C24","C25","C26"],
                        categorical_feature=["C1","C2","C3","C4","C5","C6","C7","C8","
                        C9","C10","C11","C12","C13","C14","C15","C16","C17","C18","C19
                        ","C20","C21","C22","C23","C24","C25","C26"],
                        early_stopping_rounds=5)

        print('Save model...')
        # 保存模型
        gbm.save_model('lgb_model.txt')

        print('Start predicting...')
        # 用验证集试试预测效果
        y_pred = gbm.predict(X_test, num_iteration=gbm.best_iteration)
        # 打印rmse
        print('The rmse of prediction is:', mean_squared_error(y_test, y_pred) ** 0.5)

        return gbm,y_pred,X_train,y_train

# 调用刚刚定义的训练函数开始训练
gbm,y_pred,X_train ,y_train = lgb_pred('./data/train_lgb.txt', './data/valid_lgb.txt', '\t', 256)
```

训练时打印出的日志如下：

```
Load data...
Start training...
/Applications/anaconda/envs/tensorflow1.0/lib/python3.5/site-packages/lightgbm/
engine.py:99: UserWarning: Found  num_trees  in params. Will use it instead of
argument
  warnings.warn("Found {} in params. Will use it instead of argument".format(alias
))
```

```
/Applications/anaconda/envs/tensorflow1.0/lib/python3.5/site-packages/lightgbm/
basic.py:1029: UserWarning: categorical_feature in Dataset is overrided. New
categorical_feature is ['C1', 'C10', 'C11', 'C12', 'C13', 'C14', 'C15', 'C16', 'C17
', 'C18', 'C19', 'C2', 'C20', 'C21', 'C22', 'C23', 'C24', 'C25', 'C26', 'C3', 'C4',
 'C5', 'C6', 'C7', 'C8', 'C9']
  warnings.warn('categorical_feature in Dataset is overrided. New
  categorical_feature is {}'.format(sorted(list(categorical_feature))))
/Applications/anaconda/envs/tensorflow1.0/lib/python3.5/site-packages/lightgbm/
basic.py:668: UserWarning: categorical_feature in param dict is overrided.
  warnings.warn('categorical_feature in param dict is overrided.')
[1]     valid_0's l2: 0.241954   valid_0's auc: 0.70607
Training until validation scores don't improve for 5 rounds.
[2]     valid_0's l2: 0.234704   valid_0's auc: 0.715608
[3]     valid_0's l2: 0.228139   valid_0's auc: 0.717791
[4]     valid_0's l2: 0.222168   valid_0's auc: 0.72273
[5]     valid_0's l2: 0.216728   valid_0's auc: 0.724065
[6]     valid_0's l2: 0.211819   valid_0's auc: 0.725036
[7]     valid_0's l2: 0.207316   valid_0's auc: 0.727427
[8]     valid_0's l2: 0.203296   valid_0's auc: 0.728583
[9]     valid_0's l2: 0.199582   valid_0's auc: 0.730092
[10]    valid_0's l2: 0.196185   valid_0's auc: 0.730792
[11]    valid_0's l2: 0.193063   valid_0's auc: 0.732316
[12]    valid_0's l2: 0.190268   valid_0's auc: 0.733773
[13]    valid_0's l2: 0.187697   valid_0's auc: 0.734782
[14]    valid_0's l2: 0.185351   valid_0's auc: 0.735636
[15]    valid_0's l2: 0.183215   valid_0's auc: 0.736346
[16]    valid_0's l2: 0.181241   valid_0's auc: 0.737393
[17]    valid_0's l2: 0.179468   valid_0's auc: 0.737709
[18]    valid_0's l2: 0.177829   valid_0's auc: 0.739096
[19]    valid_0's l2: 0.176326   valid_0's auc: 0.740135
[20]    valid_0's l2: 0.174948   valid_0's auc: 0.741065
[21]    valid_0's l2: 0.173675   valid_0's auc: 0.742165
[22]    valid_0's l2: 0.172499   valid_0's auc: 0.742672
[23]    valid_0's l2: 0.171471   valid_0's auc: 0.743246
[24]    valid_0's l2: 0.17045    valid_0's auc: 0.744415
[25]    valid_0's l2: 0.169582   valid_0's auc: 0.744792
```

```
[26]     valid_0's l2: 0.168746    valid_0's auc: 0.745478
[27]     valid_0's l2: 0.167966    valid_0's auc: 0.746282
[28]     valid_0's l2: 0.167264    valid_0's auc: 0.74675
[29]     valid_0's l2: 0.166582    valid_0's auc: 0.747429
[30]     valid_0's l2: 0.16594     valid_0's auc: 0.748392
[31]     valid_0's l2: 0.165364    valid_0's auc: 0.748986
[32]     valid_0's l2: 0.164844    valid_0's auc: 0.749362
Did not meet early stopping. Best iteration is:
[32]     valid_0's l2: 0.164844    valid_0's auc: 0.749362
Save model...
Start predicting...
The rmse of prediction is: 0.406009502303
```

保存 LightGBM 的变量：

```
dump = gbm.dump_model()

save_params_with_name((gbm, dump), 'gbm_dump')

gbm, dump = load_params_with_name('gbm_dump')
```

LightGBM 还提供了分析特征重要性的接口，我们来分析看看。

查看每个特征的重要程度：

```
gbm.feature_importance()
```

返回的是每一个特征的重要度：

```
array([ 15,   0,  30,  10,  12,  79,  31,  15,  14,   0,  44,   0,  29,
         0,  16,   0,  65,   0,   0,  32,   0,   0,  29,  30,  10, 120,
        30, 165,  19,  11,  69,   1,   0,   2,   0,   6,  39,   0,   5])
```

feature_importance 函数还可以传入参数 "gain"，返回的是每个特征的权重：

```
gbm.feature_importance("gain")
```

每个特征的权重：

```
array([ 69634.31561279,      0.        ,  17624.44689941,
         4734.61398315,  10529.7180481 , 199794.76257324,
        94191.14331055,  13543.23699951,  10014.74700928,
```

```
       0.        ,  191050.53414917,       0.        ,
   28020.85171509,       0.        ,    6852.7729187 ,
       0.        ,   32251.70903015,       0.        ,
       0.        ,   14341.38494873,       0.        ,
       0.        ,   11129.02203369,   12486.21105957,
    5218.96902466,   99722.85806274,   23106.2180481 ,
   79130.2718811 ,   10490.07904053,   17757.50100708,
   34302.44396973,     424.67401123,       0.        ,
     882.20599365,       0.        ,    3156.61196899,
   15901.01004028,       0.        ,    3397.2270813 ])
```

这么看实在看不出所以然来，现在我们把每个特征的重要程度排序看看：

```
def ret_feat_impt(gbm):
    # 先将特征权重变成百分比
    gain = gbm.feature_importance("gain").reshape(-1, 1) / sum(gbm.
        feature_importance("gain"))
    # 获取每个特征的名字
    col = np.array(gbm.feature_name()).reshape(-1, 1)
    # 将比值和名字组合在一起，升序排列
    return sorted(np.column_stack((col, gain)),key=lambda x: x[1],reverse=True)

ret_feat_impt(gbm)
```

这回看着直观一些了吧，当然，你也可以做个图表看着更方便：

```
[array(['I6', '0.1978774213012332'],
      dtype='<U32'), array(['I11', '0.1892171073393491'],
      dtype='<U32'), array(['C13', '0.098765862248320 32'],
      dtype='<U32'), array(['I7', '0.093287232896674 94'],
      dtype='<U32'), array(['C15', '0.078370893936512 43'],
      dtype='<U32'), array(['I1', '0.068966066127406 37'],
      dtype='<U32'), array(['C18', '0.033973258706274 91'],
      dtype='<U32'), array(['C4', '0.031942203755739 26'],
      dtype='<U32'), array(['I13', '0.027751948092299 045'],
      dtype='<U32'), array(['C14', '0.022884477973766 117'],
      dtype='<U32'), array(['C17', '0.017587090185844 79'],
      dtype='<U32'), array(['I3', '0.017455312939137 25'],
      dtype='<U32'), array(['C24', '0.015748415135270 675'],
```

```
dtype='<U32'), array(['C7', '0.014203757070472703'],
dtype='<U32'), array(['I8', '0.013413268591324624'],
dtype='<U32'), array(['C11', '0.012366386458128355'],
dtype='<U32'), array(['C10', '0.011022221770323784'],
dtype='<U32'), array(['I5', '0.010428669037920042'],
dtype='<U32'), array(['C16', '0.010389410428237439'],
dtype='<U32'), array(['I9', '0.009918639946598076'],
dtype='<U32'), array(['C2', '0.006787009911825981'],
dtype='<U32'), array(['C12', '0.005168884905437884'],
dtype='<U32'), array(['I4', '0.00468917800335175'],
dtype='<U32'), array(['C26', '0.003364625407413743'],
dtype='<U32'), array(['C23', '0.0031263193710805628'],
dtype='<U32'), array(['C21', '0.0008737398560005959'],
dtype='<U32'), array(['C19', '0.00042059860405565207'],
dtype='<U32'), array(['I2', '0.0'],
dtype='<U32'), array(['I10', '0.0'],
dtype='<U32'), array(['I12', '0.0'],
dtype='<U32'), array(['C1', '0.0'],
dtype='<U32'), array(['C3', '0.0'],
dtype='<U32'), array(['C5', '0.0'],
dtype='<U32'), array(['C6', '0.0'],
dtype='<U32'), array(['C8', '0.0'],
dtype='<U32'), array(['C9', '0.0'],
dtype='<U32'), array(['C20', '0.0'],
dtype='<U32'), array(['C22', '0.0'],
dtype='<U32'), array(['C25', '0.0'],
dtype='<U32')]
```

根据 LightGBM 训练总结出的各特征权重，我们发现 I6 和 I11 两个特征很重要，比值的和接近 40%。

5.8 用 LightGBM 的输出生成 FM 数据

GBDT 的模型训练好了，我们需要将数据输入给 GBDT 进行预测，得到输出之后再传给 FM。FM 的实现使用 LibFM，我们一起看看 LibFM 需要的数据格式是什么样的。关于数据格式的解释可以参考 libFM 1.4.2-Manual（http://www.libfm.org/libfm-1.42.manual.pdf）中的说明。

LibFM 支持文本和二进制两种数据格式,我们使用文本格式。

其数据格式与 SVMlite 和 LIBSVM 中的数据格式相同:对于每行的训练样本 (x, y),首先是 y 的值,然后是 x 的非零值。对于二分类,$y > 0$ 的情况被认为是正类,$y \leqslant 0$ 作为负类。

举例:

```
4 0:1.5 3:-7.9
2 1:1e-5 3:2
-1 6:1
...
```

这个文件包含三个样本。第一列是每个样本的目标,即第一个是 4,第二个是 2,第三个是 -1。在目标之后,每行包含 x 的非零元素,0:1.5 表示 $x_0 = 1.5$,3:-7.9 表示 $x_3 = -7.9$。这意味着 index:value 的左侧表示 x 的索引,而右侧表示 x_{index} 的值,即 $x_{index} = \text{value}$。

本例的数据描述了以下矩阵 $\boldsymbol{X}$ 和目标向量 $\boldsymbol{y}$:

$$\boldsymbol{X} = \begin{pmatrix} 1.5 & 0.0 & 0.0 & -7.9 & 0.0 & 0.0 & 0.0 \\ 0.0 & 10^{-5} & 0.0 & 2.0 & 0.0 & 0.0 & 0.0 \\ 0.0 & 0.0 & 0.0 & 0.0 & 0.0 & 0.0 & 1.0 \end{pmatrix}, \quad \boldsymbol{y} = \begin{pmatrix} 4 \\ 2 \\ -1 \end{pmatrix}$$

以上就是 LibFM 要求的数据格式,我们需要把 GBDT 输出的叶子节点值作为输入数据 $\boldsymbol{X}$ 传给 FM,数据格式是 $\boldsymbol{X}$ 中不是 0 的数据的 index:value。

一段真实数据如下:

```
0    0:31      1:61      2:93      3:108    4:149    5:182    6:212    7:242    8:277    9:310
     10:334    11:365    12:401    13:434   14:465   15:491   16:527   17:552   18:589   19:619
20:648  21:678  22:697  23:744  24:770  25:806  26:826  27:862  28:899  29:928
30:955  31:988
```

也可以将所有叶子节点的编号统一成 one-hot 编码,这样每个叶子节点值就是它在 one-hot 编码中的位置(index),值(value)是 1。这样的话,上面传给 FM 的输入数据就变成了:

```
0    31:1     61:1     93:1     108:1    149:1    182:1    212:1    242:1    277:1    310:1
     334:1    365:1    401:1    434:1    465:1    491:1    527:1    552:1    589:1    619:1
648:1  678:1  697:1  744:1  770:1  806:1  826:1  862:1  899:1  928:1
955:1  988:1
```

one-hot 编码的数据格式就留给读者来实现。

代码实现如下:

```python
def generat_lgb2fm_data(outdir, gbm, dump, tr_path, va_path, te_path, _sep = '\t'):
    # 分别生成训练集、验证集和测试集数据
    with open(os.path.join(outdir, 'train_lgb2fm.txt'), 'w') as out_train:
        with open(os.path.join(outdir, 'valid_lgb2fm.txt'), 'w') as out_valid:
            with open(os.path.join(outdir, 'test_lgb2fm.txt'), 'w') as out_test:
                # 读入lgb的数据
                df_train_ = pd.read_csv(tr_path, header=None, sep=_sep)
                df_valid_ = pd.read_csv(va_path, header=None, sep=_sep)
                df_test_= pd.read_csv(te_path, header=None, sep=_sep)

                # 先将特征X和目标y分离出来
                y_train_ = df_train_[0].values
                y_valid_ = df_valid_[0].values

                X_train_ = df_train_.drop(0, axis=1).values
                X_valid_ = df_valid_.drop(0, axis=1).values
                X_test_= df_test_.values

                # pred_leaf设为True，指示预测叶子节点的索引值
                # 针对每条数据都会有32个叶子节点索引值的输出
                # 因为是32棵树，每棵树的结果最终会落入30个叶子节点中的一个上
                # 也就是说，train_leaves的shape是[N, 32]，N是数据总数
                # train_leaves中每个值的范围都是(0~30]
                train_leaves= gbm.predict(X_train_, num_iteration=gbm.
                    best_iteration, pred_leaf=True)
                valid_leaves= gbm.predict(X_valid_, num_iteration=gbm.
                    best_iteration, pred_leaf=True)
                test_leaves= gbm.predict(X_test_, num_iteration=gbm.
                    best_iteration, pred_leaf=True)

                # 得到树的信息变量，dump['tree_info']返回含有所有树信息的数组
                tree_info = dump['tree_info']
                # 一共32棵树
                tree_counts = len(tree_info)
                # 循环遍历每棵树，修改叶子节点索引值进行统一编号
                for i in range(tree_counts):
```

```python
    # tree_info[i]['num_leaves']是每棵树的叶子节点数,其实这个数值
    # 每棵树都一样,是30
    # 此处的目的是将32棵树输出的叶子节点值统一编号。也就是说,第一
    # 棵树的叶子节点编号是0~29,第二棵树顺序编号为30~59,依此类推
    # 因为LibFM要求的数据是值不为0的x,所以给编号加1,从1开始顺序
    # 编号
    train_leaves[:, i] = train_leaves[:, i] + tree_info[i]['
        num_leaves'] * i + 1
    valid_leaves[:, i] = valid_leaves[:, i] + tree_info[i]['
        num_leaves'] * i + 1
    test_leaves[:, i] = test_leaves[:, i] + tree_info[i]['
        num_leaves'] * i + 1

    # 生成训练数据、验证数据和测试数据
    for idx in range(len(y_train_)):
        out_train.write((str(y_train_[idx]) + '\t'))
        out_train.write('\t'.join(
            ['{}:{}'.format(ii, val) for ii,val in enumerate(
            train_leaves[idx]) if float(val) != 0 ]) + '\n')

    for idx in range(len(y_valid_)):
        out_valid.write((str(y_valid_[idx]) + '\t'))
        out_valid.write('\t'.join(
            ['{}:{}'.format(ii, val) for ii,val in enumerate(
            valid_leaves[idx]) if float(val) != 0 ]) + '\n')

    for idx in range(len(X_test_)):
        out_test.write('\t'.join(
            ['{}:{}'.format(ii, val) for ii,val in enumerate(
            test_leaves[idx]) if float(val) != 0 ]) + '\n')


generat_lgb2fm_data('./data', gbm, dump, './data/train_lgb.txt', './data/valid_lgb.
txt', './data/test_lgb.txt', '\t')
```

5.9 训练 FM

训练 FM 的数据准备好了，我们调用 LibFM 进行训练。其下载地址为：https://github.com/srendle/libfm。

LibFM 的安装很简单，打开 shell，在代码目录下执行 make all 命令即可。

注意：LibFM 的代码笔者做了修改，请使用笔者的代码库中的相关代码。接下来说明都修改了哪些地方。

修改点其实跟使用 LibFFM 差不多，一是我们使用 LibFM 输出的 logits 值，就是 Sigmoid 激活函数前一层的返回值，不使用 Sigmoid 激活函数输出的概率值，所以在预测输出部分要修改；二是修改读取数据集的部分，因为我们会传入没有标签 Y 的测试集数据，所以在读取数据集的地方要有对测试集数据的处理；三是 LibFM 默认读入数据就要进行训练，而我们训练好 FM 模型后要进行预测，不使用训练的功能了，所以也要添加是否训练的判断代码。

首先修改 libfm.cpp 文件中的 main 函数，在函数开始定义命令行参数的地方，添加三个参数。

```
// 参数train_off表示是否训练，"false"表示训练，"true"表示不训练
const std::string param_train_off   = cmdline.registerParameter("train_off",
    "predict results without train");
// 参数prefix表示保存的文件名的前缀
const std::string param_prefix      = cmdline.registerParameter("prefix", "prefix
    filename for logits output");
// 参数test2predict表示传入的数据集是否有标签Y，"true"表示传入的是测试集数据
// 没有标签Y，"false"表示传入的是训练集或验证集数据
const std::string param_test2predict = cmdline.registerParameter("test2predict",
    "test file for predict into logits output");
```

继续看 main 函数，在注释行 "// (1) Load the data" 的地方（请读者自行在源代码中查找），就是加载数据集的代码。

```
// 新代码
// 判断是否加载测试集，如果是"true"，则调用我们新写的函数load_withoutY来加载
// 否则，调用原来的load函数
if(cmdline.hasParameter(param_test2predict)
&& ! cmdline.getValue(param_test2predict).compare("true")){
    train.load_withoutY(cmdline.getValue(param_train_file));
} else {
    // 原代码
```

```
        train.load(cmdline.getValue(param_train_file));
}
```

在 main 函数的最后,从注释行 "// () learn" 的地方开始,是训练 FM 和进行预测的代码。

```
// 新代码
// 添加了是否训练的判断,如果train_off是"false",则调用训练函数learn
if (! cmdline.hasParameter(param_train_off)
|| ! cmdline.getValue(param_train_off).compare("false")) {
    // 原代码
    // () learn
    fml->learn(train, test);
}


// () Prediction at the end  (not for mcmc and als)
// 新代码
// 此处添加了是否是测试集的判断,因为原代码是要训练集和测试集一起打印准确率的
// 如果只传入测试集就不调用原代码的逻辑了,参数test2predict是"true"表示只进行预测
// 不进行训练
if(cmdline.hasParameter(param_test2predict)
&& ! cmdline.getValue(param_test2predict).compare("true")){
    std::cout << "Final\t" << "Test=" << fml->evaluate(test) << std::endl;
} else {
    // 原代码
    if (cmdline.getValue(param_method).compare("mcmc")) {
        std::cout << "Final\t" << "Train=" << fml->evaluate(train) << "\tTest=" <<
            fml->evaluate(test) << std::endl;
}


// 原代码,保存预测值
// () Save prediction
if (cmdline.hasParameter(param_out)) {
    DVector<double> pred;
    pred.setSize(test.num_cases);
    fml->predict(test, pred);
    pred.save(cmdline.getValue(param_out));
}
```

```cpp
// 新代码
// 此处判断是训练还是进行预测, 参数train_off是"true"表示不是训练, 而是要进行预测
// 这部分代码主要复制自上面的预测代码, 区别是保存的预测值是不经过Sigmoid激活函数处
// 理的返回值。不再调用predict, 而是调用我们新写的sgd_logits函数
if (cmdline.hasParameter(param_train_off)
&& ! cmdline.getValue(param_train_off).compare("true")) {
    // std::cout << "--------train off--------start!" << std::endl;
    DVector<double> pred;
    pred.setSize(train.num_cases);
    ((fm_learn_sgd_element*)fml)->sgd_logits(train, pred);

    // 保存LibFM处理后的值, 判断文件名是否使用了参数给定的前缀
    if (cmdline.hasParameter(param_prefix)){
        pred.save(cmdline.getValue(param_prefix) + ".fm.logits");
    } else {
        pred.save("fm.logits");
    }
    // std::cout << "--------train off--------end!" << std::endl;
}

// 原代码, 保存模型
// () save the FM model
if (cmdline.hasParameter(param_save_model)) {
    std::cout << "Writing FM model to "<< cmdline.getValue(param_save_model) << std
    ::endl;
    fm.saveModel(cmdline.getValue(param_save_model));
}
```

接下来修改 fm_learn_sgd_element.h, 在 fm_learn_sgd_element 类中新添加一个函数 sgd_logits, 用来做预测, 不使用 Sigmoid 激活函数。该函数复制自 fm_learn_sgd.h 文件中的 predict 函数并做了少许修改。

```cpp
void sgd_logits(Data& data, DVector<double>& out) {
    assert(data.data->getNumRows() == out.dim);
    for (data.data->begin(); !data.data->end(); data.data->next()) {
        double p = predict_case(data);
```

```
        // 删除原函数中的Sigmoid激活函数处理, 主要是指TASK_CLASSIFICATION的部分
        // std::cout << p << std::endl;
        // if (task == TASK_REGRESSION ) {
        //   p = std::min(max_target, p);
        //   p = std::max(min_target, p);
        // } else if (task == TASK_CLASSIFICATION) {
        //   p = 1.0/(1.0 + exp(-p));
        // } else {
        //   throw "task not supported";
        // }
        out(data.data->getRowIndex()) = p;
    }
}
```

最后修改 Data.h 文件，添加加载测试数据的函数 load_withoutY。该函数主要复制自 Data::load 函数，只做了少许修改。

第一处修改是在注释行 "(1) determine the number of rows and the maximum feature_id" 的地方。

```
// 首先将if判断删除, 因为每行数据的第一项不再是标签Y了
// if (sscanf(pline, "%f%n", &_value, &nchar) >=1)
{
    // 同理, sscanf没有调用, 也就不用加nchar了
    // pline += nchar;
    // 设定标签Y的最小值与最大值0和1, 删除原来的min和max代码
    min_target = 0;//std::min(_value, min_target);
    max_target = 1;//std::max(_value, max_target);
    num_rows++;
    while (sscanf(pline, "%d:%f%n", &_feature, &_value, &nchar) >= 2) {
        pline += nchar;
        num_feature = std::max(_feature, num_feature);
        has_feature = true;
        num_values++;
    }
    if(*pline == 13){
        num_feature = std::max(_feature, num_feature);
        has_feature = true;
```

```
            num_values++;
        }
        //printf("%d:%f\n",_feature, _value);
        //printf("%s %c %d\n", pline, *pline, *pline);
        while ((*pline != 0) && ((*pline == ' ')||(*pline == 9))) { printf("%c %d\n",
            *pline, *pline); pline++; } // skip trailing spaces
        if ((*pline != 13) && (*pline != 0)  && (*pline != '#')) {
            throw "cannot parse line \"" + line + "\" at character " + pline[0];
        }
}
// 因为最开始的if删除了，所以也不需要else代码了
/*else {
    throw "cannot parse line \"" + line + "\" at character " + pline[0];
}*/
```

第二处修改是在注释行"(2) read the data"的地方，这里开始读取数据，修改方法跟上面基本一样。

```
// 跟上面一样，删除if判断，因为每行数据的第一项不再是标签Y了
// if (sscanf(pline, "%f%n", &_value, &nchar) >=1)
{
    // 也不需要加nchar了，因为没有调用if判断的sscanf函数
    // pline += nchar;
    assert(row_id < num_rows);
    // 删除标签Y的赋值语句
    // target.value[row_id] = _value;
    data.value[row_id].data = &(cache[cache_id]);
    data.value[row_id].size = 0;

    while (sscanf(pline, "%d:%f%n", &_feature, &_value, &nchar) >= 2) {
        pline += nchar;
        assert(cache_id < num_values);
        cache[cache_id].id = _feature;
        cache[cache_id].value = _value;
        cache_id++;
        data.value[row_id].size++;
    }
```

```
    if(*pline == 13){
        assert(cache_id < num_values);
        cache[cache_id].id = _feature;
        cache[cache_id].value = _value;
        cache_id++;
        data.value[row_id].size++;
    }
    row_id++;

    while ((*pline != 0) && ((*pline == ' ')  || (*pline == 9))) { pline++; } //
skip trailing spaces
    if ((*pline != 13) && (*pline != 0)  && (*pline != '#')) {
        throw "cannot parse line \"" + line + "\" at character " + pline[0];
    }
}
// 因为if判断删除了，所以也不需要else代码了
/*else {
    throw "cannot parse line \"" + line + "\" at character " + pline[0];
}*/
```

以上就是修改 LibFM 的代码。接下来介绍命令行参数，看看 LibFM 怎么用。

LibFM 通过参数 -train 使用训练数据集训练 FM 模型，并通过参数 -test 使用测试数据集进行预测。更多参数请参考官方文档，我们使用的参数如下：

- -dim 'k0,k1,k2'：用于指定 FM 的维度。$k0 \in \{0,1\}$ 确定是否在模型中使用全局偏置项（bias），即 w_0；$k1 \in \{0,1\}$ 确定是否在模型中使用单向交互（每个变量的偏置项），即 w；k2 给出用于成对相互作用的因子数。默认值为 1, 1, 8。

- -init_stdev：用于初始化参数 V 的正态分布的标准差。应该在这里使用非零的正值；默认值为 0.1。

- -iter：迭代次数。默认值为 100。

- -learn_rate：SGD 的学习率。默认值为 0.1。

- -method：学习方法（SGD、SGDA、ALS、MCMC）。默认值为 MCMC。

- -regular 'r0,r1,r2'：正则化参数应该具有零或正值。这里只介绍与 SGD 相关的设置，对于 SGD 可以通过以下方式指定正则化值。

    – 一个值（-regular r0）：所有模型参数都使用相同的正则化值。

- 三个值（-regular 'r0,r1,r2'）：0-way 交互（w_0）使用 r0 作为正则化值；1-way 交互（w）使用 r1 和 r2。
- 没有值：如果参数-regular 没有被指定，则相当于没有正则化，即-regular 0。
- `-task`：指定任务是分类还是回归。r＝回归，c＝二分类。[必填项]
- `-test`：测试数据文件名。[必填项]
- `-train`：训练数据文件名。[必填项]
- `-save_model`：指定保存模型的文件名。
- `-load_model`：指定加载模型的文件名。
- `-train_off`：指定是训练（"false"）还是预测（"true"）。
- `-prefix`：指定预测值保存文件名的前缀。
- `-test2predict`：指定传入的是否是没有标签 Y 的测试数据（"true"）。

我们设置参数为迭代 64 次，使用 SGD 训练，学习率是 0.00000001（说真的，笔者也觉得这个学习率很夸张），训练好的模型保存为文件 fm_model。

训练 FM 的 Python 代码如下：

```
cmd = './libfm/libfm/bin/libFM -task c -train ./data/train_lgb2fm.txt -test ./data/valid_lgb2fm.txt -dim '1,1,8' -iter 64 -method sgd -learn_rate 0.00000001 -regular '0,0,0.01' -init_stdev 0.1 -save_model fm_model'
os.popen(cmd).readlines()
```

训练的输出如下：

```
['----------------------------------------------------------------------\n',
 'libFM\n',
 '  Version: 1.4.4\n',
 '  Author:  Steffen Rendle, srendle@libfm.org\n',
 '  WWW:     http://www.libfm.org/\n',
 'This program comes with ABSOLUTELY NO WARRANTY; for details see license.txt.\n',
 'This is free software, and you are welcome to redistribute it under certain\n',
 'conditions; for details see license.txt.\n',
 '----------------------------------------------------------------------\n',
 'Loading train...\t\n',
 'has x = 1\n',
 'has xt = 0\n',
```

```
'num_rows=899991\tnum_values=28799712\tnum_features=32\tmin_target=0\tmax_target
=1\n',
'Loading test... \t\n',
'has x = 1\n',
'has xt = 0\n',
'num_rows=100009\tnum_values=3200288\tnum_features=32\tmin_target=0\tmax_target=1\
n',
'#relations: 0\n',
'Loading meta data...\t\n',
'learnrate=1e-08\n',
'learnrates=1e-08,1e-08,1e-08\n',
'#iterations=64\n',
"SGD: DON'T FORGET TO SHUFFLE THE ROWS IN TRAINING DATA TO GET THE BEST RESULTS.\n
",
'#Iter=  0\tTrain=0.625438\tTest=0.619484\n',
'#Iter=  1\tTrain=0.636596\tTest=0.632013\n',
'#Iter=  2\tTrain=0.627663\tTest=0.623114\n',
'#Iter=  3\tTrain=0.609776\tTest=0.606605\n',
'#Iter=  4\tTrain=0.563581\tTest=0.56092\n',
'#Iter=  5\tTrain=0.497907\tTest=0.495655\n',
'#Iter=  6\tTrain=0.461677\tTest=0.461408\n',
'#Iter=  7\tTrain=0.453666\tTest=0.452639\n',
'#Iter=  8\tTrain=0.454026\tTest=0.453419\n',
'#Iter=  9\tTrain=0.456836\tTest=0.455919\n',
'#Iter= 10\tTrain=0.46032\tTest=0.459339\n',
'#Iter= 11\tTrain=0.466546\tTest=0.465358\n',
'#Iter= 12\tTrain=0.473565\tTest=0.472317\n',
'#Iter= 13\tTrain=0.481726\tTest=0.480967\n',
'#Iter= 14\tTrain=0.492357\tTest=0.491216\n',
'#Iter= 15\tTrain=0.504419\tTest=0.502935\n',
'#Iter= 16\tTrain=0.517793\tTest=0.516214\n',
'#Iter= 17\tTrain=0.533604\tTest=0.532102\n',
'#Iter= 18\tTrain=0.552926\tTest=0.5515\n',
'#Iter= 19\tTrain=0.575645\tTest=0.573198\n',
'#Iter= 20\tTrain=0.59418\tTest=0.590887\n',
'#Iter= 21\tTrain=0.610691\tTest=0.607815\n',
```

```
'#Iter= 22\tTrain=0.626138\tTest=0.623384\n',
'#Iter= 23\tTrain=0.640751\tTest=0.637923\n',
'#Iter= 24\tTrain=0.65393\tTest=0.652141\n',
'#Iter= 25\tTrain=0.666099\tTest=0.6641\n',
'#Iter= 26\tTrain=0.677933\tTest=0.675419\n',
'#Iter= 27\tTrain=0.689539\tTest=0.687108\n',
'#Iter= 28\tTrain=0.700177\tTest=0.697397\n',
'#Iter= 29\tTrain=0.709265\tTest=0.706156\n',
'#Iter= 30\tTrain=0.716553\tTest=0.713266\n',
'#Iter= 31\tTrain=0.723218\tTest=0.719635\n',
'#Iter= 32\tTrain=0.729163\tTest=0.726065\n',
'#Iter= 33\tTrain=0.734428\tTest=0.731354\n',
'#Iter= 34\tTrain=0.738863\tTest=0.735844\n',
'#Iter= 35\tTrain=0.74284\tTest=0.740323\n',
'#Iter= 36\tTrain=0.746316\tTest=0.743793\n',
'#Iter= 37\tTrain=0.749123\tTest=0.746333\n',
'#Iter= 38\tTrain=0.751573\tTest=0.748493\n',
'#Iter= 39\tTrain=0.753264\tTest=0.750292\n',
'#Iter= 40\tTrain=0.754803\tTest=0.751642\n',
'#Iter= 41\tTrain=0.756011\tTest=0.753062\n',
'#Iter= 42\tTrain=0.756902\tTest=0.753892\n',
'#Iter= 43\tTrain=0.757642\tTest=0.754872\n',
'#Iter= 44\tTrain=0.758293\tTest=0.755372\n',
'#Iter= 45\tTrain=0.758855\tTest=0.755782\n',
'#Iter= 46\tTrain=0.759293\tTest=0.756322\n',
'#Iter= 47\tTrain=0.759695\tTest=0.756652\n',
'#Iter= 48\tTrain=0.760084\tTest=0.756982\n',
'#Iter= 49\tTrain=0.760343\tTest=0.757252\n',
'#Iter= 50\tTrain=0.76055\tTest=0.757332\n',
'#Iter= 51\tTrain=0.760706\tTest=0.757582\n',
'#Iter= 52\tTrain=0.760944\tTest=0.757842\n',
'#Iter= 53\tTrain=0.761035\tTest=0.757952\n',
'#Iter= 54\tTrain=0.761173\tTest=0.758152\n',
'#Iter= 55\tTrain=0.761291\tTest=0.758382\n',
'#Iter= 56\tTrain=0.76142\tTest=0.758412\n',
'#Iter= 57\tTrain=0.761541\tTest=0.758452\n',
```

```
'#Iter= 58\tTrain=0.761677\tTest=0.758572\n',
'#Iter= 59\tTrain=0.76175\tTest=0.758692\n',
'#Iter= 60\tTrain=0.761829\tTest=0.758822\n',
'#Iter= 61\tTrain=0.761855\tTest=0.758862\n',
'#Iter= 62\tTrain=0.761918\tTest=0.759002\n',
'#Iter= 63\tTrain=0.761988\tTest=0.758972\n',
'Final\tTrain=0.761988\tTest=0.758972\n',
'Writing FM model to fm_model\n']
```

现在 FM 模型训练好了，我们把训练数据、验证数据和测试数据输入给 FM，得到 FM 的输出，输出的文件名为 *.fm.logits。

训练集的输出如下：

```
cmd = './libfm/libfm/bin/libFM -task c -train ./data/train_lgb2fm.txt -test ./data/valid_lgb2fm.txt -dim ’1,1,8’ -iter 32 -method sgd -learn_rate 0.00000001 -regular ’0,0,0.01’ -init_stdev 0.1 -load_model fm_model -train_off true -prefix tr'
os.popen(cmd).readlines()

['----------------------------------------------------------------------------\n',
 'libFM\n',
 '  Version: 1.4.4\n',
 '  Author:  Steffen Rendle, srendle@libfm.org\n',
 '  WWW:     http://www.libfm.org/\n',
 'This program comes with ABSOLUTELY NO WARRANTY; for details see license.txt.\n',
 'This is free software, and you are welcome to redistribute it under certain\n',
 'conditions; for details see license.txt.\n',
 '----------------------------------------------------------------------------\n',
 'Loading train...\t\n',
 'has x = 1\n',
 'has xt = 0\n',
 'num_rows=899991\tnum_values=28799712\tnum_features=32\tmin_target=0\tmax_target=1\n',
 'Loading test... \t\n',
 'has x = 1\n',
 'has xt = 0\n',
 'num_rows=100009\tnum_values=3200288\tnum_features=32\tmin_target=0\tmax_target=1\n',
```

```
'#relations: 0\n',
'Loading meta data...\t\n',
'Reading FM model... \t\n',
'Final\tTrain=0.761987\tTest=0.758982\n']
```

验证集的输出如下:

```
cmd = './libfm/libfm/bin/libFM -task c -train ./data/valid_lgb2fm.txt -test ./data/valid_lgb2fm.txt -dim '1,1,8' -iter 32 -method sgd -learn_rate 0.00000001 -regular '0,0,0.01' -init_stdev 0.1 -load_model fm_model -train_off true -prefix va'
os.popen(cmd).readlines()
```

```
['----------------------------------------------------------------------------\n',
 'libFM\n',
 '  Version: 1.4.4\n',
 '  Author:  Steffen Rendle, srendle@libfm.org\n',
 '  WWW:     http://www.libfm.org/\n',
 'This program comes with ABSOLUTELY NO WARRANTY; for details see license.txt.\n',
 'This is free software, and you are welcome to redistribute it under certain\n',
 'conditions; for details see license.txt.\n',
 '----------------------------------------------------------------------------\n',
 'Loading train...\t\n',
 'has x = 1\n',
 'has xt = 0\n',
 'num_rows=100009\tnum_values=3200288\tnum_features=32\tmin_target=0\tmax_target=1\n',
 'Loading test... \t\n',
 'has x = 1\n',
 'has xt = 0\n',
 'num_rows=100009\tnum_values=3200288\tnum_features=32\tmin_target=0\tmax_target=1\n',
 '#relations: 0\n',
 'Loading meta data...\t\n',
 'Reading FM model... \t\n',
 'Final\tTrain=0.758982\tTest=0.758982\n']
```

测试集的输出如下:

```
cmd = './libfm/libfm/bin/libFM -task c -train ./data/test_lgb2fm.txt -test ./data/
valid_lgb2fm.txt -dim '1,1,8' -iter 32 -method sgd -learn_rate 0.00000001 -
regular '0,0,0.01' -init_stdev 0.1 -load_model fm_model -train_off true -prefix
te -test2predict true'
os.popen(cmd).readlines()
```

```
['----------------------------------------------------------------------------\n',
 'libFM\n',
 '  Version: 1.4.4\n',
 '  Author:  Steffen Rendle, srendle@libfm.org\n',
 '  WWW:     http://www.libfm.org/\n',
 'This program comes with ABSOLUTELY NO WARRANTY; for details see license.txt.\n',
 'This is free software, and you are welcome to redistribute it under certain\n',
 'conditions; for details see license.txt.\n',
 '----------------------------------------------------------------------------\n',
 'Loading train...\t\n',
 'has x = 1\n',
 'has xt = 0\n',
 'num_rows=1000000\tnum_values=32000000\tnum_features=32\tmin_target=0\tmax_target=1\n',
 'Loading test... \t\n',
 'has x = 1\n',
 'has xt = 0\n',
 'num_rows=100009\tnum_values=3200288\tnum_features=32\tmin_target=0\tmax_target=1\n',
 '#relations: 0\n',
 'Loading meta data...\t\n',
 'Reading FM model... \t\n',
 'Final\tTest=0.758982\n']
```

5.10 实现点击率预估：基于 TensorFlow 1.x

本节的完整代码请见本项目中的 ctr.ipynb 文件。

5.10.1 构建神经网络

至此，网络模型的 FFM、GBDT 和 FM 部分就完成了，下面开始构建神经网络并把这些组件融合起来。

```python
# 嵌入向量的维度
embed_dim = 32
# 分类数据的最大数据个数。我们把C1~C26中的分类数据统一编号，这个数就是编号总数，但
# 其实并不准确，因为我们只用了数据集的子集，这个数够用
sparse_max = 30000 # sparse_feature_dim = 117568

# 分类特征的个数(C1~C26)
sparse_dim = 26
# 数值特征的个数(I1~I13)
dense_dim = 13
# 全连接层的输出维度
out_dim = 400
```

定义输入占位符：

```python
import tensorflow as tf
def get_inputs():
    # 数值特征的输入
    dense_input = tf.placeholder(tf.float32, [None, dense_dim], name="dense_input")
    # 分类特征的输入
    sparse_input = tf.placeholder(tf.int32, [None, sparse_dim], name="sparse_input")
    # FFM和FM的输入(*.fm.logits和*ffm.out.logit文件内容)
    FFM_input = tf.placeholder(tf.float32, [None, 1], name="FFM_input")
    FM_input = tf.placeholder(tf.float32, [None, 1], name="FM_input")
    # 学习目标Y和学习率
    targets = tf.placeholder(tf.float32, [None, 1], name="targets")
    LearningRate = tf.placeholder(tf.float32, name = "LearningRate")
    return dense_input, sparse_input, FFM_input, FM_input, targets, LearningRate
```

输入分类特征，从嵌入层获得嵌入向量：

```python
def get_sparse_embedding(sparse_input):
    with tf.name_scope("sparse_embedding"):
```

```
    # 定义嵌入矩阵尺寸为[分类特征数量,嵌入向量维度32]
    sparse_embed_matrix = tf.Variable(tf.random_uniform([sparse_max,
        embed_dim], -1, 1), name = "sparse_embed_matrix")
    sparse_embed_layer = tf.nn.embedding_lookup(sparse_embed_matrix,
        sparse_input, name = "sparse_embed_layer")
    # 每行数据有26个分类特征(C1~C26)输入进来,所以这里将从嵌入矩阵获得的26个嵌
    # 入向量合并起来,成为26 * 32 = 832维的向量
    sparse_embed_layer = tf.reshape(sparse_embed_layer, [-1, sparse_dim *
        embed_dim])
    return sparse_embed_layer
```

将数值特征和嵌入向量连接在一起,并连接三层全连接层:

```
def get_dnn_layer(dense_input, sparse_embed_layer):
    with tf.name_scope("dnn_layer"):
        # 将数值特征和嵌入向量连接在一起,(?, 845 = 832 + 13)
        input_combine_layer = tf.concat([dense_input, sparse_embed_layer], 1)
        # 连接三层全连接层
        fc1_layer=tf.layers.dense(input_combine_layer, out_dim, name = "fc1_layer",
            activation=tf.nn.relu)
        fc2_layer = tf.layers.dense(fc1_layer, out_dim, name = "fc2_layer",
            activation=tf.nn.relu)
        fc3_layer = tf.layers.dense(fc2_layer, out_dim, name = "fc3_layer",
            activation=tf.nn.relu)
    return fc3_layer
```

1. 构建计算图

如前所述,将 FFM 和 FM 的输出经过全连接,再和数值特征、嵌入向量的三层全连接层的输出连接在一起,做 Logistics 回归,此处可以对照图 5.3 所示的网络模型结构来学习。

采用 LogLoss 损失、FtrlOptimizer 优化器。

```
tf.reset_default_graph()
train_graph = tf.Graph()
with train_graph.as_default():
    # 网络的右半部分DNN
    dense_input, sparse_input, FFM_input, FM_input, targets, lr = get_inputs()
```

```python
sparse_embed_layer = get_sparse_embedding(sparse_input)
fc3_layer = get_dnn_layer(dense_input, sparse_embed_layer)
# 网络的左半部分，当然大部分都处理完了，只需要读取*.fm.logits和*ffm.out.logit
# 文件内容即可
ffm_fc_layer = tf.layers.dense(FFM_input, 1, name = "ffm_fc_layer")
fm_fc_layer = tf.layers.dense(FM_input, 1, name = "fm_fc_layer")
# 网络的左、右两部分连接在一起
feature_combine_layer = tf.concat([ffm_fc_layer, fm_fc_layer, fc3_layer], 1)
#(?, 402)

with tf.name_scope("inference"):
    # 最后一层全连接层，并做LR（Logistics Regression）
    logits = tf.layers.dense(feature_combine_layer, 1, name = "logits_layer")
    pred = tf.nn.sigmoid(logits, name = "prediction")

with tf.name_scope("loss"):
    # LogLoss损失，Logistics回归到点击率
    logloss_cost = tf.losses.log_loss(labels=targets, predictions=pred)
    cost = logloss_cost
    loss = tf.reduce_mean(cost)
# Ftrl优化损失
global_step = tf.Variable(0, name="global_step", trainable=False)
optimizer = tf.train.FtrlOptimizer(lr)
gradients = optimizer.compute_gradients(loss)
train_op = optimizer.apply_gradients(gradients, global_step=global_step)

# 准确率（accuracy）
with tf.name_scope("score"):
    correct_prediction = tf.equal(tf.to_float(pred > 0.5), targets)
    accuracy = tf.reduce_mean(tf.to_float(correct_prediction), name="accuracy")
```

超参数设置如下：

```python
# 迭代次数，数据量太大，我们只训练了一代（1 epochs）
num_epochs = 1
# 批量大小
batch_size = 32
```

```python
# 学习率
learning_rate = 0.01
# 设置每处理N个批量数据显示一次统计信息
show_every_n_batches = 25
# 训练好的模型文件存放位置
save_dir = './save'
# FFM训练集数据位置
ffm_tr_out_path = './tr_ffm.out.logit'
# FFM验证集数据位置
ffm_va_out_path = './va_ffm.out.logit'
# FM训练集数据位置
fm_tr_out_path = './tr.fm.logits'
# FM验证集数据位置
fm_va_out_path = './va.fm.logits'
# DNN训练集数据位置
train_path = './data/train.txt'
# DNN验证集数据位置
valid_path = './data/valid.txt'
```

从文件中读取 FFM 输出的数据：

```python
ffm_train = pd.read_csv(ffm_tr_out_path, header=None)
ffm_train = ffm_train[0].values

ffm_valid = pd.read_csv(ffm_va_out_path, header=None)
ffm_valid = ffm_valid[0].values
```

从文件中读取 FM 输出的数据：

```python
fm_train = pd.read_csv(fm_tr_out_path, header=None)
fm_train = fm_train[0].values

fm_valid = pd.read_csv(fm_va_out_path, header=None)
fm_valid = fm_valid[0].values
```

将 DNN 数据和 FM、FFM 输出的数据读取出来，并连接在一起：

```python
train_data = pd.read_csv(train_path, header=None)
```

```python
train_data = train_data.values

valid_data = pd.read_csv(valid_path, header=None)
valid_data = valid_data.values
# 将DNN、FFM和FM的训练集数据连接在一起
cc_train = np.concatenate((ffm_train.reshape(-1, 1), fm_train.reshape(-1, 1),
    train_data), 1)
# 将DNN、FFM和FM的验证集数据连接在一起
cc_valid = np.concatenate((ffm_valid.reshape(-1, 1), fm_valid.reshape(-1, 1),
    valid_data), 1)
# 打乱数据
np.random.shuffle(cc_train)
np.random.shuffle(cc_valid)

# 分离出特征X和目标y
train_y = cc_train[:,-1]
test_y = cc_valid[:,-1]

train_X = cc_train[:,0:-1]
test_X = cc_valid[:,0:-1]
```

训练数据的平均点击率：

```python
np.mean(train_y)
```

0.25485810413659693

验证数据的平均点击率：

```python
np.mean(test_y)
```

0.25576698097171252

2. 下采样数据迭代器

通过上面的平均点击率能够看出，大量的数据都是没有被用户点击的，真正被点击的广告只占25%。也就是说，目标 y 有大部分都是0，这种样本不均衡的数据会导致网络更倾向于预测结果是0。举个极端的例子，假如样本里面99%的数据标签都是0，只有1%的数据标签是1，这样

的网络不需要学习,只要永远输出 0 就能达到 99% 的准确率,但这样的网络是没有用的。为了解决样本不均衡的问题,需要对训练数据做下采样。思路是,从训练数据中取正例(标签是 1)的全部数据,然后取跟正例数据量相同的负例(标签是 0),这样正例和负例的样本数是相同的,各占 50%。下采样数据主要用于训练,代码实现如下:

```python
def get_batches_downsample(Xs, ys, batch_size):
    # 得到全部负例数据的下标
    ind_0 = ys==0
    # 得到全部正例数据的下标
    ind_1 = ys==1
    # 取出负例数据
    Xs_0 = Xs[ind_0]
    ys_0 = ys[ind_0]
    # 取出正例数据
    Xs_1 = Xs[ind_1]
    ys_1 = ys[ind_1]

    # Xs_0.shape[0]是负例的数量,Xs_1.shape[0]是正例的数量
    # 从负例中随机取出正例个数的数据下标
    sampling_ind = np.random.permutation(Xs_0.shape[0])[:Xs_1.shape[0]]
    # 取出负例数据
    Xs_0_sampling = Xs_0[sampling_ind]
    ys_0_sampling = ys_0[sampling_ind]
    # 将正例和负例拼接起来,得到样本均衡的数据
    Xs_downsampled = np.concatenate((Xs_0_sampling, Xs_1))
    ys_downsampled = np.concatenate((ys_0_sampling, ys_1))
    # 随机打乱数据顺序
    downsampled_ind = np.random.permutation(Xs_downsampled.shape[0])
    Xs_downsampled = Xs_downsampled[downsampled_ind]
    ys_downsampled = ys_downsampled[downsampled_ind]
    # 迭代返回批量数据
    for start in range(0, len(Xs_downsampled), batch_size):
        end = min(start + batch_size, len(Xs_downsampled))
        yield Xs_downsampled[start:end], ys_downsampled[start:end]
```

3. 全数据迭代器

这是用所有数据进行迭代返回批量数据,用于验证集。

```python
def get_batches(Xs, ys, batch_size):
    for start in range(0, len(Xs), batch_size):
        end = min(start + batch_size, len(Xs))
        yield Xs[start:end], ys[start:end]
```

5.10.2 训练网络

训练网络,代码实现如下:

```python
%matplotlib inline
%config InlineBackend.figure_format = 'retina'
import matplotlib.pyplot as plt
from sklearn.model_selection import train_test_split
import time
import datetime
from sklearn.metrics import log_loss
from sklearn.learning_curve import learning_curve
from sklearn import metrics

# 参数downsample_flg表示是否使用下采样数据
def train_model(num_epochs, downsample_flg=True):
    losses = {'train':[], 'test':[]}
    acc_lst = {'train':[], 'test':[]}
    pred_lst = []
    
    with tf.Session(graph=train_graph) as sess:
        # 收集数据给TensorBoard用
        grad_summaries = []
        for g, v in gradients:
            if g is not None:
                grad_hist_summary = tf.summary.histogram("{}/grad/hist".format(v.
                name.replace(':', '_')), g)
```

```python
            sparsity_summary = tf.summary.scalar("{}/grad/sparsity".format(v.
                name.replace(':', '_')), tf.nn.zero_fraction(g))
            grad_summaries.append(grad_hist_summary)
            grad_summaries.append(sparsity_summary)
    grad_summaries_merged = tf.summary.merge(grad_summaries)

    # 摘要的输出目录
    timestamp = str(int(time.time()))
    out_dir = os.path.abspath(os.path.join(os.path.curdir, "runs", timestamp))
    print("Writing to {}\n".format(out_dir))

    # 损失的摘要
    loss_summary = tf.summary.scalar("loss", loss)

    # 训练摘要
    train_summary_op = tf.summary.merge([loss_summary, grad_summaries_merged])
    train_summary_dir = os.path.join(out_dir, "summaries", "train")
    train_summary_writer = tf.summary.FileWriter(train_summary_dir, sess.graph)

    # 推理摘要
    inference_summary_op = tf.summary.merge([loss_summary])
    inference_summary_dir = os.path.join(out_dir, "summaries", "inference")
    inference_summary_writer = tf.summary.FileWriter(inference_summary_dir,
        sess.graph)

    sess.run(tf.global_variables_initializer())
    sess.run(tf.local_variables_initializer())
    saver = tf.train.Saver()
    # 训练循环
    for epoch_i in range(num_epochs):

        # 将数据集分成训练集和测试集
        if downsample_flg:
            # 下采样数据
            train_batches=get_batches_downsample(train_X, train_y, batch_size)
            batch_num = len(train_y[train_y==1])*2 // batch_size
```

```python
else:
    # 全数据
    train_batches = get_batches(train_X, train_y, batch_size)
    batch_num = len(train_X) // batch_size
test_batches = get_batches(test_X, test_y, batch_size)

# 训练的迭代，保存训练损失
for batch_i in range(batch_num):
    # 取得批量数据
    x, y = next(train_batches)

    feed = {
        dense_input: x.take([2,3,4,5,6,7,8,9,10,11,12,13,14],1),
        sparse_input: x.take
        ([15,16,17,18,19,20,21,22,23,24,25,26,27,28,29,30,31,32,
        33,34,35,36,37,38,39,40],1),
        FFM_input: np.reshape(x.take(0,1), [batch_size, 1]),
        FM_input: np.reshape(x.take(1,1), [batch_size, 1]),
        targets: np.reshape(y, [batch_size, 1]),
        lr: learning_rate}
    # 开始训练
    step, train_loss, summaries, _, prediction, acc = sess.run(
        [global_step, loss, train_summary_op, train_op, pred, accuracy
        ], feed)

    # 保存损失和准确率
    prediction = prediction.reshape(y.shape)
    losses['train'].append(train_loss)

    acc_lst['train'].append(acc)
    # 保存摘要给TensorBoard用
    train_summary_writer.add_summary(summaries, step)

    # 计算auc
    if(np.mean(y) != 0):
        auc = metrics.roc_auc_score(y, prediction)
```

```python
        else:
            auc = -1

        # 每处理show_every_n_batches个批量数据显示一次统计信息
        if (epoch_i * (batch_num) + batch_i) % show_every_n_batches == 0:
            time_str = datetime.datetime.now().isoformat()
            print('{}: Epoch {:>3} Batch {:>4}/{}   train_loss = {:.3f}  accuracy = {}   auc = {}'.format(
                time_str,
                epoch_i,
                batch_i,
                (batch_num),
                train_loss,
                acc,
                auc))

    # 使用测试数据的迭代，处理同上
    for batch_i  in range(len(test_X) // batch_size):
        x, y = next(test_batches)

        feed = {
            dense_input: x.take([2,3,4,5,6,7,8,9,10,11,12,13,14],1),
            sparse_input: x.take
            ([15,16,17,18,19,20,21,22,23,24,25,26,27,28,29,30,31,32,33,34,35,
             36,37,38,39,40],1),
            FFM_input: np.reshape(x.take(0,1), [batch_size, 1]),
            FM_input: np.reshape(x.take(1,1), [batch_size, 1]),
            targets: np.reshape(y, [batch_size, 1]),
            lr: learning_rate}

        step, test_loss, summaries, prediction, acc = sess.run(
            [global_step, loss, inference_summary_op, pred, accuracy], feed
        ) #cost

        # 保存测试损失和准确率
        prediction = prediction.reshape(y.shape)
```

```python
            losses['test'].append(test_loss)

            acc_lst['test'].append(acc)
            inference_summary_writer.add_summary(summaries, step)
            pred_lst.append(prediction)

            if(np.mean(y) != 0):
                auc = metrics.roc_auc_score(y, prediction)
            else:
                auc = -1

            time_str = datetime.datetime.now().isoformat()
            if (epoch_i * (len(test_X) // batch_size) + batch_i) %
show_every_n_batches == 0:
                print('{}: Epoch {:>3} Batch {:>4}/{}   test_loss = {:.3f}
                    accuracy = {}   auc = {}'.format(
                    time_str,
                    epoch_i,
                    batch_i,
                    (len(test_X) // batch_size),
                    test_loss,
                    acc,
                    auc))
                print(metrics.classification_report(y, np.float32(prediction >
                0.5)))

    # 保存模型
    saver.save(sess, save_dir)
    print('Model Trained and Saved')
    save_params((losses, acc_lst, pred_lst, save_dir))
    return losses, acc_lst, pred_lst, save_dir
```

下采样数据训练：

```python
# 开始训练，参数是迭代次数epochs
losses, acc_lst, pred_lst, load_dir = train_model(1)
```

全数据训练：

```
losses, acc_lst, pred_lst, load_dir = train_model(1, False)
```

接下来输出验证集上的一些训练信息,看看训练的效果如何:

- 平均准确率。
- 平均损失。
- 平均 Auc。
- 预测的平均点击率。
- 精确率、召回率、F1 Score 等信息。

对于全数据,因为数据中大部分都是负例,正例较少,如果模型全部猜 0 就能有 75% 的准确率,所以准确率这个指标是不可信的。

我们需要关注正例的精确率和召回率,当然最主要还是要看 LogLoss 的值,在 Kaggle 比赛中采用的评价指标是 LogLoss。

```
def train_info():
    print("Test Mean Acc : {}".format(np.mean(acc_lst['test'])))
    print("Test Mean Loss : {}".format(np.mean(losses['test'])))
    print("Mean Auc : {}".format(metrics.roc_auc_score(test_y[:-9], np.array(
        pred_lst).reshape(-1, 1))))
    print("Mean prediction : {}".format(np.mean(np.array(pred_lst).reshape(-1,1))))
    print(metrics.classification_report(test_y[:-9], np.float32(np.array(pred_lst).
        reshape(-1, 1) > 0.5)))
```

1. 全数据训练情况

全数据训练情况如下:

```
train_info()

Test Mean Acc : 0.7814300060272217
Test Mean Loss : 0.46838584542274475
Mean Auc : 0.7792937214782675
Mean prediction : 0.2552148997783661
             precision    recall  f1-score   support

        0.0       0.81      0.93      0.86     74426
```

	precision	recall	f1-score	support
1.0	0.63	0.34	0.45	25574
avg / total	0.76	0.78	0.76	100000

正例的 recall 和 f1-score 都很小，对用户点击的预测很差。负例的 recall 已经快 100% 了，数据不均衡导致网络更倾向于预测成 0，正例的预测还需要提高。

2. 下采样数据训练情况

下采样数据训练情况如下：

```
train_info()

Test Mean Acc : 0.7068700194358826
Test Mean Loss : 0.5600311160087585
Mean Auc : 0.778957561289977
Mean prediction : 0.4345361292362213
             precision    recall  f1-score   support

        0.0       0.87      0.71      0.78     74423
        1.0       0.45      0.70      0.55     25577

avg / total       0.77      0.71      0.72    100000
```

从训练一轮的结果看，下采样数据训练使得网络能够对用户点击预测得更好，跟全数据训练比起来，下采样的数据更均衡。正例的 recall 和 f1-score 比全数据训练要高一些，说明假阴性（False Negative）变少了，提高了用户点击预测的能力。因为网络更主动预测出用户点击事件，造成了假阳性（False Positive）增加，导致正例的精确率降低了一些。采用下采样数据训练避免了数据不均衡，增加训练次数有助于提高网络的预测能力。

5.10.3 点击率预估

我们使用测试数据实际测试一下。

```
# 读取FFM、FM和DNN的测试数据集
ffm_test_out_path = './te_ffm.out.logit'
fm_test_out_path = './te.fm.logits'
```

```python
test_path = './data/test.txt'

ffm_test = pd.read_csv(ffm_test_out_path, header=None)
ffm_test = ffm_test[0].values

fm_test = pd.read_csv(fm_test_out_path, header=None)
fm_test = fm_test[0].values

test_data = pd.read_csv(test_path, header=None)
test_data = test_data.values

pred_test_X = np.concatenate((ffm_test.reshape(-1, 1), fm_test.reshape(-1, 1), test_data), 1)
```

使用函数 get_tensor_by_name() 从 loaded_graph 中获取张量，后面的点击率预估要用到。

```python
def get_tensors(loaded_graph):

    dense_input = loaded_graph.get_tensor_by_name("dense_input:0")
    sparse_input = loaded_graph.get_tensor_by_name("sparse_input:0")
    FFM_input = loaded_graph.get_tensor_by_name("FFM_input:0")
    FM_input = loaded_graph.get_tensor_by_name("FM_input:0")
    pred = loaded_graph.get_tensor_by_name("inference/prediction:0")

    targets = loaded_graph.get_tensor_by_name("targets:0")
    lr = loaded_graph.get_tensor_by_name("LearningRate:0")
    return dense_input, sparse_input, FFM_input, FM_input, targets, lr, pred
```

定义点击预测函数，这部分就是对网络正向传播做推理（Inference），预测用户是否会点击广告。

```python
def predict_click(x, axis = 0):
    loaded_graph = tf.Graph()
    with tf.Session(graph=loaded_graph) as sess:
        # 加载模型
        loader = tf.train.import_meta_graph(load_dir + '.meta')
        loader.restore(sess, load_dir)
```

```
# 取得张量
dense_input, sparse_input, FFM_input, FM_input, __, _, pred = get_tensors(
    loaded_graph)
feed = {
    dense_input: np.reshape(x.take([2,3,4,5,6,7,8,9,10,11,12,13,14],axis)
    , [1 if axis == 0 else len(x.take([2,3,4,5,6,7,8,9,10,11,12,13,14],
    axis)), 13]),
    sparse_input: np.reshape(x.take
    ([15,16,17,18,19,20,21,22,23,24,25,26,27,28,29,30,31,32,33,34,
    35,36,37,38,39,40],axis), [1 if axis == 0 else len(x.take
    ([15,16,17,18,19,20,21,22,23,24,25,26,27,28,29,30,31,32,33,34,35,36,
    37,38,39,40],axis)), 26]),
    FFM_input: np.reshape(x.take(0,axis), [1 if axis == 0 else len(x.take
    (0,axis)), 1]),
    FM_input: np.reshape(x.take(1,axis), [1 if axis == 0 else len(x.take
    (0,axis)), 1])}

# 执行预测
clicked = sess.run([pred], feed)
return (np.int32(np.array(clicked) > 0.5))[0]
```

全数据训练网络的预测：

```
# 传入测试集的前20个数据
predict_click(pred_test_X[:20], 1)
```

输出结果如下：

```
array([[0],
       [0],
       [0],
       [0],
       [0],
       [0],
       [0],
       [0],
       [0],
       [0],
       [0],
```

```
       [0],
       [0],
       [1],
       [1],
       [0],
       [0],
       [0],
       [1],
       [1],
       [0]], dtype=int32)
```

下采样数据训练网络的预测：

```
# 传入测试集的前20个数据
predict_click(pred_test_X[:20], 1)
```

输出结果如下：

```
array([[0],
       [0],
       [0],
       [1],
       [0],
       [0],
       [1],
       [0],
       [0],
       [1],
       [0],
       [0],
       [1],
       [1],
       [0],
       [0],
       [0],
       [1],
       [1],
       [0]], dtype=int32)
```

从结果上看，下采样数据训练的网络会多预测出一些用户点击事件。

定义取得测试集数据的迭代函数：

```
def get_test_batches(Xs, batch_size):
    for start in range(0, len(Xs), batch_size):
        end = min(start + batch_size, len(Xs))
        yield Xs[start:end]
```

同样是对前向传播做推理，使用测试数据集：

```
def predict_test(batch_size, axis = 1):
    loaded_graph = tf.Graph()
    with tf.Session(graph=loaded_graph) as sess:
        # 加载模型
        loader = tf.train.import_meta_graph(load_dir + '.meta')
        loader.restore(sess, load_dir)

        # 取得张量
        dense_input, sparse_input, FFM_input, FM_input, __, _, pred = get_tensors(
            loaded_graph)

        test_batches = get_test_batches(pred_test_X, batch_size)
        total_num = len(pred_test_X)

        pred_lst = []
        for batch_i in range(total_num // batch_size):
            # 取得批量数据
            x = next(test_batches)

            feed = {
                    dense_input: np.reshape(x.take([2,3,4,5,6,7,8,9,10,11,12,13,14],
                        axis), [1 if axis == 0 else len(x.take
                        ([2,3,4,5,6,7,8,9,10,11,12,13,14],axis)), 13]),
                    sparse_input: np.reshape(x.take
                        ([15,16,17,18,19,20,21,22,23,24,25,26,27,28,29,30,31,
                        32,33,34,35,36,37,38,39,40],axis), [1 if axis == 0 else len(x.
                        take
                        ([15,16,17,18,19,20,21,22,23,24,25,26,27,28,29,30,31,32,33,34,
```

```
                    35,36,37,38,39,40],axis)), 26]),
                FFM_input: np.reshape(x.take(0,axis), [1 if axis == 0 else len(x.
                    take(0,axis)), 1]),
                FM_input: np.reshape(x.take(1,axis), [1 if axis == 0 else len(x.
                    take(0,axis)), 1])}

            # 执行预测
            clicked = sess.run([pred], feed)
            pred_lst.append(clicked)

            # 每处理show_every_n_batches个批量数据显示一次统计信息
            if ((total_num // batch_size) + batch_i) % show_every_n_batches == 0:
                print('Batch {:>4}/{}   mean click = {}'.format(
                    batch_i,
                    (total_num // batch_size),
                    np.mean(np.array(clicked))))
    # 返回所有预测的结果
    return pred_lst
```

全数据训练网络对测试数据的预测：

```
pred_lst = predict_test(64)
```

打印预测结果的平均点击率：

```
np.mean(pred_lst)
```

0.27048749

下采样数据训练网络对测试数据的预测：

```
pred_lst = predict_test(64)
```

打印预测结果的平均点击率：

```
np.mean(pred_lst)
```

0.45447949

5.11 实现点击率预估：基于 TensorFlow 2.0

现在演示如何使用 TensorFlow 2.0 实现点击率预估，大部分代码跟 5.10 节的内容是一样的，这里主要介绍网络的构建和训练。其中网络模型的构建跟前几章稍有些不同，因为本次网络的输入有 4 个，但其实也没什么不同，按照网络结构的定义一层一层地实现就行，最后构建模型时将 4 个输入一起传入即可，详情请看实现代码。本节的完整代码请见本项目的 ctr_tf2.ipynb 文件。

```
import tensorflow as tf
import datetime
from tensorflow import keras
from tensorflow.python.ops import summary_ops_v2
import time
%matplotlib inline
%config InlineBackend.figure_format = 'retina'
import matplotlib.pyplot as plt
from sklearn.model_selection import train_test_split
from sklearn.metrics import log_loss
from sklearn.model_selection import learning_curve
from sklearn import metrics as sk_metrics

MODEL_DIR = "./models"

# 定义点击率预测网络类
class ctr_network(object):
    def __init__(self, batch_size=32):
        self.batch_size = batch_size
        self.best_loss = 9999
        
        self.losses = {'train': [], 'test': []}
        self.pred_lst = []
        self.test_y_lst = []
        
        # 定义输入
        dense_input = tf.keras.layers.Input(shape=(dense_dim,), name='dense_input')
        sparse_input = tf.keras.layers.Input(shape=(sparse_dim,),
            name='sparse_input')
```

```python
FFM_input = tf.keras.layers.Input(shape=(1,), name='FFM_input')
FM_input = tf.keras.layers.Input(shape=(1,), name='FM_input')

# 输入分类特征,从嵌入层获得嵌入向量
sparse_embed_layer =
    tf.keras.layers.Embedding(sparse_max,
                              embed_dim,
                              input_length=sparse_dim)(sparse_input)
sparse_embed_layer =
    tf.keras.layers.Reshape([sparse_dim * embed_dim])(sparse_embed_layer)

# 输入数值特征,和嵌入向量连接在一起经过三层全连接层,(?, 845 = 832 + 13)
input_combine_layer = tf.keras.layers.concatenate([dense_input,
    sparse_embed_layer])
fc1_layer =
tf.keras.layers.Dense(out_dim, name="fc1_layer",
    activation='relu')(input_combine_layer)
fc2_layer =
    tf.keras.layers.Dense(out_dim, name="fc2_layer", activation='relu')(
        fc1_layer)
fc3_layer =
    tf.keras.layers.Dense(out_dim, name="fc3_layer", activation='relu')(
        fc2_layer)

# 将FFM、FM和DNN的全连接层输出连接在一起
ffm_fc_layer = tf.keras.layers.Dense(1, name="ffm_fc_layer")(FFM_input)
fm_fc_layer = tf.keras.layers.Dense(1, name="fm_fc_layer")(FM_input)
feature_combine_layer =
    tf.keras.layers.concatenate([ffm_fc_layer, fm_fc_layer, fc3_layer], 1)
        # (?, 402)

# 最后一层全连接层,得到点击率
logits_output =
tf.keras.layers.Dense(1, name="logits_layer",
                      activation='sigmoid')(feature_combine_layer)
# 构建模型:一共4个输入、1个输出
```

```python
        self.model = 
        tf.keras.Model(inputs=[dense_input,
                               sparse_input,
                               FFM_input,
                               FM_input], outputs=[logits_output])
        self.model.summary()

        # Ftrl优化器在tf.compat.v1里面
        self.optimizer = tf.compat.v1.train.FtrlOptimizer(0.01)
        # LogLoss损失
        self.ComputeLoss = tf.keras.losses.LogLoss()

        # 创建保存模型的文件夹
        if tf.io.gfile.exists(MODEL_DIR):
            pass
        else:
            tf.io.gfile.makedirs(MODEL_DIR)

        train_dir = os.path.join(MODEL_DIR, 'summaries', 'train')
        test_dir = os.path.join(MODEL_DIR, 'summaries', 'eval')

        # 日志记录器
#        self.train_summary_writer = summary_ops_v2.create_file_writer(train_dir,
# flush_millis=10000)
#        self.test_summary_writer = summary_ops_v2.create_file_writer(test_dir,
# flush_millis=10000, name='test')

        checkpoint_dir = os.path.join(MODEL_DIR, 'checkpoints')
        self.checkpoint_prefix = os.path.join(checkpoint_dir, 'ckpt')
        self.checkpoint = tf.train.Checkpoint(model=self.model,
            optimizer=self.optimizer)

        # 加载保存的模型文件
        self.checkpoint.restore(tf.train.latest_checkpoint(checkpoint_dir))

    # 计算准确率
```

```python
    def compute_metrics(self, labels, pred):
        correct_prediction = tf.equal(tf.keras.backend.cast(pred > 0.5, 'float32'),
            labels)
        accuracy = tf.reduce_mean(tf.keras.backend.cast(correct_prediction,
            'float32'), name="accuracy")
        return accuracy

    # 单步训练函数
    @tf.function
    def train_step(self, x, y):
        # 使用磁带记录损失计算
        metrics = 0
        with tf.GradientTape() as tape:
            pred = self.model([x[0],
                               x[1],
                               x[2],
                               x[3]], training=True)
            loss = self.ComputeLoss(y, pred)
            metrics = self.compute_metrics(y, pred)
        # 利用磁带做优化
        grads = tape.gradient(loss, self.model.trainable_variables)
        self.optimizer.apply_gradients(zip(grads, self.model.trainable_variables))
        return loss, metrics, pred

    # 训练函数
    def training(self, train_dataset, test_dataset, downsample_flg=True, epochs=1,
        log_freq=50):

        train_X, train_y = train_dataset
        # 训练循环
        for epoch_i in range(epochs):
            # 根据参数决定是否使用下采样数据
            if downsample_flg:
                train_batches = get_batches_downsample(train_X, train_y, self.
                    batch_size)
                batch_num = (len(train_y[train_y==1])*2 // self.batch_size)
```

```python
    else:
        train_batches = get_batches(train_X, train_y, self.batch_size)
        batch_num = len(train_X) // self.batch_size

train_start = time.time()
# with self.train_summary_writer.as_default():
if True:
    start = time.time()
    # 损失、准确率和auc的指标
    avg_loss = tf.keras.metrics.Mean('loss', dtype=tf.float32)
    avg_acc = tf.keras.metrics.Mean('acc', dtype=tf.float32)
    avg_auc = tf.keras.metrics.Mean('auc', dtype=tf.float32)

    # 迭代数据集
    for batch_i in range(batch_num):
        x, y = next(train_batches)
        if len(x) < self.batch_size:
            break
        # 单步训练
        loss, metrics, pred = 
        self.train_step([
            x.take([2, 3, 4, 5, 6, 7, 8, 9, 10, 11, 12, 13, 14], 1),
            x.take([15, 16, 17, 18, 19, 20, 21, 22, 23, 24,
                    25, 26, 27, 28, 29, 30, 31, 32, 33, 34,
                    35, 36, 37, 38, 39, 40], 1),
            np.reshape(x.take(0, 1), [self.batch_size, 1]),
            np.reshape(x.take(1, 1), [self.batch_size, 1])],
            np.expand_dims(y, 1))
        avg_loss(loss)
        avg_acc(metrics)

        prediction = tf.reshape(pred, y.shape)
        self.losses['train'].append(loss)

        if (np.mean(y) != 0):
            # 计算auc指标
```

```python
                auc = sk_metrics.roc_auc_score(y, prediction)
            else:
                auc = -1

            avg_auc(auc)
            # 打印指标
            if tf.equal((epoch_i * (batch_num) + batch_i) % log_freq, 0):
            #         summary_ops_v2.scalar('loss', avg_loss.result(),
            #                             step=self.optimizer.iterations)
            #         summary_ops_v2.scalar('mae',self.ComputeMetrics.result(),
            #                             step=self.optimizer.iterations)
            #         summary_ops_v2.scalar('acc', avg_acc.result(),
            #                             step=self.optimizer.iterations)

                rate = log_freq / (time.time() - start)

                print('Epoch {:>3} Batch {:>4}/{} Loss: {:0.6f} acc: {:0.6f
                    } auc = {} ({} steps/sec)'.format(
                        epoch_i, batch_i, batch_num, avg_loss.result(),
                        (avg_acc.result()), avg_auc.result(), rate))

                avg_auc.reset_states()
                avg_loss.reset_states()

                avg_acc.reset_states()
                start = time.time()

        train_end = time.time()
        print('\nTrain time for epoch #{} : {}'.format(epoch_i + 1, train_end -
            train_start))
        # 使用验证数据集测试训练情况
        # with self.test_summary_writer.as_default():
        self.testing(test_dataset)

    self.export_path = os.path.join(MODEL_DIR, 'export')
    tf.saved_model.save(self.model, self.export_path)
```

```python
# 测试函数，跟训练函数差不多
def testing(self, test_dataset):
    test_X, test_y = test_dataset
    test_batches = get_batches(test_X, test_y, self.batch_size)

    # 各类指标
    avg_loss = tf.keras.metrics.Mean('loss', dtype=tf.float32)
    avg_acc = tf.keras.metrics.Mean('acc', dtype=tf.float32)
    avg_auc = tf.keras.metrics.Mean('auc', dtype=tf.float32)
    avg_prediction = tf.keras.metrics.Mean('prediction', dtype=tf.float32)

    self.pred_lst=[]
    self.test_y_lst=[]
    # 迭代数据集
    batch_num = (len(test_X) // self.batch_size)
    for batch_i in range(batch_num):
        x, y = next(test_batches)
        if len(x) < self.batch_size:
            break
        # 调用模型得到点击率
        pred=self.model([x.take([2,3,4,5,6,7,8,9,10,11,12,13,14],1),
                        x.take(
                            [15, 16, 17, 18, 19, 20, 21, 22, 23, 24, 25,
                            26, 27, 28, 29, 30, 31, 32, 33, 34, 35,
                            36, 37, 38, 39, 40], 1),
                        np.reshape(x.take(0, 1), [self.batch_size, 1]),
                        np.reshape(x.take(1, 1), [self.batch_size, 1])],
                        training=False)
        # 计算损失和准确率
        test_loss = self.ComputeLoss(np.expand_dims(y, 1), pred)
        avg_loss(test_loss)
        acc = self.compute_metrics(np.expand_dims(y, 1), pred)
        avg_acc(acc)

        # 保存测试损失、准确率和预测的点击率
```

```python
            prediction = tf.reshape(pred, y.shape)
            avg_prediction(prediction)
            self.losses['test'].append(test_loss)

            self.pred_lst.append(prediction)
            self.test_y_lst.append(y)

            if (np.mean(y) != 0):
                # 计算auc
                auc = sk_metrics.roc_auc_score(y, prediction)
            else:
                auc = -1
            avg_auc(auc)

    self.pred_lst = np.concatenate([val for val in self.pred_lst])
    self.test_y_lst = np.concatenate([val for val in self.test_y_lst])
    # 打印各类指标
    print('Model test set loss: {:0.6f}  acc: {:0.6f}  auc = {} prediction = 
        {}'.format(
            avg_loss.result(), avg_acc.result(),
            avg_auc.result(), avg_prediction.result()))
    print(sk_metrics.classification_report(self.test_y_lst, tf.keras.backend.
        cast((self.pred_lst) > 0.5, 'float32')))
    # summary_ops_v2.scalar('loss', avg_loss.result(), step=step_num)
    # summary_ops_v2.scalar('mae', self.ComputeMetrics.result(), step=step_num)
    # summary_ops_v2.scalar('acc', avg_acc.result(), step=step_num)

    # 如果当前的损失是最小的,则保存损失和模型
    if avg_loss.result() < self.best_loss:
        self.best_loss = avg_loss.result()
        print("best loss = {}".format(self.best_loss))
        self.checkpoint.save(self.checkpoint_prefix)

# 点击预测函数,跟上一节的实现是一样的
def predict_click(self, x, axis = 0):
    clicked = self.model(
```

```
             [np.reshape(x.take([2,3,4,5,6,7,8,9,10,11,12,13,14],axis),
                        [1 if axis == 0
                         else len(x.take([2,3,4,5,6,7,8,9,10,11,12,13,14],axis)),
                        13]),
              np.reshape(x.take([15,16,17,18,19,20,21,22,23,24,25,26,27,28,
                                29,30,31,32,33,34,35,36,37,38,39,40],axis),
                        [1 if axis == 0 else len(x.take
                         ([15,16,17,18,19,20,21,22,23,24,
                         25,26,27,28,29,30,31,32,33,
                         34,35,36,37,38,39,40],axis)), 26]),
              np.reshape(x.take(0,axis),
                        [1 if axis == 0
                         else len(x.take(0,axis)), 1]),
              np.reshape(x.take(1,axis), [1 if axis == 0 else len(x.take(0,axis)),
                  1])])

          return (np.int32(np.array(clicked) > 0.5))
```

现在用 TensorFlow 2.0 构建模型和训练网络大家应该都不陌生了吧，主要变化的就是构建模型时由于网络结构的不同而使用了不同的层和参数，本章网络最大的不同就在于有 4 个输入，仅此而已。

5.12　本章小结

本章给出了点击率预估的完整过程。为了演示如何实现点击率预估模型，这里没有进行完整数据的训练，并且有很多超参数可以调整。从只训练了一代（1 epochs）的结果来看，验证集上的 LogLoss 损失是 0.46（全数据训练）～0.56（下采样数据训练），其他数据都在 70%～80% 之间，这跟 FFM、GBDT 和 FM 网络训练的准确率差不多，仍然有改进调整的空间。

从网络的训练结果上看，全数据训练的网络更倾向于预测成 0，因为全数据存在严重的数据不均衡，而采用下采样的数据训练方式能够改善网络的训练效果。

6 人脸识别

6.1 概述

随着深度学习和卷积神经网络的发展，很多基于人脸识别的应用和产品已经进入了大众的生活中，比如 iPhone X 的 Face ID、人脸打卡、地铁站和火车站的人脸刷票、人脸支付等。还有一类应用就是通过人脸识别的技术找人，百度已经在使用该技术帮助寻找失踪儿童了，甚至有机场和车站可以通过人脸识别寻找可疑人员。相信未来人脸识别的应用场景会越来越多，越来越深入大家的生活。

限于篇幅，本章重要代码放在书里，其他代码读者可以去 GitHub 上下载，地址：https://github.com/chengstone/Face_Recognizer，TensorFlow 2.0 的实现在 Face_Recognizer-upgradedToTF2.0（https://github.com/chengstone/Face_Recognizer/tree/master/Face_Recognizer-upgradedToTF2.0）文件夹中。

先看看在图片中找人的效果，如图 6.1 所示。

图 6.1 在图片中找人的效果

视频找人的识别结果请参见：https://raw.githubusercontent.com/chengstone/Face_Recognizer/master/out.mp4。

从视频中找人其实跟从图片中找人是一样的，视频是图片的集合，现在我们介绍从图片中找人的实现思路。基本上就是三步：一是人脸检测（从图片中识别出人脸的位置），本章将介绍通过 OpenCV、dlib 和多任务级联卷积神经网络（Multi-Task Convolutional Neural Networks，MTCNN）等方式检测人脸；二是提取人脸特征，本章讲解通过 dlib、facenet 和 VGG16 网络三种方式提取人脸特征；三是通过特征的比较决定是否找到目标人物。本章将围绕这三个步骤讲解实现方法，最后实现了一个现实场景下的视频找人应用。

6.2 人脸检测

本节我们讲解人脸检测，一共讨论三种方法，先讨论 OpenCV 人脸检测的方法，再讨论使用 dlib 做人脸检测的方法，最后详细讨论 MTCNN 人脸检测的原理和实现。

6.2.1 OpenCV 人脸检测

OpenCV 提供了人脸检测的模型和方法。其使用方法很简单，首先要确保电脑中安装了 OpenCV。可以使用如下命令安装 OpenCV：

```
conda install --channel https://conda.anaconda.org/anaconda opencv=3.4.1
```

或者

```
pip install opencv-python
```

接下来，要找到 haarcascade_frontalface_alt.xml 文件在电脑中的位置，它是 OpenCV 提供的人脸检测器。该文件在笔者电脑中的位置是：

/Applications/anaconda/pkgs/libopencv-3.4.1-he076b03_1/share/OpenCV/haarcascades/

你会看到在该文件夹下有很多检测器文件，其中我们要使用的与人脸检测相关的文件如表 6.1 所示。

默认的人脸检测器经常会出现误差，不太准确；而 alt_tree 用时较长，alt 和 alt2 的识别效果还可以，我们选择使用 alt 人脸检测器。

通过调用 `cv2.CascadeClassifier.detectMultiScale` 函数，即可完成人脸检测，函数返回的是人脸框的位置。

表 6.1　与人脸检测相关的文件

文件名	分类器类型
haarcascade_frontalface_alt_tree.xml	人脸检测器（Tree）
haarcascade_frontalface_alt.xml	人脸检测器（Haar_1）
haarcascade_frontalface_alt2.xml	人脸检测器（快速 Haar）
haarcascade_frontalface_default.xml	人脸检测器（默认）

函数原型如下：

```
detectMultiScale(self, image, scaleFactor=None, minNeighbors=None, flags=None,
    minSize=None, maxSize=None)
```

- image：待识别人脸的图像，通常传入灰度图以加快检测速度。
- scaleFactor：用来指定每次扫描图像时图像尺寸缩放的比例，通常传入 1.1，表示每次扫描的图像尺寸都会扩大 10%。
- minNeighbors：指定每个候选框应保留多少个邻居，也就是指定候选框的相邻矩形框的最小个数，如果相邻矩形框的个数小于该参数值，那么它们都会被舍弃。如果将该参数设置成 0 的话，则不舍弃任何矩形框，全部返回。通常传入 3。
- flags：可以传入 cv2.CASCADE_SCALE_IMAGE 表示按比例正常检测；也可以传入 cv2.CASCADE_DO_CANNY_PRUNING，表示用 Canny 边缘检测过滤边缘太多或者太少的区域，这些区域通常不是人脸。
- minSize 和 maxSize 用来限制目标区域的大小。

接下来，我们来实践如何使用 OpenCV 检测人脸，代码文件是 face_detect_main.py。

```python
import tensorflow as tf
from scipy import misc
import numpy as np
import os
import cv2


# 4个人脸检测器的文件名
# 人脸检测器（默认）
haarcascade_frontalface_default = "haarcascade_frontalface_default.xml"
# 人脸检测器（快速Haar）
haarcascade_frontalface_alt2 = "haarcascade_frontalface_alt2.xml"
```

```python
# 人脸检测器（Tree）
haarcascade_frontalface_alt_tree = "haarcascade_frontalface_alt_tree.xml"
# 人脸检测器（Haar_1）
haarcascade_frontalface_alt = "haarcascade_frontalface_alt.xml"

# 人脸检测器所在路径
CAS_PATH = "/Applications/anaconda/pkgs/libopencv-3.4.1-he076b03_1/share/OpenCV/haarcascades/"

# 检测函数，传入灰度图和级联分类器
def detect(img, cascade):
    # 调用级联分类器的人脸检测函数，返回人脸框
    rects = cascade.detectMultiScale(img, scaleFactor=1.1, minNeighbors=3, flags=cv2.CASCADE_SCALE_IMAGE)
    if len(rects) == 0:
        return []
    rects[:,2:] += rects[:,:2]
    return rects

# 在图像上描画出人脸框
def draw_rects(img, rects, color):
    for x1, y1, x2, y2 in rects:
        cv2.rectangle(img, (x1, y1), (x2, y2), color, 2)

# 人脸检测主函数
def cv_findFace(img, image_size=160):

    # 将图像转换成灰度图，并通过直方图均衡化来提高图像质量
    gray = cv2.cvtColor(img, cv2.COLOR_RGB2GRAY)
    gray = cv2.equalizeHist(gray)

    # 使用人脸检测器创建级联分类器
    cascade_fn = os.path.join(CAS_PATH, haarcascade_frontalface_alt)
    cascade = cv2.CascadeClassifier(cascade_fn)
    # 使用级联分类器检测人脸
    rects = detect(gray, cascade)
```

```python
    if len(rects) != 0:
        img_list = []
        # 将框出的人脸保存起来
        for rect in rects:
            vis = img[rect[1]:rect[3], rect[0]:rect[2], :]
            aligned=misc.imresize(vis, (image_size, image_size), interp='bilinear')
            img_list.append(aligned)
        # 在图像上画出检测到的每张人脸
        draw_rects(img, rects, (0, 255, 0))

        # 将结果保存到本地文件中
        misc.imsave("./cv_face_detect.png", img)

        # 返回检测到的人脸和人脸框
        images = np.stack(img_list)
        return images, rects
    else:
        return [], []

if __name__ == '__main__':
    image_path = "./frame_tmp.jpg"
    img = misc.imread(os.path.expanduser(image_path), mode='RGB')
    cv_findFace(img)
```

检测效果如图 6.2 所示。

图 6.2　OpenCV 人脸检测效果

6.2.2 dlib 人脸检测

dlib 是含有机器学习算法的函数库，同样提供了与人脸识别相关的接口。

dlib 官网地址：http://dlib.net。

在终端输入如下命令安装 dlib：

```
pip install dlib
```

如果提示缺少 cython，则输入如下命令安装它，然后再安装 dlib：

```
pip install cython
```

dlib 人脸检测有两种方式，一种是使用 dlib.get_frontal_face_detector 函数获取到人脸检测器进行检测：

```
face_detector = dlib.get_frontal_face_detector()
```

另一种是通过 dlib.cnn_face_detection_model_v1 函数加载基于 CNN 的人脸检测模型进行检测，dlib 官方提供了训练好的人脸检测模型，官方文件地址：http://dlib.net/files/，下载 mmod_human_face_detector.dat.bz2 文件，解压缩后得到 mmod_human_face_detector.dat 文件用于人脸检测：

```
cnn_face_detector = dlib.cnn_face_detection_model_v1(mmod_human_face_detector)
```

代码中 mmod_human_face_detector 是 mmod_human_face_detector.dat 文件的路径。

无论使用以上哪种人脸检测器，检测方式都是一样的，实际上获取到的是函数指针，我们直接调用即可。调用人脸检测器函数需要传入两个参数，第一个参数是原图像，第二个参数通常传入 1，表示应该对图像进行一次上采样，这将允许我们检测更多的人脸。函数返回检测到的人脸框。

现在我们用 dlib 实现人脸检测，代码同样被写在 face_detect_main.py 中，请将 mmod_human_face_detector.dat 文件放在与代码同级的目录下。

```
import tensorflow as tf
from scipy import misc
import numpy as np
import os
import cv2
import dlib


# dlib官方提供的人脸检测器模型
```

```python
mmod_human_face_detector = "./mmod_human_face_detector.dat"

# 用来将人脸检测器返回的人脸框位置转换成[x1, y1, x2, y2]的形式
def convert_rect(rect):
    return [rect.left(), rect.top(), rect.right(), rect.bottom()]

# 人脸检测主函数,mode用来指定使用普通人脸检测器还是CNN人脸检测器
def dlib_findFace(img, mode="cnn", image_size=160):

    if mode == "cnn":
        # 传入人脸检测器模型,获得CNN人脸检测器
        cnn_face_detector = dlib.cnn_face_detection_model_v1(
            mmod_human_face_detector)
        # 检测人脸,返回人脸对象
        faces = cnn_face_detector(img, 1)
        rects = []
        # 转换人脸对象中的人脸框并保存
        for face in faces:
            rects.append(convert_rect(face.rect))
    else:
        # 获得人脸检测器
        face_detector = dlib.get_frontal_face_detector()
        # 检测人脸,返回人脸框
        boxs = face_detector(img, 1)
        rects = []
        # 转换人脸框并保存
        for rect in boxs:
            rects.append(convert_rect(rect))

    if len(rects) != 0:
        img_list = []
        # 将框出的人脸保存起来
        for rect in rects:
            vis = img[rect[1]:rect[3], rect[0]:rect[2], :]
            aligned=misc.imresize(vis, (image_size, image_size), interp='bilinear')
            img_list.append(aligned)
```

```python
        # 在图像上画出检测到的每张人脸
        draw_rects(img, rects, (0, 255, 0))

        # 将结果保存到本地文件中
        misc.imsave("./dlib_face_detect_{}.png".format(mode), img)

        # 返回检测到的人脸和人脸框
        images = np.stack(img_list)
        return images, rects
    else:
        return [], []

if __name__ == '__main__':
    image_path = "./frame_tmp.jpg"

    img = misc.imread(os.path.expanduser(image_path), mode='RGB')
    dlib_findFace(img, mode="cnn", image_size=160)

    img = misc.imread(os.path.expanduser(image_path), mode='RGB')
    dlib_findFace(img, mode="normal", image_size=160)
```

普通人脸检测器和 CNN 人脸检测器的检测效果分别如图 6.3 和图 6.4 所示。

图 6.3 普通人脸检测器的检测效果

图 6.4 CNN 人脸检测器的检测效果

6.2.3 MTCNN 人脸检测

1. MTCNN 人脸检测的原理和训练

MTCNN 是论文 *Joint Face Detection and Alignment using Multi-task Cascaded Convolutional Networks* 中提出的一种人脸检测算法,论文首页地址:https://kpzhang93.github.io/MTCNN_face_detection_alignment/,开源实现地址:https://github.com/kpzhang93/MTCNN_face_detection_alignment。

论文中提出的框架思想是,使用多个 CNN 的级联架构整合人脸检测和人脸关键点两个任务。该框架分为三个阶段,每个阶段对应一个 CNN。第一阶段,通过浅层 CNN 快速生成候选窗口;第二阶段,通过更复杂的 CNN 过滤掉大量非人脸窗口来精选窗口;第三阶段,使用更强大的 CNN 再次从候选窗口中丢弃部分窗口,细化窗口结果(Bounding Box)并输出 5 个面部关键点位置(Landmark Position)。

MTCNN 人脸检测的总体流程如图 6.5 所示(图片出自论文原文)。

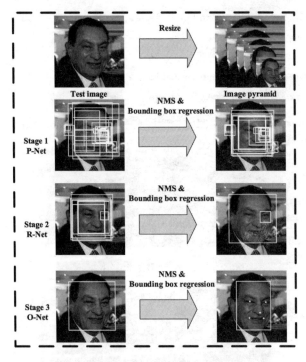

图 6.5　MTCNN 人脸检测的总体流程

将给定图像调整到不同的比例,作为级联框架的输入。阶段一:使用称为 Proposal Network (P-Net)的全卷积网络来获得候选面部窗口及其边界框回归向量,然后使用边界框回归向量校准候选窗口。接下来，采用非极大值抑制（NMS）去除高度重叠的候选窗口。P-Net 的网络结构如图 6.6 所示（图片出自论文原文）。

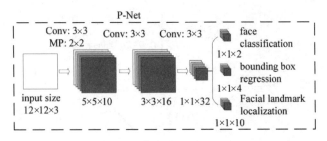

图 6.6　P-Net 的网络结构

阶段二：将 P-Net 输出的所有候选窗口送到另一个 CNN 中进行训练，称为优化网络（Refine Network，R-Net）。同样使用边界框回归优化候选窗口和 NMS，进一步丢弃大量错误候选窗口（False-positive）。R-Net 在网络最后使用了全连接层，其网络结构如图 6.7 所示（图片出自论文原文）。

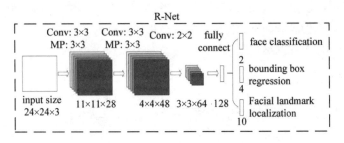

图 6.7　R-Net 的网络结构

阶段三：使用 O-Net（Output Network）输出最终的人脸窗口和 5 个面部关键点位置，作用和 R-Net 一样，其网络结构比 R-Net 多了一层卷积层，这使得网络获得了更好的表达能力。O-Net 的网络结构如图 6.8 所示（图片出自论文原文）。

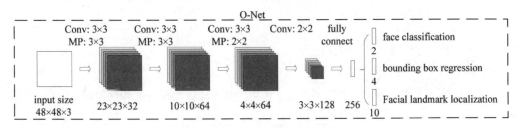

图 6.8　O-Net 的网络结构

论文指出在三个网络中使用了 PReLU（Parametric Rectified Linear Unit）作为非线性激活函数。为什么使用 PReLU 而不是 ReLU 呢？我们看看两者的比较就知道原因了。ReLU 的定义：

$$y = \text{ReLU}(x) = \begin{cases} x, & \text{如果 } x \geqslant 0 \\ 0, & \text{如果 } x < 0 \end{cases}$$

ReLU 图形如图 6.9 所示。

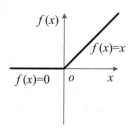

图 6.9　ReLU 图形

与 Sigmoid 激活函数相比，ReLU 能够有效缓解梯度消失的问题。当 $x > 0$ 时导数为 1，ReLU 可以保持梯度不衰减，但随着训练的持续，部分结果可能会落入 $x < 0$ 的区域，导致权重无法更新，这会影响参数的学习和网络的收敛性。

PReLU 的定义：

$$\text{PReLU}(x_i) = f(y_i) = \begin{cases} y_i, & \text{如果 } y_i > 0 \\ a_i y_i, & \text{如果 } y_i \leqslant 0 \end{cases}$$

PReLU 图形如图 6.10 所示（图片出自前述论文）。

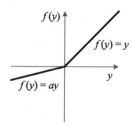

图 6.10　PReLU 图形

PReLU 中的 a 是可学习的，并且不固定，可以避免 ReLU 出现的饱和性问题。参数 a 的更新公式：

$$\Delta a := \mu \Delta a + \epsilon \frac{\partial \mathcal{E}}{\partial a_i}$$

其中 μ 是动量，ϵ 是学习率。这样 PReLU 引入了少量的参数，理论上增加了一些计算量和过拟合风险，但实验结果表明 PReLU 具有更好的性能，与 ReLU 相比，收敛速度更快。关于 PReLU 的详细信息，读者可以参考论文 *Delving Deep into Rectifiers: Surpassing Human-Level Performance on ImageNet Classification*（地址：https://arxiv.org/pdf/1502.01852.pdf）中的描述。

从三个网络的结构来看，网络都很小、很简单，包括层数和卷积核大小，这保证了算法的效率，减少了计算和训练时间，同时也保证了有较好的识别效果。每个网络都包含三个部分的训练：人脸分类、边界框回归和人脸关键点定位。其中人脸分类部分的主要任务是判断是否为人脸的二分类问题；边界框回归是指根据训练集标记好的人脸框位置做回归训练；人脸关键点定位同样也是回归问题。我们主要关注的是人脸边界框回归，不关注 5 个人脸关键点的位置。

那么对于这三个部分的训练，可以将训练数据分成 4 部分。

- Negatives（IoU < 0.3）：非人脸。
- Positives（IoU > 0.65）：人脸。
- Part faces（0.4 < IoU < 0.65）：部分人脸、不完整的人脸。
- Landmark faces：标记好关键点的人脸。

其中 Negatives 和 Positives 用于训练人脸分类，Positives 和 Part faces 用于人脸边界框回归（Bounding Box Regression），Landmark faces 用于人脸关键点定位。

训练数据的比例是，Negatives:Positives:Part faces:Landmark faces=3:1:1:2。

IoU（Intersection over Union）用来度量两个边界框的重合程度，计算公式是 IoU $= \frac{\text{边界框的交集}}{\text{边界框的并集}}$，可以用来表示预测的候选框和标记的真实人脸边界框的相似程度，IoU 值越高，说明越相似。最好的情况就是完全重叠，IoU 等于 1。IoU 的图示如图 6.11 所示。

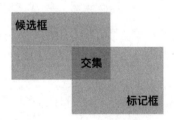

图 6.11　IoU 的图示

这样就可以理解在上面训练数据 4 部分的划分当中使用 IoU 的原因了。IoU < 0.3，说明候选框跟人脸标记框重叠的部分很少，可以认为不是人脸；IoU > 0.65，说明重叠部分很多，可以认为是人脸；0.4 < IoU < 0.65，在是与非之间，可以认为是一部分人脸。

我们需要两类数据集，一类是带有人脸框标记的数据集；另一类就是带有人脸关键点标记的数据集。

通过这两类数据集做数据预处理，生成 Negatives、Positives、Part faces 和 Landmark faces 4 类训练数据。其做法就是在每张图片上进行随机选框，然后跟标记的选框计算 IoU，再按照之前说过的小于 0.3、大于 0.65，在 0.4 和 0.65 之间的 IoU 分类到不同的训练数据中，这样训练数据集的预处理就做好了，可以开始训练了。

关于训练，前文已经说过了，一共是三个部分的训练：人脸分类、边界框回归和人脸关键点定位。其实每个部分单拿出来都是简单的训练问题，比如人脸分类是二分类问题，边界框回归和人脸关键点定位都是回归问题。

这里说说边界框的回归训练。边界框是一个具有 4 个元素的一维向量 (x_left, y_top, x_right, y_bottom)，现在假设有通过神经网络预测出的边界框 bbox_pred 和目标边界框 bbox_target，跟其他回归训练一样，我们可以使用 MSE（Mean Square Error，均方误差）作为训练的损失函数，代入 MSE 公式就是：MSE(bbox_pred) = E(bbox_pred – bbox_target)^2。这样三个网络分别进行训练即可。

2. 使用 FaceNet 做人脸检测

在 MTCNN 的开源代码中，作者已经提供了训练好的参数，在 https://github.com/kpzhang93/MTCNN_face_detection_alignment/tree/master/code/codes/MTCNNv1/model 地址中我们可以看到作者训练好的参数 det1、det2 和 det3，但都是用于 Caffe 的，所以需要转换。当然，我们不用自己去做转换，在开源社区中已经有人转换好了，这就是接下来项目中要使用的 FaceNet，地址是：https://github.com/davidsandberg/facenet。

FaceNet 是用 TensorFlow 实现的人脸识别项目，作者使用 Inception ResNet 网络来实现人脸识别，所以实际上我们完全可以使用 FaceNet 来做人脸特征提取和人脸识别功能。我们只使用 FaceNet 作者实现的 MTCNN 和 MTCNN 作者提供的网络参数来做人脸检测。

首先把 FaceNet 克隆到本地，执行命令：`git clone https://github.com/davidsandberg/facenet`。

我们主要使用 facenet/src/align/ 目录下的文件，目录结构如图 6.12 所示。

其中 det1.npy、det2.npy 和 det3.npy 就是作者转换好的参数文件；在 detect_face.py 中定义了 MTCNN 的网络；align_dataset_mtcnn.py 就是使用 MTCNN 做人脸检测的主程序。

我们一起看看 detect_face.py 中 MTCNN 的实现，完全按照论文中三个网络结构的描述来定义即可。在 MTCNN 的开源地址 https://github.com/kpzhang93/MTCNN_face_detection_alignment/tree/master/code/codes/MTCNNv1/model 中，*.prototxt 文件描述了详细的网络定义。

图 6.12　FaceNet 的 MTCNN 代码目录结构

PNet 的定义如下：

```
class PNet(Network):
    def setup(self):
        # feed('data')的作用是先找到网络的输入层（Input Layer），然后一层层堆叠
        (self.feed('data')
            # 定义卷积层conv1，3×3的卷积核，输出维度是10，步幅是1
            .conv(3, 3, 10, 1, 1, padding='VALID', relu=False, name='conv1')
            # PReLU激活函数
            .prelu(name='PReLU1')
            # 池化窗口的大小和步幅都是1×2×2×1
            .max_pool(2, 2, 2, 2, name='pool1')
            # 定义卷积层conv2，3×3的卷积核，输出维度是16，步幅是1
            .conv(3, 3, 16, 1, 1, padding='VALID', relu=False, name='conv2')
            .prelu(name='PReLU2')
            # 定义卷积层conv3，3×3的卷积核，输出维度是32，步幅是1
            .conv(3, 3, 32, 1, 1, padding='VALID', relu=False, name='conv3')
            .prelu(name='PReLU3')
            # 定义卷积层conv4-1，1×1的卷积核，输出维度是2，步幅是1
            .conv(1, 1, 2, 1, 1, relu=False, name='conv4-1')
            # softmax层，这层输出的就是"是人脸"或"不是人脸"的概率
            .softmax(3,name='prob1'))
        # 找到PReLU3层，在后面堆叠新的层
        (self.feed('PReLU3')
            # 定义卷积层conv4-2，1×1的卷积核，输出维度是4，步幅是1，这层的输出
            # 是预测的人脸框
```

```
            .conv(1, 1, 4, 1, 1, relu=False, name='conv4-2'))
```

RNet 和 ONet 的定义方式相同,如下所示。

```
class RNet(Network):
    def setup(self):
        # 先找到输入层开始堆叠网络
        (self.feed('data')
            # 定义卷积层conv1,3×3的卷积核,输出维度是28,步幅是1
            .conv(3, 3, 28, 1, 1, padding='VALID', relu=False, name='conv1')
            # PReLU激活函数
            .prelu(name='prelu1')
            # 池化窗口的大小是1×3×3×1,步幅是1×2×2×1
            .max_pool(3, 3, 2, 2, name='pool1')
            # 定义卷积层conv2,3×3的卷积核,输出维度是48,步幅是1
            .conv(3, 3, 48, 1, 1, padding='VALID', relu=False, name='conv2')
            .prelu(name='prelu2')
            .max_pool(3, 3, 2, 2, padding='VALID', name='pool2')
            # 定义卷积层conv3,2×2的卷积核,输出维度是64,步幅是1
            .conv(2, 2, 64, 1, 1, padding='VALID', relu=False, name='conv3')
            .prelu(name='prelu3')
            # 从RNet开始网络加入了全连接层,输出维度是128
            # 看起来这层给的名字conv4好奇怪,感觉起名叫fc1更合适
            .fc(128, relu=False, name='conv4')
            .prelu(name='prelu4')
            # 全连接层,输出维度是2
            .fc(2, relu=False, name='conv5-1')
            # softmax层,得到"是人脸"或"不是人脸"的概率
            .softmax(1,name='prob1'))
        # 找到prelu4层,在后面加入新层
        (self.feed('prelu4')
            # 全连接层,输出维度是4,作为预测的人脸框
            .fc(4, relu=False, name='conv5-2'))

class ONet(Network):
    def setup(self):
        (self.feed('data')
```

```
        .conv(3, 3, 32, 1, 1, padding='VALID', relu=False, name='conv1')
        .prelu(name='prelu1')
        .max_pool(3, 3, 2, 2, name='pool1')
        .conv(3, 3, 64, 1, 1, padding='VALID', relu=False, name='conv2')
        .prelu(name='prelu2')
        .max_pool(3, 3, 2, 2, padding='VALID', name='pool2')
        .conv(3, 3, 64, 1, 1, padding='VALID', relu=False, name='conv3')
        .prelu(name='prelu3')
        .max_pool(2, 2, 2, 2, name='pool3')
        # ONet多了一层卷积层
        .conv(2, 2, 128, 1, 1, padding='VALID', relu=False, name='conv4')
        .prelu(name='prelu4')
        .fc(256, relu=False, name='conv5')
        .prelu(name='prelu5')
        .fc(2, relu=False, name='conv6-1')
        .softmax(1, name='prob1'))

    (self.feed('prelu5')
        .fc(4, relu=False, name='conv6-2'))
    # ONet除输出"是人脸"或"不是人脸"的概率,以及人脸框外,还会输出人脸关键点
    # 找到prelu5层,在后面追加新层
    (self.feed('prelu5')
        # 定义全连接层,输出维度是10,作为预测的人脸关键点
        .fc(10, relu=False, name='conv6-3'))
```

上面三个类 PNet、RNet 和 ONet 都继承自父类 Network，调用的各种层的生成函数都来自父类，具体的实现这里就不展开解释了，感兴趣的读者可以继续阅读源码。

在 detect_face.py 中最重要的两个函数是 create_mtcnn 和 detect_face，其中前者用来创建 MTCNN 的三个网络；后者用来检测人脸。

接下来，我们来实践如何检测人脸，主要使用的就是 create_mtcnn 和 detect_face 函数。

其代码同样写在 face_detect_main.py 中，主要参照的是 FaceNet 中 compare.py 的代码。

```
# 将FaceNet项目中的align文件夹复制到我们的代码同级目录下,引入align.detect_face
import align.detect_face
import tensorflow as tf
from scipy import misc
```

```python
import numpy as np
import os
import cv2


# 调用create_mtcnn创建MTCNN的三个网络
mtcnn_graph = tf.Graph()
with mtcnn_graph.as_default():
    gpu_options = tf.GPUOptions(per_process_gpu_memory_fraction=1)
    mtcnn_sess = tf.Session(graph=mtcnn_graph,
                        config=tf.ConfigProto(gpu_options=gpu_options,
                        log_device_placement=False))
    mtcnn_sess.run(tf.global_variables_initializer())
    with mtcnn_sess.as_default():
        pnet, rnet, onet = align.detect_face.create_mtcnn(mtcnn_sess, None)


# 画矩形框
def draw_single_rect(img, rect, color):
    cv2.rectangle(img, (int(rect[0]), int(rect[1])), (int(rect[2]), int(rect[3])),
        color, 2)


# 人脸检测
def mtcnn_findFace(img, image_size=160, margin=44):
    minsize = 20    # 人脸的最小尺寸
    threshold = [0.6, 0.7, 0.7]    # 三个网络判断是否是人脸的阈值
    factor = 0.709    # 缩放因子

    # 调用detect_face得到人脸框
    img_size = np.asarray(img.shape)[0:2]
    bounding_boxes, _ = align.detect_face.detect_face(img, minsize, pnet, rnet,
        onet, threshold, factor)

    img_list = []
    rects = []

    for i in range(len(bounding_boxes)):
        det = np.squeeze(bounding_boxes[i, 0:4])
```

```python
    bb = np.zeros(4, dtype=np.int32)
    bb[0] = np.maximum(det[0] - margin / 2, 0)
    bb[1] = np.maximum(det[1] - margin / 2, 0)
    bb[2] = np.minimum(det[2] + margin / 2, img_size[1])
    bb[3] = np.minimum(det[3] + margin / 2, img_size[0])

    rects.append([bb[0], bb[1], bb[2], bb[3]])
    # 将检测到的人脸框抠出来并保存
    cropped = img[bb[1]:bb[3], bb[0]:bb[2], :]
    aligned=misc.imresize(cropped, (image_size, image_size), interp='bilinear')
    img_list.append(aligned)

    # 将检测到的人脸框描画出来
    draw_single_rect(img, bb, (0, 255, 0))

# 将结果保存到本地文件中
misc.imsave("./mtcnn_face_detect.png", img)

# 返回检测到的人脸和人脸框
images = np.stack(img_list)
return images, rects

if __name__ == '__main__':
    image_path = "./frame_tmp.jpg"
    img = misc.imread(os.path.expanduser(image_path), mode='RGB')
    mtcnn_findFace(img)
```

MTCNN 的人脸检测效果如图 6.13 所示。

图 6.13　MTCNN 的人脸检测效果

至此，第一步人脸检测就完成了，下一步我们来研究人脸特征的提取。

6.3 提取人脸特征

本节我们讲解人脸特征的提取，一共讨论三种方法，首先讨论 FaceNet 的实现，然后讨论使用 VGG 网络提取人脸特征的方法，最后讨论使用 dlib 提供的方法实现人脸特征的提取。

6.3.1 使用 FaceNet 提取人脸特征

既然使用了 FaceNet，那么就先简单说说如何使用 FaceNet 提供的方法提取人脸特征，有关 FaceNet 的网络定义和训练等方法这里不做讲解。同样，本实现参照的是 FaceNet 中 compare.py 的代码。本实现需要在 FaceNet 首页上下载最新的参数模型，名字是 "20180402-114759"。FaceNet 的 download_and_extract.py 提供了以代码方式下载的方法，调用 download_and_extract_file 函数，提供模型名和保存数据的地址即可。我们将下载的模型文件保存到 model 文件夹下，代码文件是 facenet_feature_extraction.py。

```
import facenet
import tensorflow as tf
from scipy import misc
import numpy as np
import os
import cv2
# 引入上一节定义的MTCNN人脸检测的函数mtcnn_findFace
from face_detect_main import mtcnn_findFace

facenet_graph = tf.Graph()

with facenet_graph.as_default():
    facenet_sess = tf.Session(graph=facenet_graph)
    facenet_sess.run(tf.global_variables_initializer())

    with facenet_sess.as_default():
        # 加载模型
        facenet.load_model('./model/20180402-114759/')
```

```python
    # 获得输入输出的张量，images_placeholder是输入图像，embeddings是输出的特征
    images_placeholder = tf.get_default_graph().get_tensor_by_name("input:0")
    embeddings = tf.get_default_graph().get_tensor_by_name("embeddings:0")
    phase_train_placeholder = tf.get_default_graph().get_tensor_by_name(
        "phase_train:0")

# 传入findFace返回的人脸图像
def feature_extraction(tmp_img):
    # 图像预处理
    prewhitened = facenet.prewhiten(tmp_img)

    if prewhitened.ndim == 3:
        prewhitened = np.expand_dims(prewhitened, 0)

    with facenet_graph.as_default():

        # 执行前向传播计算人脸特征
        feed_dict = {images_placeholder: prewhitened, phase_train_placeholder:
            False}
        emb = facenet_sess.run(embeddings, feed_dict=feed_dict)
    # 返回人脸特征
    return emb

if __name__ == '__main__':
    image_path = "./frame_tmp.jpg"
    img = misc.imread(os.path.expanduser(image_path), mode='RGB')
    images, rects = mtcnn_findFace(img)
    feature_extraction(images)
```

因为FaceNet是通过人脸数据集进行训练的，所以使用FaceNet提取的人脸特征是网络经过若干次训练学习得到的。接下来要讲解的使用VGG网络进行人脸特征的提取就稍稍有些不同了。

6.3.2　使用VGG网络提取人脸特征

从0到1构建一个好用的网络并不容易，有很多细节需要去实验和调整。最好的方式就是站在巨人的肩膀上解决问题，利用迁移学习的思想，在现成网络的基础上进行改造往往是一个

不错的开始。VGGNet 是在 ILSVRC 比赛中被证明过的很好的分类网络。当然，除 VGG（Visual Geometry Group）网络以外，分类网络还有很多其他网络模型，这里不选择过于庞大、复杂的网络，使用 VGG16 足够学习人脸特征，符合我们的需求。

关于 VGG 网络的论文请参见：http://x-algo.cn/wp-content/uploads/2017/01/VERY-DEEP-CONVOLUTIONAL-NETWORK-SFOR-LARGE-SCALE-IMAGE-RECOGNITION.pdf。

VGG 网络结构如图 6.14 所示（图片出自该论文）。

ConvNet Configuration					
A	A-LRN	B	C	D	E
11 weight layers	11 weight layers	13 weight layers	16 weight layers	16 weight layers	19 weight layers
input (224×224 RGB image)					
conv3-64	conv3-64 LRN	conv3-64 conv3-64	conv3-64 conv3-64	conv3-64 conv3-64	conv3-64 conv3-64
maxpool					
conv3-128	conv3-128	conv3-128 conv3-128	conv3-128 conv3-128	conv3-128 conv3-128	conv3-128 conv3-128
maxpool					
conv3-256 conv3-256	conv3-256 conv3-256	conv3-256 conv3-256	conv3-256 conv3-256 conv1-256	conv3-256 conv3-256 conv3-256	conv3-256 conv3-256 conv3-256 conv3-256
maxpool					
conv3-512 conv3-512	conv3-512 conv3-512	conv3-512 conv3-512	conv3-512 conv3-512 conv1-512	conv3-512 conv3-512 conv3-512	conv3-512 conv3-512 conv3-512 conv3-512
maxpool					
conv3-512 conv3-512	conv3-512 conv3-512	conv3-512 conv3-512	conv3-512 conv3-512 conv1-512	conv3-512 conv3-512 conv3-512	conv3-512 conv3-512 conv3-512 conv3-512
maxpool					
FC-4096					
FC-4096					
FC-1000					
soft-max					

图 6.14　VGG 网络结构

VGGNet 定义了 6 种不同的网络结构，都有 5 组卷积，每组都使用 3×3 的卷积核（C 列中的 conv1 除外），每组卷积之后是 2×2 的最大池化，最后是 3 层全连接。我们使用的 VGG16 就是 13 层卷积加 3 层全连接（D 列）。

VGG16 网络的实现我们使用 tensorflow-vgg 的开源代码（地址：https://github.com/machrisaa/tensorflow-vgg），需要在首页下载训练好的参数文件 vgg16.npy。

VGG16 网络的 vgg16.py 代码实现如下：

```
import inspect
```

```python
import os

import numpy as np
import tensorflow as tf
import time

VGG_MEAN = [103.939, 116.779, 123.68]

class Vgg16:
    def __init__(self, vgg16_npy_path=None):
        if vgg16_npy_path is None:
            path = inspect.getfile(Vgg16)
            path = os.path.abspath(os.path.join(path, os.pardir))
            path = os.path.join(path, "vgg16.npy")
            vgg16_npy_path = path
            print(path)
        # 加载网络参数
        self.data_dict = np.load(vgg16_npy_path, encoding='latin1').item()
        print("npy file loaded")

    def build(self, rgb):
        """
        从npy加载变量以构建VGG网络
        :参数 rgb: 值被缩放至 [0, 1]的RGB图像 [batch, height, width, 3]
        """

        start_time = time.time()
        print("build model started")
        rgb_scaled = rgb * 255.0

        # 将RGB转换为BGR
        red, green, blue = tf.split(axis=3, num_or_size_splits=3, value=rgb_scaled)
        assert red.get_shape().as_list()[1:] == [224, 224, 1]
        assert green.get_shape().as_list()[1:] == [224, 224, 1]
        assert blue.get_shape().as_list()[1:] == [224, 224, 1]
        bgr = tf.concat(axis=3, values=[
```

```python
            blue - VGG_MEAN[0],
            green - VGG_MEAN[1],
            red - VGG_MEAN[2],
        ])
        assert bgr.get_shape().as_list()[1:] == [224, 224, 3]
        # 第1组卷积
        self.conv1_1 = self.conv_layer(bgr, "conv1_1")
        self.conv1_2 = self.conv_layer(self.conv1_1, "conv1_2")
        self.pool1 = self.max_pool(self.conv1_2, 'pool1')
        # 第2组卷积
        self.conv2_1 = self.conv_layer(self.pool1, "conv2_1")
        self.conv2_2 = self.conv_layer(self.conv2_1, "conv2_2")
        self.pool2 = self.max_pool(self.conv2_2, 'pool2')
        # 第3组卷积
        self.conv3_1 = self.conv_layer(self.pool2, "conv3_1")
        self.conv3_2 = self.conv_layer(self.conv3_1, "conv3_2")
        self.conv3_3 = self.conv_layer(self.conv3_2, "conv3_3")
        self.pool3 = self.max_pool(self.conv3_3, 'pool3')
        # 第4组卷积
        self.conv4_1 = self.conv_layer(self.pool3, "conv4_1")
        self.conv4_2 = self.conv_layer(self.conv4_1, "conv4_2")
        self.conv4_3 = self.conv_layer(self.conv4_2, "conv4_3")
        self.pool4 = self.max_pool(self.conv4_3, 'pool4')
        # 第5组卷积
        self.conv5_1 = self.conv_layer(self.pool4, "conv5_1")
        self.conv5_2 = self.conv_layer(self.conv5_1, "conv5_2")
        self.conv5_3 = self.conv_layer(self.conv5_2, "conv5_3")
        self.pool5 = self.max_pool(self.conv5_3, 'pool5')
        # 全连接1
        self.fc6 = self.fc_layer(self.pool5, "fc6")
        assert self.fc6.get_shape().as_list()[1:] == [4096]
        self.relu6 = tf.nn.relu(self.fc6)
        # 全连接2
        self.fc7 = self.fc_layer(self.relu6, "fc7")
        self.relu7 = tf.nn.relu(self.fc7)
        # 全连接3
```

```python
        self.fc8 = self.fc_layer(self.relu7, "fc8")

        self.prob = tf.nn.softmax(self.fc8, name="prob")

        self.data_dict = None
        print(("build model finished: %ds" % (time.time() - start_time)))

    def avg_pool(self, bottom, name):
        return tf.nn.avg_pool(bottom, ksize=[1, 2, 2, 1], strides=[1, 2, 2, 1], padding='SAME', name=name)

    def max_pool(self, bottom, name):
        return tf.nn.max_pool(bottom, ksize=[1, 2, 2, 1], strides=[1, 2, 2, 1], padding='SAME', name=name)

    def conv_layer(self, bottom, name):
        with tf.variable_scope(name):
            filt = self.get_conv_filter(name)

            conv = tf.nn.conv2d(bottom, filt, [1, 1, 1, 1], padding='SAME')

            conv_biases = self.get_bias(name)
            bias = tf.nn.bias_add(conv, conv_biases)

            relu = tf.nn.relu(bias)
            return relu

    def fc_layer(self, bottom, name):
        with tf.variable_scope(name):
            shape = bottom.get_shape().as_list()
            dim = 1
            for d in shape[1:]:
                dim *= d
            x = tf.reshape(bottom, [-1, dim])

            weights = self.get_fc_weight(name)
```

```
            biases = self.get_bias(name)

            fc = tf.nn.bias_add(tf.matmul(x, weights), biases)

            return fc

    def get_conv_filter(self, name):
        return tf.constant(self.data_dict[name][0], name="filter")

    def get_bias(self, name):
        return tf.constant(self.data_dict[name][1], name="biases")

    def get_fc_weight(self, name):
        return tf.constant(self.data_dict[name][0], name="weights")
```

关于使用 VGG16，这里有两种方法：一种是迁移学习，就是使用 VGG16 的网络和人脸数据集继续训练，使得 VGG16 的参数学习到人脸特征；另一种就是使用 VGG16 的网络参数直接计算出的特征作为人脸特征。

先说说迁移学习的方法。迁移学习的目的是让网络学习到人脸特征的参数，VGG16 的参数文件是固定的，我们不想改变 VGG16 的参数，所以需要修改 VGG16 的网络。我们可以去掉 VGG16 的最后一层全连接，然后加上两层新的全连接。新加的第一层全连接可以跟上一层全连接一样，输出是 4096；新加的第二层全连接：假如我们使用的人脸数据集有 1000 张人脸，那么这一层全连接就可以跟 VGG16 的最后一层全连接一样，输出是 1000。最后是 softmax 层。这样进行训练时，从第一组卷积开始到 VGG16 的倒数第二层全连接参数都是 VGG16 网络的，参数不变化；而新加的最后两层全连接的参数是随着训练不断变化的，网络结构看起来如图 6.15 所示。

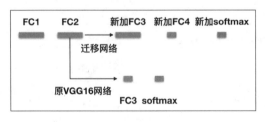

图 6.15　迁移网络结构

这样对迁移网络进行训练之后，使用新 FC3 层的输出作为人脸图像的特征，不使用新 FC4 层和 softmax 层的输出结果。

关于迁移学习方法的实现请感兴趣的读者自己尝试吧。

接下来介绍如何使用 VGG16 计算人脸特征。该方法很简单，就是使用网络的第二层全连接的输出作为人脸特征，不使用 FC3 层和 softmax 层的输出。因为 FC3 层的输出是学习了分类结果的参数，FC2 层的输出是学习了分类数据抽象特征的参数，适合我们使用。在 tensorflow-vgg 的开源代码实现中，第二层全连接是 self.fc7，我们可以通过前向传播得到 self.fc7 的输出。

以下代码是使用 VGG16 计算人脸特征的实践，请确保下载了 tensorflow-vgg 的代码和参数文件 vgg16.npy。

vgg16_feature_extraction.py 代码实现如下：

```python
# 引入VGG16网络
from tensorflow_vgg import vgg16
from tensorflow_vgg import utils
# 引入上一节定义的MTCNN人脸检测的函数mtcnn_findFace
from face_detect_main import mtcnn_findFace
import tensorflow as tf
from scipy import misc
import numpy as np
import os
import cv2
# 构建VGG16网络并加载参数
vgg = vgg16.Vgg16()
input_ = tf.placeholder(tf.float32, [None, 224, 224, 3])
with tf.name_scope("content_vgg"):
    vgg.build(input_)


def feature_extraction(imgs):
    path = "./face_feature.png"
    batch = []

    if len(imgs) > 0:
        if imgs.ndim > 3:
            for i in range(imgs.shape[0]):
                cv2.imwrite("./face_feature.png", imgs[i])
                # 调用开源代码实现的函数load_image加载人脸图像
                img = utils.load_image(path)
                batch.append(img)
```

```
        else:
            
            cv2.imwrite("./face_feature.png", imgs)
            img = utils.load_image(path)
            batch.append(img.reshape((1, 224, 224, 3)))
            batch = np.concatenate(batch)

        with tf.Session() as sess:
            feed_dict = {input_: batch}
            # 前向传播获得vgg.fc7(就是第二个全连接层)的输出作为人脸特征
            feature = sess.run(vgg.fc7, feed_dict=feed_dict)
            if imgs.ndim == 3:
                feature = np.reshape(feature, (1, 4096))
        # 返回人脸特征
        return feature

if __name__ == '__main__':
    image_path = "./frame_tmp.jpg"
    img = misc.imread(os.path.expanduser(image_path), mode='RGB')
    images, rects = mtcnn_findFace(img)
    feature_extraction(images)
```

FaceNet 输出的人脸特征长度是 512,VGG16 网络输出的人脸特征长度是 4096,可以根据不同的应用场景来选择不同的特征提取方式,因为网络越复杂,处理时间会越长。比如这里 VGG16 的处理速度就要慢于 FaceNet。接下来介绍另外一种人脸特征的提取方法:dlib。

6.3.3 使用 dlib 提取人脸特征

使用 dlib 提取人脸特征需要以下几个步骤,参考自 dlib 官方的例子程序(https://github.com/davisking/dlib/blob/a18e72e51685e0306f889642da15ff7994362dc3/python_examples/face_recognition.py)。

(1)使用人脸检测器检测人脸,得到人脸框。当然,这一步使用任何人脸检测方法都可以,不一定要用 dlib 的方法,但是要注意传给 dlib 的人脸框需要符合 dlib 要求的格式。

```
face_detector = dlib.get_frontal_face_detector()
boxs = face_detector(img, 1)
```

（2）定义 dlib 的特征点预测器 shape_predictor。dlib 提供了两种特征点预测器，即 shape_predictor_68_face_landmarks.dat 和 shape_predictor_5_face_landmarks.dat，分别能够计算出人脸的 68 个特征点和 5 个特征点。同样，这两个文件需要在 http://dlib.net/files/ 中下载，请事先下载好并放在代码同级目录下。

```
shape_predictor_68_face_landmarks = "./shape_predictor_68_face_landmarks.dat"
shape_predictor_5_face_landmarks = "./shape_predictor_5_face_landmarks.dat"

face_68_landmarks = dlib.shape_predictor(shape_predictor_68_face_landmarks)
face_5_landmarks = dlib.shape_predictor(shape_predictor_5_face_landmarks)
```

（3）将原始图像和检测到的每个人脸框传给预测器，得到每张人脸的多个关键点。这里要注意，如果没有使用 dlib 的人脸检测方法，那么传给 dlib 的人脸框 rect 要符合 dlib 要求的格式。实际上传入的是 dlib.rectangle 对象，人脸框坐标位置顺序是左、上、右、下：dlib.rectangle(left, top, right, bottom)。

```
shape_68 = face_68_landmarks(img, rect)
shape_5 = face_5_landmarks(img, rect)
```

（4）定义 dlib 的人脸识别模型，需要使用官方训练好的模型文件 dlib_face_recognition_resnet_model_v1.dat，请在官网下载并解压缩到代码同级目录下。

```
face_rec_model_path = "./dlib_face_recognition_resnet_model_v1.dat"
facerec = dlib.face_recognition_model_v1(face_rec_model_path)
```

（5）调用 dlib 人脸识别模型的 facerec.compute_face_descriptor 方法，传入原始图像和每张人脸的关键点位置，计算人脸特征。

```
face_descriptor_68 = facerec.compute_face_descriptor(img, shape_68)
face_descriptor_5 = facerec.compute_face_descriptor(img, shape_5)
```

经过以上步骤，人脸特征就被计算出来了。

我们实践一下，dlib_feature_extraction.py 代码实现如下：

```
from scipy import misc
import numpy as np
import os
import cv2
import dlib
# 引入上一节定义的MTCNN人脸检测的函数mtcnn_findFace
```

```python
from face_detect_main import mtcnn_findFace

# 特征点预测器文件和dlib人脸识别模型文件
shape_predictor_68_face_landmarks = "./shape_predictor_68_face_landmarks.dat"
shape_predictor_5_face_landmarks = "./shape_predictor_5_face_landmarks.dat"
face_rec_model_path = "./dlib_face_recognition_resnet_model_v1.dat"

# 转换人脸框为dlib的rectangle对象
def convert_to_rect(rect):
    return dlib.rectangle(rect[0], rect[1], rect[2], rect[3])

# 提取人脸特征
def feature_extraction(img, rects, mode="68"):

    feature = []

    if len(rects) > 0:

        show_img = img.copy()

        # 定义特征点预测器
        if mode == "68":
            sp = dlib.shape_predictor(shape_predictor_68_face_landmarks)
        else:
            sp = dlib.shape_predictor(shape_predictor_5_face_landmarks)

        # 定义dlib人脸识别模型
        facerec = dlib.face_recognition_model_v1(face_rec_model_path)
        for rect in rects:
            # 计算每张人脸的特征点
            shape = sp(img, convert_to_rect(rect))

            # 把特征点描画出来，检验程序是否正确
            for pt in shape.parts():
                pt_pos = (pt.x, pt.y)
                cv2.circle(show_img, pt_pos, 2, (0, 255, 0), 1)
```

```python
        # 将原始图像和特征点传入人脸识别模型，得到人脸特征
        face_descriptor = facerec.compute_face_descriptor(img, shape)
        feature.append(face_descriptor)

    feature = np.array(feature)
    # 将结果保存到本地文件中
    misc.imsave("./dlib_feature_extraction_{}.png".format(mode), show_img)

    # 返回人脸特征
    return feature

if __name__ == '__main__':
    # 通过MTCNN获得人脸框位置
    image_path = "./frame_tmp.jpg"
    img = misc.imread(os.path.expanduser(image_path), mode='RGB')
    images, rects = mtcnn_findFace(img)

    img = misc.imread(os.path.expanduser(image_path), mode='RGB')
    feature_extraction(img, rects, "68")

    img = misc.imread(os.path.expanduser(image_path), mode='RGB')
    feature_extraction(img, rects, "5")
```

68个特征点和5个特征点的描画图像分别如图 6.16 和图 6.17 所示。

我们可以根据实际情况来选择使用 68 个特征点还是 5 个特征点的人脸特征。

图 6.16　68 个特征点的描画图像

图 6.17　5 个特征点的描画图像

6.4　人脸特征的比较

关于人脸特征的比较，这里提供两种方法，其中一种方法是计算特征间的欧氏距离，距离越近，说明特征越相近；另一种方法是计算余弦相似度，相似度越高，说明越有可能是同一个人。

第一种方法是计算欧氏距离，n 维空间的公式是：

$$d(\boldsymbol{x}, \boldsymbol{y}) = \sqrt{\sum_{i=1}^{n}(x_i - y_i)^2}$$

在代码中可以通过调用 Numpy 的 np.linalg.norm 函数计算欧氏距离。

第二种方法是计算余弦相似度。通过计算两个向量间夹角的余弦值来度量两者的相似度，余弦值的范围在 $[-1, 1]$ 之间，夹角越小，余弦值就越接近于 1，这意味着两个向量的方向越接近，说明两者更相似；越接近于 -1，说明两者的方向相反；接近于 0，表示两个向量趋于正交。n 维空间的公式是：

$$\cos\theta = \frac{\sum_{i=1}^{n}(A_i \times B_i)}{\sqrt{\sum_{i=1}^{n}(A_i)^2} \times \sqrt{\sum_{i=1}^{n}(B_i)^2}}$$

从二维空间的角度来看，如图 6.18 所示。

在代码中可以使用 sklearn.metrics.pairwise.cosine_similarity 函数计算余弦相似度。

现在我们实践一下，通过比较特征间的距离和相似度，实现人脸识别。首先准备两张图片，其中一张是目标图片，就是要找的人的图片，图 6.19 所示。另一张是源图片，就是要从哪张图片中找人，如图 6.20 所示。

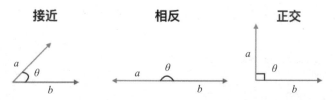

图 6.18　余弦相似度二维图示

图 6.19　目标：要找的人是谁

图 6.20　源：在哪张图片中找人

首先提取出目标人脸的特征，然后从源图片中将找到的所有人脸的特征也提取出来，对这些特征进行比较，在所有的余弦相似度结果中相似度最高并且大于预设的阈值时，对应的人脸就是我们要找的目标。在代码中预设的阈值是 0.8。最后把人脸框、相似度和距离描画出来。

这次实践我们使用 dlib 做人脸检测和特征提取，感兴趣的读者也可以试试其他组合。

代码文件是 main_recognition.py，实现如下：

```
# 引入前几节完成的特征提取和人脸检测函数
import dlib_feature_extraction
import vgg16_feature_extraction
import facenet_feature_extraction
from face_detect_main import dlib_findFace
from face_detect_main import cv_findFace
from face_detect_main import mtcnn_findFace
from argparse import ArgumentParser
from scipy import misc
```

```python
import numpy as np
import os
import sklearn.metrics.pairwise as pw
import cv2

# 定义人脸识别类
class Recognizer():
    # 初始化函数，传入的是两张图片的路径、人脸检测类型、特征提取类型和阈值
    # 函数的主要目的是提取目标人物的人脸特征
    def __init__(self, src_path, target_path, detect, feat_ext, threshold):
        self.src_path = src_path
        self.detect = detect
        self.feat_ext = feat_ext
        self.threshold = threshold

        # 先检测目标图片，提取出人脸特征
        img = misc.imread(os.path.expanduser(target_path), mode='RGB')
        self.target_images, self.target_rects = self.findFace(detect, img.copy())
        self.target_features=self.feature_extraction(feat_ext, self.target_images,
            self.target_rects, img.copy())

    # 人脸检测函数，根据传入的人脸检测类型选择不同的检测方法
    def findFace(self, detect, img):
        if detect == 'mtcnn':
            return mtcnn_findFace(img)
        elif detect == 'cv':
            return cv_findFace(img)
        elif detect == 'dlib_cnn':
            return dlib_findFace(img, mode="cnn", image_size=160)
        elif detect == 'dlib':
            return dlib_findFace(img, mode="normal", image_size=160)
        else:
            return [], []

    # 特征提取函数，根据传入的特征提取类型选择不同的特征提取方法
    def feature_extraction(self, feat_ext, images, rects, img):
```

```python
        if feat_ext == 'facenet':
            return facenet_feature_extraction.feature_extraction(images)
        elif feat_ext == 'vgg16':
            return vgg16_feature_extraction.feature_extraction(images)
        elif feat_ext == 'dlib_68':
            return dlib_feature_extraction.feature_extraction(img, rects, "68")
        else:
            return []

    # 余弦相似度函数，计算两个特征间的相似度
    def cosine_similarity(self, src_feature, target_feature):
        if len(src_feature) == 0 or len(target_feature) == 0:
            return np.empty((0))

        predicts = pw.cosine_similarity(src_feature, target_feature)
        return predicts

    # 欧氏距离函数，计算两个特征间的欧氏距离
    def euclidean_distance(self, src_feature, target_feature):
        if len(src_feature) == 0 or len(target_feature) == 0:
            return np.empty((0))

        return np.linalg.norm(src_feature - target_feature, axis=1)

    # 在图片上描画矩形框
    def draw_single_rect(self, img, rect, color):
        cv2.rectangle(img, (rect[0], rect[1]), (rect[2], rect[3]), color, 2)

    # 人脸识别主函数，从image图片中识别目标人物
    def process(self, image):
        # 先将图片中的所有人脸特征提取出来
        self.src_images, self.src_rects = self.findFace(self.detect, image.copy())
        self.src_features = self.feature_extraction(self.feat_ext, self.src_images,
            self.src_rects, image.copy())

        # 计算特征间的欧氏距离
```

```python
        distances=self.euclidean_distance(self.src_features, self.target_features)
        # 计算特征间的余弦相似度
        cosine_distances = self.cosine_similarity(self.src_features,
            self.target_features)

        # 如果没有找到，则退出函数
        if len(cosine_distances) == 0 or len(distances) == 0:
            return image

        # 得到相似度最大的下标
        index_x, index_y = np.where(cosine_distances == np.max(cosine_distances))
        # 循环描画人脸框、欧氏距离和相似度
        for i in range(len(cosine_distances)):
            # 如果当前下标是相似度最大值的下标，并且相似度大于阈值，则说明找到目标
            # 人物了，用绿色表示
            if i == index_x and cosine_distances[i] >= self.threshold:
                pen = (0, 255, 0)
            else:
                # 否则不是目标人物，用红色表示
                pen = (255, 0, 0)

            # 描画出矩形框、欧氏距离和相似度
            self.draw_single_rect(image, self.src_rects[i], pen)
            cv2.putText(image, "similarity : " + str(np.round(cosine_distances[i],
                2)), (self.src_rects[i][0], self.src_rects[i][1] - 7),
                    cv2.FONT_HERSHEY_DUPLEX, 0.8, pen)
            cv2.putText(image, "distance : " + str(round(distances[i], 2)),
                    (self.src_rects[i][0], self.src_rects[i][1] - 28),
                    cv2.FONT_HERSHEY_DUPLEX, 0.8, pen)

        # 将结果保存到本地文件中
        misc.imsave("./recognition_result.png", image)
        return image

if __name__ == '__main__':
    # 设定参数，一共5组参数
```

```python
# detect: 表示人脸检测类型。一共4种类型, 'mtcnn', 'cv', 'dlib_cnn', 'dlib'
# feat_ext: 表示特征提取类型。一共3种类型, 'facenet', 'vgg16', 'dlib_68'
# src: 源图片路径,从该图片中识别目标人物
# target: 目标图片路径,指定要找的人是谁
# threshold: 决定是否找到人的阈值。如果相似度大于或等于该值,则说明成功找到目标
parser = ArgumentParser()
parser.add_argument('--detect', default='dlib', choices=['mtcnn', 'cv',
    'dlib_cnn', 'dlib'], type=str, help='mtcnn, cv, dlib_cnn, dlib')
parser.add_argument('--feat_ext', default='dlib_68', choices=['facenet',
    'vgg16', 'dlib_68'], type=str, help='facenet, vgg16, dlib_68')
parser.add_argument('--src',  dest='src', help='image/video path',
    required=True)
parser.add_argument('--target', default="", dest='target',
    help='image/video path', required=True)
parser.add_argument('--threshold', type=float, default=0.8,
        dest='threshold', help='the videos and pictures threshold',
        metavar='THRESHOLD')

options = parser.parse_args()

# 定义人脸识别对象
cls = Recognizer(options.src, options.target, options.detect, options.feat_ext,
    options.threshold)
src_img = misc.imread(os.path.expanduser(options.src), mode='RGB')
# 开始人脸识别
cls.process(src_img)
```

在终端输入如下命令运行程序:

```
python main_recognition.py --src src.jpg --target me.jpg
```

人脸识别结果如图6.21所示。

从结果来看,成功找到了目标。目标人脸与两张人脸的相似度很接近,目标之间的相似度是0.94;目标与另一张人脸的相似度是0.91,因为是父子嘛,像一些很正常。

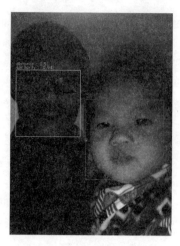

图 6.21 人脸识别结果

6.5 从视频中找人的实现

从视频中找人跟在图片中找人是一样的,因为视频是图片的集合,只要稍微修改上一节的代码,将视频中一帧一帧的图片传给识别函数去处理即可。

我们使用 MoviePy 来处理视频。在终端输入如下命令安装 MoviePy:

```
pip install moviepy
```

另外,还需要安装 requests。在终端输入如下命令安装 requests:

```
conda install requests
```

在代码中引入 MoviePy 的 VideoFileClip 类:

```
from moviepy.editor import VideoFileClip
```

然后定义 VideoFileClip 对象,传入视频文件全路径:

```
video_clip = VideoFileClip(fullPathName)
```

接下来设置回调函数,用来接收每帧图像:

```
video_out_clip = video_clip.fl_image(cls.process)
```

最后开始视频处理,并将结果保存到输出路径下:

```
video_out_clip.write_videofile(outpath, audio=False)
```

经过上面几行代码的处理，MoviePy 会将输入视频中的每一帧图像传给人脸识别函数，该函数会将处理好的图像再返回给 MoviePy，最后将所有处理过的图像输出到视频文件中。

复用上一节的代码，下面给出修改后的代码。

```python
if __name__ == '__main__':
    # 设定参数，一共5组参数
    # detect：表示人脸检测类型。一共4种类型，'mtcnn', 'cv', 'dlib_cnn', 'dlib'
    # feat_ext：表示特征提取类型。一共3种类型，'facenet', 'vgg16', 'dlib_68'
    # src：源图片路径，从该图片中识别目标人物
    # target：目标图片路径，指定要找的人是谁
    # threshold：决定是否找到人的阈值。如果相似度大于或等于该值，则说明成功找到目标
    parser = ArgumentParser()
    parser.add_argument('--detect', default='dlib', choices=['mtcnn', 'cv',
        'dlib_cnn', 'dlib'], type=str, help='mtcnn, cv, dlib_cnn, dlib')
    parser.add_argument('--feat_ext', default='dlib_68', choices=['facenet',
        'vgg16', 'dlib_68'], type=str, help='facenet, vgg16, dlib_68')
    parser.add_argument('--src', dest='src', help='image/video path',
        required=True)
    parser.add_argument('--target', default="", dest='target',
        help='image/video path', required=True)
    parser.add_argument('--threshold', type=float, default=0.8,
                        dest='threshold', help='the videos and pictures threshold',
                        metavar='THRESHOLD')

    options = parser.parse_args()

    # 定义人脸识别对象
    cls = Recognizer(options.src, options.target, options.detect, options.feat_ext,
     options.threshold)

    # 判断是否从图片中找人
    if options.src.split('.')[-1] in ['jpg', 'JPG', 'jpeg', 'bmp', 'png']:
        src_img = misc.imread(os.path.expanduser(options.src), mode='RGB')
        # 开始人脸识别
        cls.process(src_img)
    # 判断是否从视频中找人
    elif options.src.split('.')[-1] in \
```

```
            ['mov', 'MOV', 'rm', 'RM', \
             'rmvb', 'RMVB', 'mp4', 'MP4', \
             'avi', 'AVI', 'wmv', 'WMV', \
             '3gp', '3GP', 'mpeg', 'MPEG', \
             'mkv', 'MKV']:
        # 输出文件路径
        outpath = './result_out.mp4'
        # 定义VideoFileClip对象，传入视频文件路径
        video_clip = VideoFileClip(options.src)
        # 设置回调函数，用来接收每帧图像
        video_out_clip = video_clip.fl_image(cls.process)
        # 开始视频处理，并将结果保存到输出文件中
        video_out_clip.write_videofile(outpath, audio=False)
```

在终端输入如下命令运行程序：

```
python main_recognition.py --src ./chengshd/IMG_3170.mp4 --target me.jpg
```

从视频中找人的识别结果请参见：https://raw.githubusercontent.com/chengstone/Face_Recognizer/master/out.mp4。

6.6　视频找人的案例实践

经过前面几节的学习，我们已经掌握了如何从视频中找人。但是如果读者亲自实践了上述代码之后，可能会想，每次找人都经过完整的一遍视频处理，然后再生成一个新视频，这跟实际的应用场景相符吗？在本节中我们就考虑一个实际的应用场景并去实现它。假如有这样一个需求，要从视频中找到某人，因为视频有可能很大，时间很长，需要找到目标人物在视频中哪个时间点出现过，并且输出在该时间段首次出现的图片即可。比如有一个1小时长的视频，在12分30秒至15分12秒、45分3秒至45分12秒两次连续出现目标人物，那么只输出12分30秒和45分3秒的两张图片即可，而且要明确记录这两次出现的时间，就是12分30秒和45分3秒。

先说说实现思路。在上一节中我们是对视频中一帧一帧的图片做处理的，通常来说这没有批量处理有效率。所以，这次我们首先将视频的所有帧解析出来，同样对每帧图片识别出人脸，然后对所有的人脸分批提取人脸特征，按照时间线保存人脸特征。最后将所有的人脸特征与目标人脸特征做比较，这样会得到一个评分向量，通过阈值能够知道向量中哪些是找到的目标人物，对应的位置就是相应的时间线，最终得到时间和目标图片。

这里是通过 FFmpeg 实现视频流操作的，所以需要事先下载好并放在代码同级目录下，下载地址是：http://ffmpeg.org/download.html。

需要下载两个文件，一个是 ffmpeg；另一个是 ffprobe。

先介绍要用到的命令：

`ffmpeg -y -i 文件路径/chengshd_1.avi -ss 00:00:0.0 -t 00:00:31.70 -q:v 2 -f image2 输出路径\%6d.jpg`

这个命令的作用是将视频文件输出成 JPG 文件，从 -ss 指定的位置开始，共处理 -t 指定的时间。

- -y：表示覆盖输出文件，如果某文件已经存在，则不经提示直接覆盖。
- -i：指定输入文件。
- -ss：从指定的时间开始处理，时间单位是秒。
- -t：表示持续时间（共要多少时间），即从 -ss 指定的时间开始处理多少秒。
- -q:v：表示存储为 JPEG 的图像质量，2 是高质量。
- -f：指定输出的文件格式。
- 最后是输出的文件名规则。

另一个命令是：

`ffprobe -loglevel quiet -print_format json -show_format -show_streams -i 文件路径\chengshd_1.avi`

这个命令的作用是查看视频文件的信息，得到 JSON 形式的字符串。

- -loglevel quiet：设置日志级别，quiet 表示什么都不显示，静默模式。
- -print_format json：表示输出的格式是 JSON。
- -show_format：查看视频文件的 format 信息。
- show_streams：查看视频中的流信息。
- -i：指定输入文件。

最终的输出看起来是这样的：

```
{
    "streams": [
        {
            "index": 0,
            "codec_name": "h264",
```

```
"codec_long_name": "H.264 / AVC / MPEG-4 AVC / MPEG-4 part 10",
"profile": "High",
"codec_type": "video",
"codec_time_base": "1837/108000",
"codec_tag_string": "avc1",
"codec_tag": "0x31637661",
"width": 1440,
"height": 1080,
"coded_width": 1440,
"coded_height": 1088,
"has_b_frames": 0,
"pix_fmt": "yuvj420p",
"level": 40,
"color_range": "pc",
"color_space": "smpte170m",
"color_transfer": "bt709",
"color_primaries": "bt709",
"chroma_location": "center",
"refs": 1,
"is_avc": "true",
"nal_length_size": "4",
"r_frame_rate": "30/1",
"avg_frame_rate": "54000/1837",
"time_base": "1/600",
"start_pts": 0,
"start_time": "0.000000",
"duration_ts": 1837,
"duration": "3.061667",
"bit_rate": "8375142",
"bits_per_raw_sample": "8",
"nb_frames": "90",
"disposition": {
    "default": 1,
    "dub": 0,
    "original": 0,
    "comment": 0,
```

```
            "lyrics": 0,
            "karaoke": 0,
            "forced": 0,
            "hearing_impaired": 0,
            "visual_impaired": 0,
            "clean_effects": 0,
            "attached_pic": 0,
            "timed_thumbnails": 0
        },
        "tags": {
            "rotate": "90",
            "creation_time": "2017-02-01T06:56:47.000000Z",
            "language": "und",
            "handler_name": "Core Media Data Handler",
            "encoder": "H.264"
        },
        "side_data_list": [
            {
                "side_data_type": "Display Matrix",
                "displaymatrix": "\n00000000:           0      65536           0\n00000001:      -65536          0           0\n00000002:    70778880          0  1073741824\n",
                "rotation": -90
            }
        ]
    },
    {
        "index": 1,
        "codec_name": "pcm_s16le",
        "codec_long_name": "PCM signed 16-bit little-endian",
        "codec_type": "audio",
        "codec_time_base": "1/44100",
        "codec_tag_string": "lpcm",
        "codec_tag": "0x6d63706c",
        "sample_fmt": "s16",
        "sample_rate": "44100",
```

```
            "channels": 1,
            "bits_per_sample": 16,
            "r_frame_rate": "0/0",
            "avg_frame_rate": "0/0",
            "time_base": "1/44100",
            "start_pts": 0,
            "start_time": "0.000000",
            "duration_ts": 135020,
            "duration": "3.061678",
            "bit_rate": "705600",
            "nb_frames": "135044",
            "disposition": {
                "default": 1,
                "dub": 0,
                "original": 0,
                "comment": 0,
                "lyrics": 0,
                "karaoke": 0,
                "forced": 0,
                "hearing_impaired": 0,
                "visual_impaired": 0,
                "clean_effects": 0,
                "attached_pic": 0,
                "timed_thumbnails": 0
            },
            "tags": {
                "creation_time": "2017-02-01T06:56:47.000000Z",
                "language": "und",
                "handler_name": "Core Media Data Handler"
            }
        },
        {
            "index": 2,
            "codec_type": "data",
            "codec_tag_string": "mebx",
            "codec_tag": "0x7862656d",
```

```
                "r_frame_rate": "0/0",
                "avg_frame_rate": "0/0",
                "time_base": "1/600",
                "start_pts": 0,
                "start_time": "0.000000",
                "duration_ts": 1837,
                "duration": "3.061667",
                "bit_rate": "43200",
                "nb_frames": "1",
                "disposition": {
                    "default": 1,
                    "dub": 0,
                    "original": 0,
                    "comment": 0,
                    "lyrics": 0,
                    "karaoke": 0,
                    "forced": 0,
                    "hearing_impaired": 0,
                    "visual_impaired": 0,
                    "clean_effects": 0,
                    "attached_pic": 0,
                    "timed_thumbnails": 0
                },
                "tags": {
                    "creation_time": "2017-02-01T06:56:47.000000Z",
                    "language": "und",
                    "handler_name": "Core Media Data Handler"
                }
            }
        ],
        "format": {
            "filename": "./chengshd/IMG_3224.MOV",
            "nb_streams": 3,
            "nb_programs": 0,
            "format_name": "mov,mp4,m4a,3gp,3g2,mj2",
            "format_long_name": "QuickTime / MOV",
```

```
            "start_time": "0.000000",
            "duration": "3.061667",
            "size": "3484678",
            "bit_rate": "9105308",
            "probe_score": 100,
            "tags": {
                "major_brand": "qt  ",
                "minor_version": "0",
                "compatible_brands": "qt  ",
                "creation_time": "2017-02-01T06:56:47.000000Z",
                "com.apple.quicktime.content.identifier": "386A2282-0FFB-470C-A8DE-2
F1D605D6ED6",
                "com.apple.quicktime.location.ISO6709": "+41.7742+123.3850+055.467/",
                "com.apple.quicktime.make": "Apple",
                "com.apple.quicktime.model": "iPhone 6s Plus",
                "com.apple.quicktime.software": "10.2",
                "com.apple.quicktime.creationdate": "2017-02-01T14:56:47+0800"
            }
        }
}
```

这里有很多有用的信息，这次我们只使用 format 格式中的 duration 值，这个值表示该视频文件的时长。我们用这个时长作为将视频输出成图片命令中的-t 参数。

现在定义几个函数完成上面两个功能，代码实现在 action.py 文件中。

```python
# 定义两个程序的全路径
ffprobe_path = os.path.join(os.getcwd(), "ffprobe")
ffmpeg_path = os.path.join(os.getcwd(), "ffmpeg")

# 定义获取视频信息的函数，参数是视频文件
def getVideoProbeInfo(filename):
    # 获取视频信息的命令行
    command = [ffprobe_path, "-loglevel", "quiet", "-print_format", "json",
               "-show_format", "-show_streams", "-i",
               filename]
    # 执行命令行
    result = subprocess.Popen(command, shell=False, stdout=subprocess.PIPE,
```

```python
        stderr=subprocess.STDOUT)
    out = result.stdout.read()
    # 返回JSON结果
    return str(out.decode('utf-8'))

# 从传入的JSON信息中返回format下的duration值，单位是秒
def getDuration(VIDEO_PROBE):
    data = json.loads(VIDEO_PROBE)["format"]['duration']
    return data

# 通过FFmpeg将视频的每帧保存成图片
def fullVideoProc(filename, output_dir, sec_idx, end_idx, allFrames = True,
framesPerSec = 1):
    if allFrames == True:
        # 这个命令行就是上面介绍的
        command = [ffmpeg_path,"-y","-i",filename, "-ss", str(sec_idx), "-t",
                   str(end_idx), "-q:v", "2", "-f",
                   "image2",output_dir+"%6d.jpg"]
    else:
        # 这个命令行多了一个-r，传入的framesPerSec值是1，目的是每秒只取一帧画面，可
        # 以加速处理，但是由于抛弃了很多帧画面，结果会有遗漏
        command = [ffmpeg_path, "-y", "-i", filename, "-ss", str(sec_idx), "-t",
                   str(end_idx), "-r", str(framesPerSec), "-q:v", "2", "-f",
                   "image2", output_dir + "%6d.jpg"]
    # 执行命令行
    result = subprocess.Popen(command,shell=False,stdout = subprocess.PIPE,
        stderr = subprocess.STDOUT)
    out = result.stdout.read()
```

继续看主函数的实现：

```python
# 开始跟之前一样，有同样的参数
if __name__ == '__main__':
    # 设定参数，一共5组参数
    # detect：表示人脸检测类型。一共4种类型，'mtcnn', 'cv', 'dlib_cnn', 'dlib'
    # feat_ext：表示特征提取类型。一共3种类型，'facenet', 'vgg16', 'dlib_68'
    # src：源图片路径，从该图片中识别目标人物
```

```python
# target: 目标图片路径，指定要找的人是谁
# threshold: 决定是否找到人的阈值。如果相似度大于或等于该值，则说明成功找到目标
parser = ArgumentParser()
parser.add_argument('--detect', default='dlib', choices=['mtcnn', 'cv',
                    'dlib_cnn', 'dlib'], type=str,
                    help='mtcnn, cv, dlib_cnn, dlib')
parser.add_argument('--feat_ext', default='dlib_68', choices=['facenet',
                    'vgg16', 'dlib_68'], type=str,
                    help='facenet, vgg16, dlib_68')
parser.add_argument('--src', dest='src', help='image/video path',
    required=True)
parser.add_argument('--target', default="", dest='target',
    help='image/video path', required=True)
parser.add_argument('--threshold', type=float, default=0.8,
                    dest='threshold', help='the videos and pictures threshold',
                    metavar='THRESHOLD')

options = parser.parse_args()

# 参数src指定要解析的视频
videoPath = options.src
# 目标人物的图片
target_arg = options.target

total_time = []

# 先获取视频文件的信息，JSON格式
VIDEO_PROBE = getVideoProbeInfo(videoPath)
# 得到视频文件的时长，单位为秒
sec = float(getDuration(VIDEO_PROBE))

print(sec)

dstpath = os.path.join(os.getcwd(), "target_output/")

# 创建target_output文件夹，用来保存视频的每帧图片
```

```
checkFile(dstpath)
shutil.rmtree(dstpath)
os.mkdir(dstpath)

print("[0] now processing the video, this will take a several minutes...")
time_start = time.time()

# 通过FFmpeg将视频每帧解析出来
fullVideoProc(videoPath, dstpath, 0, sec, True)

time_end = time.time()
print(time_end - time_start, "s")
total_time.append(time_end - time_start)

# 定义人脸识别对象，并先把目标人物的人脸特征保存起来
cls = Recognizer(options.src, options.target, options.detect, options.feat_ext,
    options.threshold)

# 开始在众多画面中找目标人物
startFindFace(dstpath, videoPath, cls)

print("done.")
```

我们一起看 startFindFace 函数的实现，因为事先把视频的所有帧都保存下来了，并且也提取了目标人物的人脸特征，接下来要做的事情分如下几步。

（1）在每帧图片中找人脸，并把人脸图片保存起来。这里要注意，之前通过 FFmpeg 解析视频时，保存的文件名格式是 %6d.jpg，6 位数字按顺序命名，如图 6.22 所示。文件名中的数字相当于视频中的每一帧画面所在的帧，如果知道视频的 fps（每秒传输帧数），我们就可以在找到人脸时用来计算出现的时间了。

保存人脸图片的文件名也要能够区分出某张人脸是哪帧画面的，所以文件名是人脸所在的帧加上人脸序号，如图 6.23 所示。比如图 6.23 中的例子，前 5 张人脸是第一帧画面的，一共 5 张人脸，从 0 到 4 排列。

（2）从保存人脸的文件夹中批量读取图片提取人脸特征，需要设定批量大小（Batch Size）。

（3）将视频中的人脸特征与目标人物的特征进行比较。

（4）通过算法，把找到的目标在视频中出现的时间计算出来，并保存图片。

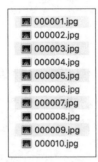

图 6.22　文件名

图 6.23　人脸图片的文件名

下面是 action.py 中 startFindFace 函数的代码。

```
def startFindFace(dstpath, videoPath, cls):

    result_output = "result_output/"
    # 创建保存最终结果的文件夹
    checkFile(os.path.join(os.getcwd(), result_output))
    shutil.rmtree(result_output)
    os.mkdir(result_output)
    print ("[1] now finding the face, this will take a while...")

    time_start = time.time()

    # 用于保存人脸图片的文件夹
    face_dst_path = os.path.join(os.getcwd(), "./face_output/")

    # 得到视频的fps
    videoCapture = cv2.VideoCapture(videoPath)
    fps = videoCapture.get(cv2.CAP_PROP_FPS)
    print("fps = ", fps)

    # 识别每帧画面中的人脸并保存起来
    cls.getDstFaceFileName(dstpath, face_dst_path)

    time_end = time.time()
    print("finding face using time : ", time_end - time_start, "s")
    total_time.append(time_end - time_start)
```

```python
print("[2] now running the tensorflow, this will take a while...")
time_start = time.time()

# 批量提取人脸特征
cls.batch_feature_extraction(face_dst_path, batch_size=64)

# print "file counts : ", len(dstFilenameArr)
print("tensorflow ouput shape : ", cls.src_features.shape)
time_end = time.time()
print("tensorflow using time : ", time_end - time_start, "s")
total_time.append(time_end - time_start)

time_start = time.time()
print("[3] now calculate similar ...")

# 计算视频中的人脸和目标人脸的相似度
results = cls.get_cosine_similarity_results()

print("results = ", results)
print("results length : ", len(results))
print("labels length : ", len(cls.labels))
print("dst_rects_lst length : ", len(cls.dst_rects_lst))
print("results shape : ", np.array(results).shape)

# bool_results中只有"真"和"假"两个值，表示每个相似度代表的人脸是不是目标人
# 物。结果按照时间顺序保存，当目标人物连续出现在视频中时，bool_results中的结果
# 也是连续的。在后面的算法中，只保留第一次出现的结果，其余结果都设置成"假"
bool_results = results >= cls.threshold
if (np.array(results).shape[1] > 1):
    print("np.array(results).shape[1] > 1")
    index_results = bool_results.take(0, 1)
    for i in range(np.array(results).shape[1]):
        index_results = np.logical_or(index_results, bool_results.take(i, 1))
else:
    print("np.array(results).shape[1] !> 1")
```

```python
        index_results = bool_results

print(index_results)
print(index_results.shape)

print("================================")

# ****  下面算法就是用来处理目标人物连续出现时，只保留第一次出现的
# 结果，其余结果都设置成False   ****
find_flag = False
prev_find_flag = False
loop_index = 0
# cls.frame_face_num是字典，key是每帧文件名，value是该帧画面中的人脸数
# 因为results是按照每帧画面中的人脸顺序排列的，我们需要知道每帧画面中有多少人脸
# 才能准确处理好results
# 循环处理每帧画面，frame_face_num_vals是每帧画面中的人脸数
for frame_face_num_vals in cls.frame_face_num.values():
    # 循环每帧画面中的所有人脸
    for i in range(frame_face_num_vals):
        # 如果某张人脸是目标人物，则设置成找到目标了
        if (index_results[i + loop_index] == True):
            find_flag = True

    # 判断在前一帧画面中是否找到了目标
    if prev_find_flag == True:
        # 如果前一帧找到了目标，并且当前帧也找到了目标，则说明是连续出现的
        # 那么把当前帧的结果都设置成false，只保留前一帧找到目标的结果就可以了
        if find_flag == True:
            for i in range(frame_face_num_vals):
                index_results[i + loop_index] = False
        # 如果前一帧找到了目标，但是当前帧没有找到目标，则设置前一帧没有找到目
        # 标，继续下一帧的循环
        else:
            prev_find_flag = False
    else:
        # 如果前一帧没有找到目标，但是当前帧找到了目标，则设置前一帧为找到目
```

```python
            # 标，继续下一帧的循环
            if find_flag == True:
                prev_find_flag = True
        find_flag = False
        loop_index += frame_face_num_vals

# 经过上面算法的处理，results中值为true的帧都是目标每次连续出现时
# 第一次出现的画面
print(index_results)

print("dst_rects_lst size : ", len(cls.dst_rects_lst))
# 对所有帧的结果循环
for i in range(len(index_results)):
    # 如果在该帧画面中找到了目标
    if (index_results[i] == True):
        # print labels[i]

        # labels中保存的是人脸文件名，比如000001_2.jpg
        # 通过文件名的方式得到帧号，比如000001，取整以后就是1
        name = int(cls.labels[i].split("_")[0])

        # 在函数的开始我们已经得到了视频的fps，将帧号除以fps就是该帧在视频中的
        # 秒数
        seconds = name // int(fps)
        # print seconds

        # 把秒数换算成时分秒的形式
        m, s = divmod(seconds, 60)
        h, m = divmod(m, 60)

        # 以时分秒的形式保存的文件名
        new_file_name = str(h) + "_" + str(m) + "_" + str(s) + ".jpg"
        print("find a face : " + new_file_name)

        print(i, int(cls.labels[i].split("_")[1]))
        print(dstpath + cls.labels[i].split("_")[0] + ".jpg"
```

```python
            # 根据人脸文件名能够知道它所在帧的文件名，比如000001_2.jpg所在帧的
            # 文件名是000001.jpg
            # 读取目标人脸所在帧的图片
            img = cv2.imread(dstpath + cls.labels[i].split("_")[0] + ".jpg")

            # 在图片上描画人脸框和相似度
            cls.draw_single_rect(img, cls.dst_rects_lst[i], (0, 255, 0))
            pen = (0, 255, 0)
            cv2.putText(img, str(np.round(results[i], 2)),
                        (cls.dst_rects_lst[i][0], cls.dst_rects_lst[i][1] - 7),
                        cv2.FONT_HERSHEY_DUPLEX, 0.8, pen)
            # 将该图片以时分秒的形式保存到结果文件夹中
            cv2.imwrite(result_output + new_file_name, img)

time_end = time.time()
print (time_end - time_start, "s")
total_time.append(time_end - time_start)
print ("-----------------------------------")
print ("processed the video time : ", total_time[0])
print ("finded face using time : ", total_time[1])
print ("tensorflow using time : ", total_time[2])
print ("calculate similar using time : ", total_time[3])

print ("total time : ", sum(total_time), "s")
print ("results picture in the ", result_output, " directory!")
```

上面函数调用了 Recognizer 对象中的功能函数，我们继续看 Recognizer 的实现，其代码在 main_recognition.py 中。首先是保存每帧画面中的所有人脸。

```python
# 在每帧画面中找出人脸，并保存
def getDstFaceFileName(self, dstpath, face_dst_path):
    # 创建用于保存人脸的文件夹
    checkFile(face_dst_path)
    shutil.rmtree(face_dst_path)
    os.mkdir(face_dst_path)
    self.dstpath = dstpath
```

```python
        filenames = os.listdir(dstpath)
        # 每帧画面的循环
        for ii, file in enumerate(filenames, 1):
            image_path = os.path.join(dstpath, file)
            # 读取每帧画面
            image = misc.imread(image_path, mode='RGB')
            # 识别出人脸，得到人脸图片和人脸框
            self.src_images, self.src_rects = self.findFace(self.detect,
                image.copy())
            # 保存每帧画面中的人脸数
            self.frame_face_num[image_path.split("/")[-1].split(".")[0]] = \
                len(self.src_rects)
            # 调用下面函数保存每张人脸
            self.getDstFaceFileName_proc(image_path, face_dst_path)

    # 循环保存每张人脸
    def getDstFaceFileName_proc(self, image_path, face_dst_path):
        # 每张人脸的循环
        for idx, rect in enumerate(self.src_rects):
            vis = self.src_images[idx]
            # 保存人脸，文件名命名规则是帧号加人脸序号，比如000001_2.jpg
            cv2.imwrite(os.path.join(face_dst_path, image_path.split("/")[-1].split
                (".")[0] + "_" + str(idx) + ".jpg"), vis)
            # 保存人脸文件名
            self.filenames.append(image_path.split("/")[-1].split(".")[0] + "_" +
                str(idx) + ".jpg")
            # 保存对应的人脸框
            self.dst_rects_lst.append(rect)
```

接下来是批量提取人脸特征。

```python
    def batch_feature_extraction(self, face_dst_path, batch_size=64):
        batch = []
        self.labels = []
        self.src_features = None
```

```python
        dlib_src_img = []
        dlib_src_rect = []

        # 所有人脸文件的循环
        for ii, file in enumerate(self.filenames, 1):
            # 保存人脸文件名，不包含后缀，例如：000001_2
            self.labels.append(file.split("/")[-1].split(".")[0])

            if self.feat_ext == 'dlib_68':
                # dlib比较特殊，需要提供人脸所在帧的图片和人脸框
                # 读取人脸所在帧的图片并保存到dlib_src_img中，例如：000001.jpg
                dlib_img=misc.imread(os.path.expanduser(os.path.join(self.dstpath,
                    file.split("_")[0] + ".jpg")), mode='RGB')
                dlib_src_img.append(dlib_img)
                # 保存人脸框
                dlib_src_rect.append(self.dst_rects_lst[ii - 1])
            else:
                # 读取人脸图片并保存到batch中
                img = misc.imread(os.path.expanduser(os.path.join(face_dst_path,
                    file)), mode='RGB')
                if self.feat_ext == 'facenet':
                    batch.append(img.reshape((160, 160, 3)))
                else:
                    img = misc.imresize(img, (224, 224, 3), interp='bilinear')
                    batch.append(img.reshape((1, 224, 224, 3)))

            # 如果读取的人脸数达到一个batch（批次）了，或者所有人脸都加载进来了，则
            # 开始批量提取人脸特征
            if ii % batch_size == 0 or ii == len(self.filenames):

                if self.feat_ext == 'facenet':
                    images = np.stack(batch)
                    codes_batch = facenet_feature_extraction.feature_extraction(
                        images)
                elif self.feat_ext == 'vgg16':
                    images = np.concatenate(batch)
```

```python
                codes_batch = vgg16_feature_extraction.feature_extraction(
                    images)
            elif self.feat_ext == 'dlib_68':
                images = np.stack(dlib_src_img)  # concatenate
                if images.ndim > 3:
                    codes_batch = []
                    for i in range(images.shape[0]):
                        feature = dlib_feature_extraction.
                        feature_extraction_single(images[i], dlib_src_rect[i],
                        "68")
                        codes_batch.append(feature)
                else:
                    codes_batch = dlib_feature_extraction.
                    feature_extraction_single(images, dlib_src_rect[0], "68")

            # 保存人脸特征
            if self.src_features is None:
                self.src_features = codes_batch
            else:
                self.src_features = np.concatenate((self.src_features,
                codes_batch))

            # 重置变量,开始下一批次的循环
            batch = []
            dlib_src_img = []
            dlib_src_rect = []
            print('{} images processed'.format(ii))
```

当得到所有的人脸特征后,开始与目标人脸比较相似度。

```python
def get_cosine_similarity_results(self):
    # 因为目标人脸有可能是一个,在维度上跟批量人脸不同,比如目标特征维度可能是
    # (1, 128),而批量人脸特征的维度可能是(batch_size, 1, 128)
    # 所以需要将批量特征的维度变成(batch_size, 128)
    if self.src_features.ndim != self.target_features.ndim:
        self.src_features = np.reshape(self.src_features, (-1,
            self.target_features.shape[self.target_features.ndim - 1]))
```

```
# 得到余弦相似度
return self.cosine_similarity(self.src_features, self.target_features)
```

经过上述代码，在视频中找人并给出目标人物所在视频的时间就实现了。在终端输入如下命令：

```
python action.py --src ./chengshd/IMG_3663.MOV --target me.jpg --feat_ext facenet
```

在 result_output 文件夹中可以看到识别的结果。

6.7 人脸识别：基于 TensorFlow 2.0

本章项目的神经网络部分使用的都是开源代码，找人的部分没有使用 TensorFlow，所以只需要将开源代码的 TensorFlow 实现修改成 TensorFlow 2.0 的代码即可。

这次修改，我们使用官方提供的升级脚本来实现。

在终端输入如下命令：

```
tf_upgrade_v2 --intree Face_Recognizer --outtree Face_Recognizer-upgradedToTF2.0
```

这样，所有的代码就转换完成了。

基本上所有 TensorFlow 1.x 的代码都会被加上 compat.v1，以保证 TensorFlow 1.x 的代码能够在 TensorFlow 2.0 的环境下成功运行。但是有的代码不会被成功转换，就是 tf.placeholder，我们需要手动修改这句代码。

这句代码是定义占位符，还记得前几章我们定义输入占位符的方法吧，使用 tf.keras.layers.Input 来替换。

举个例子，比如原代码是：

```
data = tf.placeholder(tf.float32, (None,24,24,3), 'input')
```

修改后的代码是：

```
data = tf.keras.layers.Input([24,24,3], name='input')
```

注意，在 TensorFlow 2.0 的输入占位符中，只定义该输入的形状即可（比如在上面的例子中，形状是 [24, 24, 3]），不需要加上 batch_size 的维度，所以去掉 None。

6.8 本章小结

以上就是人脸识别的实现，从人脸检测、特征提取和相似度比较三个方面做了讲解，包括对神经网络的训练和使用，以及人脸识别库提供的方法都做了探讨，最后使用上述知识实现了一个现实应用场景中的人脸识别功能。

人脸识别只是目标检测的一个子分类，上述方法可以用于识别其他目标，而不限于人脸。传统的目标检测可以通过 Hog 特征加 SVM（支持向量机）来实现，包括人脸识别也可以这样做，本章没有对传统算法进行讲解，感兴趣的读者可以自己尝试。

另外，经过尝试以上人脸识别代码后，会发现不同的模型、不同的特征，人脸识别的效果差别很大，有些能够正确识别，有些则不能，不能说哪些方法特别好，哪些方法不好。在这种情况下，可以采用投票机制，应用上述每种方法进行人脸识别，然后对结果进行投票，可以加入权重，将得票最高的结果作为最终可信的结果，得票最高的目标人脸即是找到的目标。同时，目标人脸最好多提供一些，而不像本章只有一个目标人脸特征，同样可以采用投票机制，采用集成投票的方法通常能够得到较好的结果。

7

AlphaZero / AlphaGo 实践：中国象棋

7.1 概述

随着 2016 年 AlphaGo 以 4 : 1 打败韩国围棋棋手李世乭，AlphaGo 和人工智能技术被人们所熟知。一年后，AlphaGo 的升级版又以 3 : 0 的比分完胜世界排名第一的柯洁。围棋的复杂度很高，AI 战胜人类曾被认为是很难实现的任务。可以说，AlphaGo 通过掌握围棋对弈的能力，并且战胜人类，是人工智能领域具有里程碑意义的事件。DeepMind 通过 AlphaGo 在人工智能领域的探索，证明了人工智能是可以通过"白板"训练，掌握解决复杂事件的能力的。在 AlphaGo 战胜柯洁之后，DeepMind 宣布 AlphaGo 不会再参加比赛了，以后的精力将投入到利用 AlphaGo 技术创建更通用的人工智能上，毕竟一个只会下围棋的 AI 并不是研究人工智能最终的目的。

在围棋领域成功挑战人类之后，人工智能专家们又将视线瞄准了更复杂的领域，就是《星际争霸》、*DOTA* 这种即时战略游戏。以 *DOTA2* 为例，每局比赛平均要 45 分钟，具有高维度且连续的动作和状态空间，属于不完全信息博弈。只能看到单位和建筑物周围的区域，地图的其他部分覆盖着迷雾，AI 需要根据不完整的信息做出推断，猜测对手可能采取的行动。游戏中含有数十个英雄、建筑物，几十个 NPC，上百件物品，多种技能和装备，人类能够观察到的这些游戏状态（可以从游戏中获取的信息）大约有 2 万个，而国际象棋大概有 70 个，围棋大概有 400 个。每个英雄都可以针对另一个单位或地面的位置采取很多行动，平均可以产生 1000 个可能的有效行动，与之相比，国际象棋中的平均动作是 35 个，围棋是 250 个。关于 *DOTA2* 的技术分析可以参考 OpenAI 发表的技术博客，地址：https://blog.openai.com/openai-five/。DeepMind 关于《星际争霸》的研究也已经开源，可以参考项目地址：https://github.com/deepmind/pysc2。

可以看出，即时战略游戏的复杂度要比完全信息博弈的棋类游戏高得多，以人类的智能可以驾驭这种复杂度，一旦人工智能达到或者超越了这种能力，一定会给人类社会带来重大影响，这类人工智能就可以用于处理现实世界中更复杂的事务，帮助人类做决策。

本章我们从 AlphaGo 的论文分析开始，并结合代码一起讨论 AlphaZero 在中国象棋上的实践。实际上，在 GitHub 上能够看到很多关于 AlphaGo 的实践项目，包括国际象棋、围棋、五子棋、黑白棋等。

在 DeepMind 的两篇论文发表之后，很多人都在研究该论文并尝试复现，其中比较有影响力的可能就是 LeelaZero 了，项目采用 C++ 语言编写，并且实现了客户端和服务端的分布式训练形式，这样感兴趣的人就可以使用自己的电脑生成下棋数据（棋谱），提交给服务端。围棋训练的瓶颈在于，棋谱要尽可能多，只有多下棋，AI 的棋力才会增加。根据 LeelaZero 的计算，如果以较小规模进行单机训练，通常需要 1700 年的时间才能训练出一个良好的网络模型，可见训练围棋 AI 需要多么大的计算能力。后来 Facebook AI Research（FAIR）开源了自己的围棋 AI：ELF OpenGo（https://github.com/pytorch/ELF），并跟 LeelaZero 进行了一系列比赛（198 胜 2 负）。

围棋 AI 的发展大家都有目共睹了，那么如何让 AlphaGo 的技术也用在中国象棋上，很多人都展开了尝试，也包括笔者在内。国内实现了 AlphaGo 算法在做分布式训练中国象棋的是 cc-zero，跟 LeelaZero 一样，也需要大家参与生成棋谱。网友 Icybee 实现的中国象棋程序 icyChessZero，也同样采用了分布式训练方法，感兴趣的读者可以去了解一下。

限于篇幅，本章重要代码放在书里，完整代码可以在 GitHub 上下载，地址：https://github.com/chengstone/cchess-zero，TensorFlow 2.0 的实现也在该项目下，文件名的后缀是.tf2。

7.2 论文解析

我们要参考的是 AlphaGo Zero 的论文 *Mastering the Game of Go without Human Knowledge*（地址：https://deepmind.com/documents/119/agz_unformatted_nature.pdf）和 AlphaZero 的论文 *Mastering Chess and Shogi by Self-Play with a General Reinforcement Learning Algorithm*（地址：https://arxiv.org/pdf/1712.01815.pdf）。

先从 *Mastering the Game of Go without Human Knowledge* 论文说起，算法根据这篇论文来实现，AlphaZero 只有几点不同而已。

AlphaGo Zero 在几个重要方面与 AlphaGo Fan 和 AlphaGo Lee 不同。首先，它是完全由自我对弈强化学习训练的，从随机下棋开始，没有使用人类棋谱做监督学习。其次，它只使用黑子和白子作为输入特征。再次，它使用单一的神经网络，而没有分别使用策略网络和价值网络。最后，它使用更简单的树搜索算法，依靠这个单一的神经网络来评估局面和选择走子，而不需要执行任何 rollouts（通过快速走子，得到一个终局后的胜负结果）。

AlphaGo Zero 中的神经网络 f_θ 是通过强化学习算法从自我对弈中训练出来的。这个神经网络把棋局的原始表示形式 s（棋盘状态）和它的历史走子记录作为输入，并输出走子概率和一个评估值（胜率），$(p, v) = f_\theta(s)$。向量走子概率 p 表示选择每个走子的概率，$p_a = Pr(a|s)$。评估值 v 是标量，估计当前玩家从局面 s 获胜的概率。这个神经网络将策略网络和价值网络的功能整合到统一的体系结构中，即只有一个单一的网络。

对于每个局面 s，神经网络 f_θ 的预测结果作为执行蒙特卡罗树搜索（Monte Carlo Tree Search, MCTS）的参考。MCTS 输出每个可能走子的概率 π 和胜负结果 z，这个概率通常比神经网络 $f_\theta(s)$ 输出的概率 p 更准确，选择的走子更强力。强化学习算法的主要思想是在策略迭代过程中重复以下过程：不断更新神经网络的参数 θ，使得网络输出的走子概率和胜率 $(p, v) = f_\theta(s)$ 与 MCTS 输出的走子概率和胜负结果 (π, z) 更接近；这些新的参数被用在下一次自我对弈中，使得 MCTS 更加强大。

总结起来，AlphaGo Zero 分为两个部分：一部分是蒙特卡罗树搜索；一部分是神经网络。我们是要抛弃人类棋谱的，学会如何下棋完全是通过自我对弈来完成的。过程如下：通过自我对弈生成棋谱，然后将棋谱作为输入训练神经网络，训练好的神经网络用在下一次自我对弈中预测落子和胜率，反复迭代训练，使得神经网络预测的结果越来越接近 MCTS 的预测，而 MCTS 的预测则越来越强大。AlphaGo Zero 自我对弈强化学习过程如图 7.1 所示（图片出自原论文）。

（1）自我对弈局面 $s_1, \cdots, s_T$。在每个局面 s_t，使用最新的神经网络 f_θ 执行蒙特卡罗树搜索（MCTS）。根据由 MCTS 计算的概率选择走子，$a_t \sim \pi_t$。按照游戏规则对棋局结束状态 s_T 进行评分，得到胜负结果 z。

（2）训练 AlphaGo Zero 神经网络。神经网络将棋局的原始走子状态 s_t 作为输入数据，输入给具有参数 θ 的多个卷积层，并且输出表示走子概率分布的向量 p_t 和表示当前玩家在局面 s_t 的获胜概率标量值 v_t。更新神经网络的参数 θ，以使策略向量 p_t 与搜索概率 π_t 的相似度最大化，并使得预测的获胜者 v_t 和游戏胜者 z 之间的误差最小化。新参数用于下一次迭代的自我对弈 a 中。

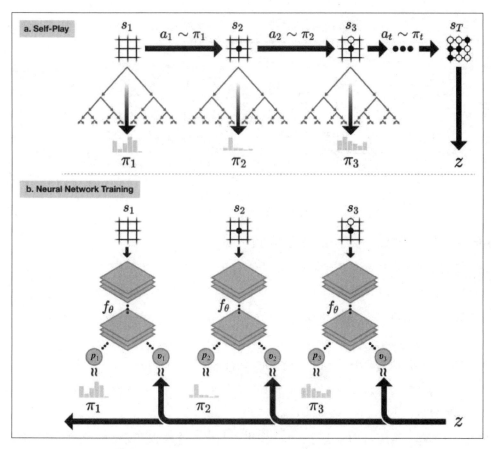

图 7.1　AlphaGo Zero 的自我对弈强化学习过程

7.2.1 蒙特卡罗树搜索算法

MCTS 用来自我对弈生成棋谱，搜索树中的每个边 (s,a) 存储先验概率 $P(s,a)$、访问计数 $N(s,a)$ 和行动价值 $Q(s,a)$。每次搜索都从根节点开始，迭代选择使置信区间上限 $Q(s,a)+U(s,a)$ 最大化的走子，其中 $U(s,a) \propto P(s,a)/(1+N(s,a))$，直到遇到叶子节点 s'。叶子节点只会被神经网络扩展和评估一次，以产生先验概率和胜率评估值，$(P(s',\cdot), V(s')) = f_\theta(s')$。更新搜索中遍历的每个边 (s,a)，增加其访问次数 $N(s,a)$，将它的行动价值更新为在这些搜索上的平均估计值，即 $Q(s,a) = 1/N(s,a)\Sigma_{s'|s,a \to s'}V(s')$，其中 $s,a \to s'$ 表示从局面 s 选择走子 a 后搜索最终到达 s'。下面结合论文中的图示（如图 7.2 所示）进行说明。

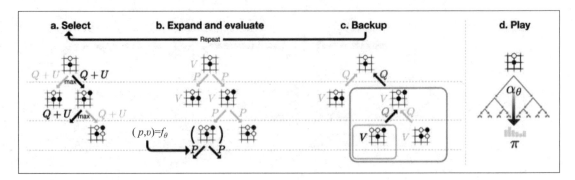

图 7.2　MCTS 过程

AlphaGo Zero 中的蒙特卡罗树搜索过程如下：

（1）每次模拟都通过选择具有最大行动价值 Q 的边，加上取决于所存储的先验概率 P 和该边的访问计数 N（每次访问都被增加一次）的上限置信区间 U 来遍历树。

（2）展开叶子节点，通过神经网络 $(P(s,\cdot), V(s)) = f_\theta(s)$ 来评估局面 s；将 P 的值存储在叶子节点扩展的边上。

（3）更新行动价值 Q 为该行动下子树中的所有评估值 V 的均值。

（4）一旦 MCTS 完成，返回局面 s 下的落子概率 π，与 $N^{1/\tau}$ 成正比，其中 N 是从根节点每次移动的访问计数，τ 是控制温度的参数。

按照论文所述，每次 MCTS 执行 1600 次模拟。过程是这样的：现在 AI 从"白板"开始自己跟自己下棋，只知道规则，不知道套路，那只好乱下。每下一步棋，都要通过 MCTS 模拟 1600 次图 7.2 中的 a~c，从而得出这次要怎么走子的结果。

现在来说说 a~c。本质上，MCTS 需要我们来维护一棵树，这棵树的每个节点都保存了每一个局面所有合法动作 $a \in A(s)$ 的边 (s,a) 的信息。这些信息包括：$N(s,a)$，是访问次数；$W(s,a)$，是总行动价值；$Q(s,a)$，是平均行动价值；$P(s,a)$，是被选择的概率。

1. 选择（Select）

每次模拟的过程都一样，从父节点的局面开始，选择一个走子。比如开局时，所有合法的走子都是可能的选择，那么该选择哪个走子呢？MCTS 选择 $Q(s,a) + U(s,a)$ 最大的那个走子（action）。Q 的公式一会在回传（Backup）中描述。U 的公式如下：

$$U(s,a) = c_{\text{puct}} P(s,a) \frac{\sqrt{\sum_b N(s,b)}}{1 + N(s,a)}$$

这个公式可以理解成：$U(s,a) = c_{\text{puct}} \times$ 概率 $P(s,a) \times$ np.sqrt（父节点访问次数 N）/（1 + 某子节点 action 的访问次数 $N(s,a)$）。

用论文中的话说，c_{puct} 是一个决定探索水平的常数；这种搜索控制策略最初倾向于具有高先验概率和低访问次数的行为，但是渐进地倾向于具有高行动价值的行为。

计算过后，我们就知道在当前局面下哪个 action 的 $Q + U$ 值最大，那么这个走子之后的局面就是第二次模拟的当前局面。比如开局时，$Q + U$ 最大的是当头炮，就使用当头炮这个 action，在下一次选择时就从当头炮的这个棋局选择下一个走子。

2. 展开和评估（Expand and evaluate）

现在开始第二次模拟。假如之前的 action 是当头炮，我们要接着这个局面选择 action，但是这个局面是一个叶子节点。也就是说，在当头炮之后可以选择哪些 action 不知道，这样就需要展开了，通过展开得到一系列可能的 action 节点。这实际上就是在扩展这棵树，从根节点开始，一点一点地扩展。

在"展开和评估"部分有一个需要关注的地方。论文中说："在队列中局面由神经网络使用 mini-batch（最小批量，大小为 8）进行评估；搜索线程被锁定，直到评估完成。叶子节点被展开，每个边 (s_L, a) 被初始化为 $N(s_L, a) = 0$，$W(s_L, a) = 0$，$Q(s_L, a) = 0$，$P(s_L, a) = p_a$，然后值 v 被回传。"

如果当前局面没有被展开过，则不知道下一步该怎么走，所以要展开，这时该神经网络出马了。把当前局面作为输入传给神经网络，神经网络会返回给我们一个 action 向量 p 和当前胜率 v。其中向量 p 是当前局面每个合法 action 的走子概率。当然，因为神经网络还没有训练好，所以将输出作为参考添加到蒙特卡罗树上。这样在当前局面下，所有可走的 action 以及对应的概率 p 就都有了，每个新增的 action 节点都按照论文中所说的对若干信息赋值，$N(s_L, a) = 0$，$W(s_L, a) = 0$，$Q(s_L, a) = 0$，$P(s_L, a) = p_a$。这些新增的节点作为当前局面节点的子节点。

3. 回传（Backup）

接下来是重点，评估和回传一起说，先看看回传做什么事。"边的统计数据在每一步 $t \leqslant L$ 中反向更新。访问计数递增，$N(s_t, a_t) = N(s_t, a_t) + 1$，并且行动价值更新为平均值，$W(s_t, a_t) = W(s_t, a_t) + v$，$Q(s_t, a_t) = \frac{W(s_t, a_t)}{N(s_t, a_t)}$。"我们使用**虚拟损失**（Virtual loss）来确保每个线程评估不同的节点。

整理思路：对于任意一个局面（就是节点），要么被展开过，要么没有被展开过（就是叶子节点）。展开过的节点可以通过选择动作进入下一个局面，下一个局面仍然是这个过程；如果展开过节点，还是可以通过选择动作进入下下一个局面；这个过程一直持续下去，直到这盘棋分出胜平负，或者遇到某个局面没有被展开过为止。

如果没有展开过节点，那么执行展开操作，通过神经网络得到每个动作的概率和胜率 v，把这些动作添加到树上，最后回传胜率 v，那么回传给谁？

我们知道这其实是一路递归的过程，一直在选择，递归必须要有结束条件，不然就是死循环了。所以分出胜负和遇到叶子节点就是递归结束条件，把胜率 v 或者分出的胜平负 value 作为返回值回传给上一层。

这个过程就是评估（Evaluate），是为回传步骤做准备的。因为在回传步骤中，我们要用 v 来更新 W 和 Q，但是如果只做了一次选择，棋局还没有结束，此时的 v 是不明确的，必须要等到一盘棋完整地下完才能知道 v 到底是多少。也就是说，现在下了一步棋，不管这步棋是好棋还是臭棋，只有下完整盘棋分出胜负，才能给这步棋评分。不管这步棋的得失，即使这步棋丢了一个"车"，但最后赢了，那么这步棋也是积极的。同样，即使这步棋吃了对方一个子，但最后输了，也不能认为这步棋就是好棋。

我们用图示来概括这个过程，如图 7.3 所示。

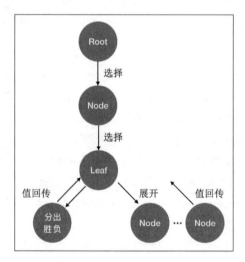

图 7.3　选择、展开、回传过程图示

当值被回传后，就要做回传动作了，这里很关键。因为是多线程同时在做 MCTS 的，由于选择算法都一样，都是选择 $Q+U$ 最大的节点，所以很有可能所有的线程最终选择的是同一个节点，这就尴尬了。我们的目的是尽可能在树上搜索出各种不同的着法，最终选择一步好棋，怎么办呢？论文中已经给出了办法，"使用**虚拟损失**来确保每个线程评估不同的节点"。

也就是说，通过选择选出某节点后，人为增大这个节点的访问次数 N，并减小节点的总行动

价值 W，因为平均行动价值 $Q = W/N$，分子减小，分母增大，就减小了 Q 值，这样递归进行时，此节点的 $Q + U$ 不是最大的，避免被选中，让其他线程尝试选择别的节点进行树搜索。这个人为增加和减少的量就是虚拟损失。

现在 MCTS 的过程越来越清晰了，通过选择选出节点后，对当前节点使用虚拟损失，通过递归继续选择，直到分出胜负或展开节点，得到返回值 value。接下来就可以使用 value 进行回传了，但首先要还原 W 和 N，之前 N 增加了虚拟损失，这次要减回去，之前减少了虚拟损失的 W 也要加回来。

然后开始做回传，"边的统计数据在每一步 $t \leqslant L$ 中反向更新。访问计数递增，$N(s_t, a_t) = N(s_t, a_t) + 1$，并且行动价值更新为平均值，$W(s_t, a_t) = W(s_t, a_t) + v$，$Q(s_t, a_t) = \frac{W(s_t, a_t)}{N(s_t, a_t)}$。"同时还要更新 U，U 的公式上面给出过。这个反向更新其实就是递归地把值返回去。这里有一点一定要**注意，就是返回值一定要符号反转**。怎么理解？就是对于当前节点是胜的，那么对于上一个节点一定是负的，所以返回的是 $-$value。

4. 走子（Play）

按照上述过程执行图 7.2 中的 a~c，论文中是每步棋执行 1600 次模拟，那就是执行 1600 次 a~c，MCTS 的过程就是模拟自我对弈的过程。当模拟结束后，基本上能覆盖大多数的棋局和着法，每步棋该怎么下、下完以后胜率是多少、得到什么样的局面都能在树上找到。然后从树上选择当前局面应该下哪一步棋，这就是走子步骤（见图 7.2 中的 d）："在搜索结束时，AlphaGo Zero 在根节点 s_0 选择一个走子 a，与其访问计数幂指数成正比，$\pi(a|s_0) = \frac{N(s_0, a)^{1/\tau}}{\sum_b N(s_0, b)^{1/\tau}}$，其中 τ 是控制探索水平的温度参数。在随后的时间步中重新使用搜索树：与所走子的动作对应的子节点成为新的根节点；保留这个节点下面的子树所有的统计信息，而树的其余部分被丢弃。如果根节点的价值和最好的子节点价值低于阈值 v_{resign}，则 AlphaGo Zero 会认输。"

当模拟结束后，对于当前局面（就是树的根节点）下的所有子节点（就是每一步对应的 action 节点），选择哪一个 action 呢？按照论文中所说的，是通过访问计数 N 来确定的。这个好理解，在实现上也容易，当前节点下的所有子节点是可以获得的，每个子节点的信息 N 都可以获得，然后从多个 action 中选择一个，这其实是多分类问题。我们使用 softmax 来得到选择某个 action 的概率，传给 softmax 的是每个 action 的 logits（$N(s_0, a)^{1/\tau}$），这其实可以改成 $1/\tau \times \log(N(s_0, a))$。这样就得到了当前局面下所有可选 action 的概率向量，最终选择概率最大的那个 action 作为要下的一步棋，并且将这个选择的节点作为树的根节点。

按照图 7.1 中 "a. Self-Play" 的说法就是，从局面 s_t 进行自我对弈的树搜索（模拟），得到 $a_t \sim \pi_t$，a_t 就是动作 action，π_t 就是所有动作的概率向量。最终在局面 s_T 下得到胜平负的结果 z，就是上面所说的 value。

至此，MCTS 算法就分析完了。

7.2.2 神经网络

上面说过，通过 MCTS 算出该下哪一步棋，然后经过 1600 次模拟算出下一步棋，如此循环直到分出胜负，这样一整盘棋就下完了。这就是一次完整的自我对弈过程，MCTS 就相当于在人大脑中的思考。我们把每步棋的局面 s_t、算出的 action 概率向量 π_t 和胜率 z_t（就是返回值 value）保存下来，作为棋谱数据训练神经网络。

神经网络的输入是局面 s，输出是预测的 action 概率向量 p 和胜率 v，$(p, v) = f_{\theta_i}(s)$。训练目标是最小化预测胜率 v 和自我对弈的胜率 z 之间的误差，并使神经网络走子概率 p 与搜索概率 π 的相似度最大化。按照论文中所说的，"参数 θ 通过梯度下降分别在均方误差和交叉熵损失之和上的损失函数 l 上进行调整，$l = (z - v)^2 - \pi^T \log p + c\|\theta\|^2$，其中 c 是控制 L2 权重正则化水平的参数（防止过拟合）。"简单点说，就是让神经网络的预测跟 MCTS 的搜索结果尽量接近。

胜率是回归问题，优化自然用 MSE 损失，概率向量的优化要用 softmax 交叉熵损失，目标就是最小化这个联合损失。

接下来，我们看看神经网络结构。神经网络结构按照论文中的描述实现即可，如图 7.4 所示。

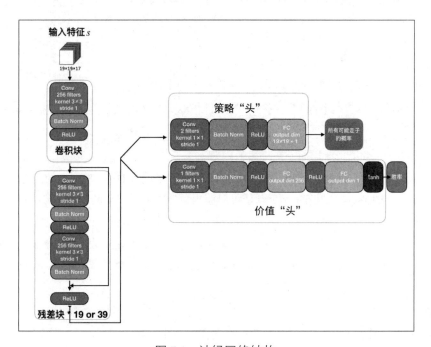

图 7.4　神经网络结构

输入特征 s 由一个残差塔处理，该塔由单个卷积块组成，后跟 19 层或 39 层残差块。

卷积块应用以下模块。

- 卷积层：滤波器的个数为 256，卷积核尺寸为 3×3，步幅为 1。
- 批次归一化（Batch Normalisation）层。
- 整流非线性 ReLU。

每个残差块将下列模块按顺序应用于其输入中。

- 卷积层：滤波器的个数为 256，卷积核尺寸为 3×3，步幅为 1。
- 批次归一化层。
- 整流非线性 ReLU。
- 卷积层：滤波器的个数为 256，卷积核尺寸为 3×3，步幅为 1。
- 批次归一化层。
- 将残差块的输入直接连接过来。
- 对第二个批次归一化层的输出和上面连接过来的残差块的输入一起做整流非线性 ReLU。

残差塔的输出分别传递到两个单独的"头"，分别计算策略和价值。

策略"头"应用以下模块。

- 卷积层：滤波器个数为 2，卷积核尺寸为 1×1，步幅为 1。
- 批次归一化层。
- 整流非线性 ReLU。
- 一个全连接的线性层，输出一个大小为 $19^2 + 1 = 362$ 的向量，对应于所有交叉点的 logits 概率（每个点的走子概率）。

价值"头"应用以下模块。

- 卷积层：滤波器个数为 1，卷积核尺寸为 1×1，步幅为 1。
- 批次归一化层。
- 整流非线性 ReLU。
- 从一个全连接的线性层到一个大小为 256 的隐层。
- 整流非线性 ReLU。
- 从一个全连接的线性层到一个标量。
- tanh 激活函数：输出范围为 $[-1, 1]$ 的标量。

至此，这篇论文基本介绍完了，有些训练和优化方面的细节这里就不介绍了。过程就是神经网络先随机初始化权重，使用 MCTS 下每一步棋，当树中节点没有被展开时通过神经网络预测出走子概率和胜率添加到树上，然后使用自我对弈的数据训练神经网络，在下一次自我对弈中使用新的训练过的神经网络进行预测，MCTS 和神经网络你中有我、我中有你，如此反复迭代，网络预测得更准确，MCTS 的结果更强大。实际上，神经网络的预测可以理解为人的直觉。

7.2.3 AlphaZero 论文解析

接下来，我们一起看看 AlphaZero 的论文 *Mastering Chess and Shogi by Self-Play with a General Reinforcement Learning Algorithm*。

AlphaZero 算法是 AlphaGo Zero 算法的更通用的版本，论文讲述了 AlphaZero 算法分别在围棋、国际象棋和将棋上的表现。

AlphaGo Zero 算法通过使用深度卷积神经网络来表达围棋知识，只通过自我对弈的强化学习来训练，在围棋中实现了超人的表现。在论文中，应用了一个类似的但完全通用的算法，该算法称之为 AlphaZero，用来像下围棋一样下国际象棋和将棋，除游戏规则外没有给予任何额外的领域知识，这个算法表明通过强化学习算法可以实现以"白板"方式学习，在多个具有挑战性的领域获得超人的表现。

在计算复杂性方面，相比国际象棋，将棋更难：它在一个更大的棋盘上玩，任何被俘获的对手棋子都会改变方向，随后可能被放置在棋盘上的任何位置。最好的将棋程序，如电脑将棋协会（CSA）的世界冠军 Elmo，直到最近才击败人类冠军。将棋程序使用与计算机国际象棋程序类似的算法，基于高度优化的 Alpha-Beta 搜索，在很多特定领域具有适应性。

围棋非常适合使用 AlphaGo 的神经网络架构，因为游戏规则是平移不变的（匹配卷积神经网络的权值共享结构），是根据棋盘上的走子点之间的相邻点的自由度来定义的（匹配卷积神经网络的局部结构），并且是旋转和反射对称的（允许数据增强和合成）。而且，行动空间很简单（一颗棋子可以放在任何可能的位置），游戏结果只有二元结果赢或输，这些都有助于神经网络的训练。

国际象棋和将棋不太适合使用 AlphaGo 的神经网络架构，因为游戏规则是与位置有关（例如，兵可以从第二横线前进两步，在第八横线上升变）和不对称的（例如，兵只能向前移动，在王翼和后翼的王车易位是不同的）。规则包括远程交互（例如，皇后可以一次穿过整个棋盘，或者从棋盘的另一边将军）。国际象棋的行动空间包括棋盘上所有棋手棋子的符合规则的位置；将棋允许将被吃掉的棋子放回棋盘上。国际象棋和将棋都可能造成平局。事实上，人们认为国际象棋最佳的解决方案是平局。

AlphaZero 使用参数为 θ 的深度神经网络，$(\boldsymbol{p}, v) = f_\theta(s)$。这个神经网络使用局面（棋盘状态）$s$ 作为输入，输出走子概率向量 $\boldsymbol{p}$，它包含每一个走子动作 a 的概率分量 $p_a = Pr(a|s)$，同时输出一个标量值 v（胜率）——从局面 s 估算预期结果 z，$v \approx [z|s]$。AlphaZero 完全从自我对弈中学习这些走子概率和价值估计，然后将学到的知识指导其搜索。

为取代具有特定领域增强的 Alpha-Beta 搜索，AlphaZero 使用通用的蒙特卡罗树搜索（MCTS）算法。每次搜索都包含一系列从根节点 s_{root} 到叶子节点遍历树的自我对弈模拟。每次模拟都是通过在每个状态 s 下，根据当前的神经网络 f_θ，选择一个访问次数低、走子概率高和价值高的走子走法 a 的。搜索返回一个表示走子概率分布的向量 $\boldsymbol{\pi}$，是在根节点状态下关于访问计数的概率分布。

AlphaZero 深度神经网络的参数 θ，从随机初始化参数开始，通过自我对弈强化学习进行训练；通过 MCTS（$a_t \sim \pi_t$）轮流为两个棋手选择走子进行下棋。在棋局结束时，根据游戏规则计算游戏结果 z 作为结束位置 s_T 的评分：-1 代表失败，0 代表平局，+1 代表胜利。更新深度神经网络的参数 θ 以使预测结果 v_t 与游戏结果 z 之间的误差最小，并且使策略向量 $\boldsymbol{p}_t$ 与搜索概率 $\boldsymbol{\pi}_t$ 的相似度最大。具体而言，参数 θ 通过在均方误差和交叉熵损失之和上的损失函数 l 上做梯度下降进行调整，$(\boldsymbol{p}, v) = f_\theta(s), l = (z-v)^2 - \boldsymbol{\pi}^{\text{T}} \log \boldsymbol{p} + c||\theta||^2$，其中 c 是控制 L2 正则化水平的参数。更新的参数被用于随后的自我对弈中。

AlphaZero 算法在如下几个方面与原始的 AlphaGo Zero 算法不同：

- AlphaGo Zero 在假设只有赢或输二元结果的情况下，对获胜概率进行估计和优化。AlphaZero 会考虑平局或潜在的其他结果，对预期的结果进行估计和优化。

- 围棋的规则是旋转和反转不变的。对此，在 AlphaGo 和 AlphaGo Zero 中有两种使用方式。首先，训练数据通过为每个局面生成 8 张对称图像来增强。其次，在 MCTS 期间，棋盘位置在被神经网络评估前，会使用随机选择的旋转或反转变换进行转换，以便蒙特卡罗评估在不同的偏差上进行平均。国际象棋和将棋的规则是不对称的。AlphaZero 不会增强训练数据，也不会在 MCTS 期间转换棋盘位置。

- 在 AlphaGo Zero 中，自我对弈是由以前所有迭代中最好的玩家生成的。在每次训练迭代之后，与最好玩家对弈测量新玩家的能力；如果以 55% 的优势获胜，那么它将取代最好的玩家，而自我对弈将由这个新玩家产生。相反，AlphaZero 只维护一个不断更新的单个神经网络，而不是等待迭代完成。自我对弈是通过使用这个神经网络的最新参数生成的，省略了评估步骤和选择最佳玩家的过程。

- AlphaGo Zero 通过贝叶斯优化调整搜索的超参数。而在 AlphaZero 中，我们为所有棋局重复使用相同的超参数，而无须进行特定于某种游戏的调整。唯一的例外是为保证探索而添加到先验策略中的噪声，这与棋局类型的典型合法走子的数量成比例。

除了上面说的几点区别，在算法上没有任何区别。

接下来，我们看看输入特征的表示。输入特征如图 7.5 所示。

围棋		国际象棋		将棋	
特征	平面	特征	平面	特征	平面
P1棋子	1	P1棋子	6	P1棋子	14
P2棋子	1	P2棋子	6	P2棋子	14
		重复局面	2	重复局面	3
				P1持驹数	7
				P2持驹数	7
颜色	1	颜色	1	颜色	1
		总走子数（回合数）	1	总走子数（回合数）	1
		P1王车易位	2		
		P2王车易位	2		
		没有进展回合数	1		
Total	17	Total	119	Total	362

图 7.5　输入特征

AlphaGo Zero 神经网络的输入是 $19 \times 19 \times 17$ 维度的图像栈，包含 17 个二值（只有两个值 0 和 1）特征平面，8 个特征平面 X_t 由二进制值组成，表示当前玩家存在的棋子（如果交叉点 i 在时间步 t 中包含玩家的棋子，那么 $X_t^i = 1$；如果交叉点是空的，包含对手的棋子，或者 $t < 0$，那么 $X_t^i = 0$）。另外 8 个特征平面 Y_t 表示对手的棋子的相应特征。为什么每个玩家 8 个特征平面呢？因为这是 8 步历史走子记录，也就是说，最近走的 8 步棋作为输入特征。最后的特征平面 C 表示棋子颜色（当前的棋盘状态），是常量，如果是黑色棋子，则为 1；如果是白色棋子，则为 0。这些平面连接在一起，给出输入特征 $s_t = [X_t, Y_t, X_{t-1}, Y_{t-1}, \cdots, X_{t-7}, Y_{t-7}, C]$。

AlphaZero 神经网络的输入是 $N \times N \times (MT + L)$ 图像栈，其表示状态使用大小为 $N \times N$ 的 T 组 M 个平面级联组成。其中 $N \times N$ 就是棋盘的大小，T 是时间步，M 是棋子类型的数量。每一组平面代表时间步 $t - T + 1, \cdots, t$ 的棋盘位置，在小于 1 的时间步中设置为 0。棋盘朝向当前玩家的角度。M 特征平面由棋手存在的棋子的二值特征平面组成，每种棋子类型都具有一个平面，第二组平面表示对手存在的棋子。对于将棋，还有额外的平面显示每种类型的持驹数。还有一个额外的 L 个常值输入平面，表示玩家的颜色、总的回合数量和特殊规则的状态。特殊规则比如：国际象棋中的王车易位、局面的重复次数（在国际象棋中 3 次重复将被判为平局；在将棋中是 4 次）、在国际象棋中没有进展的走子次数（50 次没有进展的走子将被判为平局）。

国际象棋的输入跟围棋不同，加入了各种特征平面，用来表示不同的情况，如王车易位、多少回合没有进展（没有吃子）、重复的局面（多次重复会被判为平局）等，这些特征可以根据不同的棋种去设计，我们重点关注的是棋子的输入特征。

对于围棋而言，每颗棋子都是一样的，它们都属于一类。而国际象棋分为 6 种棋子：车、马、象、后、王、兵，那么在特征平面上怎么表示呢？总不能使用 0~5 吧？还是使用 0 和 1 来表示棋盘上有子还是没子，然后，既然是 6 种棋子，那么想当然地使用 one-hot 编码，所以将特征平面分成了 6 个平面，每一个平面用来表示不同种类棋子在棋盘上的位置。

以上就是介绍的全部内容，更多的细节，比如优化参数设为多少、学习率退火设为多少等，请阅读论文。

7.3 实现中国象棋：基于 TensorFlow 1.x

本节根据论文中的算法实现一个中国象棋程序。

7.3.1 中国象棋着法表示和 FEN 格式

这里有些关于中国象棋通用引擎协议（Universal Chinese Chess Protocol，UCCI）、FEN 文件格式和象棋着法表示的知识需要讲解一下，更详细的内容可以参考象棋百科全书网，下面引用部分描述进行说明。

- FEN 文件格式（http://www.xqbase.com/protocol/cchess_fen.htm）。
- 着法表示（http://www.xqbase.com/protocol/cchess_move.htm）。
- 中国象棋通用引擎协议，版本为 3.0（http://www.xqbase.com/protocol/cchess_ucci.htm）。

中国象棋通用引擎协议（UCCI）是一种象棋界面和象棋引擎之间的基于文本的通信协议，是模仿国际象棋的 UCI 来制定的。其作用是定义了一套通信协议规范，可以使各种不同的中国象棋引擎通过同一种语言（协议）进行交流（对弈下棋）。

当然，本项目并没有完整实现 UCCI 协议，只是使用了其中的概念，主要关注的是着法的表示和 FEN 格式。

简而言之，中国象棋的着法表示，就是指某棋子从什么位置走到什么位置。中国象棋规定，对于红方来说，纵线从右到左依次用"一"到"九"表示，黑方则用"1"到"9"表示（如图 7.6 所示），每个数字表示某方的一条纵线。

着法的表示使用炮二平五、士 4 进 5 等，这是中国象棋传统的表示方法。

如图 7.7 所示的棋盘表示则借鉴了国际象棋的表示方法，按照坐标的方式来表示着法，比如炮二平五可以表示成 h2e2。

另外，棋子的表示也参照了国际象棋的方式，在协议中不使用汉字而使用字母来表示某棋子，如表 7.1 所示。

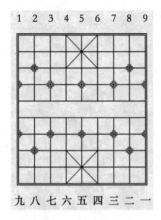

图 7.6 中国象棋传统棋盘表示方式（图片出自象棋百科全书网）

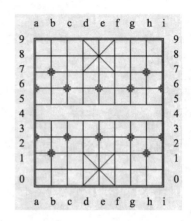

图 7.7 中国象棋棋盘坐标表示方式（图片出自象棋百科全书网）

表 7.1 中国象棋棋子代号（该表出自象棋百科全书网）

红方	黑方	字母	相当于国际象棋中的棋子
帅	将	K	King（王）
仕	士	A	Advisor（没有可比较的棋子）
相	象	B[1]	Bishop（象）
马	马	N[2]	Knight（马）
车	车	R	Rook（车）
炮	炮	C	Cannon（没有可比较的棋子）
兵	卒	P	Pawn（兵）

[1] 世界象棋联合会推荐的字母代号为 E（Elephant）。
[2] 世界象棋联合会推荐的字母代号为 H（Horse）。

这样有了代表棋子的字母，就可以用来按照某种格式表示棋局了，这种格式叫作 FEN（Forsyth-Edwards Notation）。下面解释出自象棋百科全书网。

FEN 是专门用来表示象棋局面的记录规范，在这个规范下，一个局面可以简单地用一行"FEN 格式串"来表示。

国际象棋的 FEN 格式串是由 6 段 ASCII 字符串组成的代码（各段之间用空格隔开），这 6 段代码的含义依次是：

（1）棋盘上的棋子，这是 FEN 格式串的主要部分。

（2）轮到哪一方走子。

（3）每方及该方的王翼和后翼是否还存在"王车易位"的可能。

（4）是否存在吃过路兵的可能，过路兵是经过哪个格子的。

（5）最近一次吃子或者进兵后棋局进行的步数（半回合数），用来判断"50回合自然限着"。

（6）棋局的回合数。

中国象棋没有"王车易位"和"吃过路兵"的着法，所以FEN格式串的（3）和（4）这两项空缺。下面以最初局面为例进行说明。

`rnbakabnr/9/1c5c1/p1p1p1p1p/9/9/P1P1P1P1P/1C5C1/9/RNBAKABNR w - - 0 1`

上述最初局面的6个部分分别是：

（1）rnbakabnr/9/1c5c1/p1p1p1p1p/9/9/P1P1P1P1P/1C5C1/9/RNBAKABNR

（2）w

（3）-

（4）-

（5）0

（6）1

其中：

（1）表示棋盘布局，小写字母表示黑方，大写字母表示红方。这里要注意两点，一是中国象棋棋盘有10行，所以要用9个"/"把每一行隔开；二是棋子名称用英文字母表示，在国际象棋中没有的棋子是仕（士）和炮，这里分别用字母A（a）和C（c）表示。

（2）表示轮到哪一方走子，"w"表示红方，"b"表示黑方。（有人认为红方应该用"r"表示，很多象棋软件确实是这样表示的。）

（3）空缺，始终用"-"表示。

（4）空缺，始终用"-"表示。

（5）表示双方没有吃子的走棋步数（半回合数），通常该值达到120就要判和（60回合自然限着），一旦形成局面的上一步是吃子，这里就标记为"0"。

（6）表示当前的回合数。

现在我们有了着法表示、棋子表示和棋局表示三个工具了，那么在程序中就会使用这三个工具来实现象棋对弈。

其中这里FEN格式的实现稍稍有些不同，在FEN格式串的第一部分（1）中，格式规定的最初局面表示是：

`rnbakabnr/9/1c5c1/p1p1p1p1p/9/9/P1P1P1P1P/1C5C1/9/RNBAKABNR`

因为小写字母代表黑方，大写字母代表红方，每个"/"表示一行，拆分开就是这样的：

rnbakabnr
/9
/1c5c1
/p1p1p1p1p
/9
/9
/P1P1P1P1P
/1C5C1
/9
/RNBAKABNR

每个数字代表几个空白，即没有棋子，比如第二行"/9"表示这一行没有任何棋子。

换成棋盘表示的话，就像下面这样（红方在下，黑方在上），如图7.8所示。

	a	b	c	d	e	f	g	h	i
9	r（车）	n（马）	b（象）	a（士）	k（将）	a（士）	b（象）	n（马）	r（车）
8									
7		c（炮）						c（炮）	
6	p（卒）		p（卒）		p（卒）		p（卒）		p（卒）
5									
4									
3	P（兵）		P（兵）		P（兵）		P（兵）		P（兵）
2		C（炮）						C（炮）	
1									
0	R（车）	N（马）	B（相）	A（仕）	K（帅）	A（仕）	B（相）	N（马）	R（车）
	a	b	c	d	e	f	g	h	i

图7.8 棋盘表示

本项目在实现上不同的地方在于棋局表示跟上面是相反的，最初局面表示是这样的：

RNBAKABNR/9/1C5C1/P1P1P1P1P/9/9/p1p1p1p1p/1c5c1/9/rnbakabnr

变成大写字母在前、小写字母在后了，就是先表示红方棋子，后面才是黑方棋子，就是这一点区别。

7.3.2 输入特征的设计

本章中与象棋相关的实现参考了国际象棋开源项目，地址：https://github.com/Zeta36/chess-alpha-zero/tree/master/src/chess_zero。

首先实现神经网络部分，需要先设计输入特征。其实跟国际象棋差不多，棋子分为：车、马、炮、象、士、将、兵，共7种，即每个玩家7个特征平面，一共14个特征平面。至于论文中说的其他特征平面，比如颜色、回合数、重复局面、历史走子记录等，这里没有实现，只使用了当前棋盘上每个玩家每颗棋子的位置特征作为输入，一共14个特征平面。当然，论文中说的其他特征平面你也可以尝试着实现。中国象棋棋盘表示如图7.7所示，棋盘大小是 9×10，所以输入占位符就是：

```
self.inputs_ = tf.placeholder(tf.float32, [None, 9, 10, 14], name='inputs')
```

接下来定义输入的概率向量 pi（π），需要确定向量的维度，这意味着需要确定所有合法走子集合的数量，通过打印最终数量，得到走子集合数量的结果一共是2086。

上一节我们说过着法表示使用坐标的方式，这些着法就是走子。无论是MCTS还是神经网络的输出都有走子概率，所以要先定义好所有的合法走子，这样就能够得到所有合法走子的数量。

实现思路是穷举。把每一颗棋子在每一个位置上能够达到的每一个位置（所有位置）都穷举出来，最终得到的就是所有合法走子的集合。

函数如下：

```
# 创建所有走子，走子数量为2086
def create_uci_labels():
    # 所有走子集合
    labels_array = []
    # 横坐标和纵坐标，用来组合走子着法
    letters = ['a', 'b', 'c', 'd', 'e', 'f', 'g', 'h', 'i']
    numbers = ['0', '1', '2', '3', '4', '5', '6', '7', '8', '9']

    # 因为士（仕）和象（相）的合法走子比较少，所以直接硬写进来了，没有用程序计算
    # 士（仕）的所有合法走子
    Advisor_labels=['d7e8', 'e8d7', 'e8f9', 'f9e8', 'd0e1', 'e1d0', 'e1f2', 'f2e1',
                    'd2e1', 'e1d2', 'e1f0', 'f0e1', 'd9e8', 'e8d9', 'e8f7', 'f7e8']
    # 象（相）的所有合法走子
    Bishop_labels=['a2c4', 'c4a2', 'c0e2', 'e2c0', 'e2g4', 'g4e2', 'g0i2', 'i2g0',
                   'a7c9', 'c9a7', 'c5e7', 'e7c5', 'e7g9', 'g9e7', 'g5i7', 'i7g5',
```

```
                    'a2c0', 'c0a2', 'c4e2', 'e2c4', 'e2g0', 'g0e2', 'g4i2', 'i2g4',
                    'a7c5', 'c5a7', 'c9e7', 'e7c9', 'e7g5', 'g5e7', 'g9i7', 'i7g9']

# 因为将（帅）、车、炮、兵（卒）的走法都是可以横着走和竖着走的，只是各自的规则
# 不同，但起码用坐标表示起来都一样。也就是说，如果把这4种棋子的所有走子穷举出
# 来肯定会有重复的，所以这里就统一在一起了。想象一下e2e1这步棋，炮、车、帅和
# 卒都可以这么走
# 以下循环穷举将（帅）、车、炮、兵（卒）和马的所有走子
for l1 in range(9):
    for n1 in range(10):
        destinations = [(t, n1) for t in range(9)] + \    # 横着走
                       [(l1, t) for t in range(10)] + \   # 竖着走
                       [(l1 + a, n1 + b) for (a, b) in
                        [(-2, -1), (-1, -2), (-2, 1), (1, -2), (2, -1), (-1, 2),
                         (2, 1), (1, 2)]]   # 马走日
        for (l2, n2) in destinations:
            if (l1, n1) != (l2, n2) and l2 in range(9) and n2 in range(10):
                move = letters[l1] + numbers[n1] + letters[l2] + numbers[n2]
                labels_array.append(move)

# 把士（仕）和象（相）的合法走子添加进来
for p in Advisor_labels:
    labels_array.append(p)

for p in Bishop_labels:
    labels_array.append(p)

# 返回所有合法走子
return labels_array
```

这样就有了所有走子的集合，也得到了走子数量2086。

概率向量pi的定义：

```
self.pi_ = tf.placeholder(tf.float32, [None, 2086], name='pi')
```

胜率z的定义：

```
self.z_ = tf.placeholder(tf.float32, [None, 1], name='z')
```

学习率的定义：

```python
self.learning_rate = tf.placeholder(tf.float32, name='learning_rate')
```

优化器使用 Momentum：

```python
self.momentum = 0.9
optimizer = tf.train.MomentumOptimizer(learning_rate=self.learning_rate,
    momentum=self.momentum, use_nesterov=True)
```

7.3.3 实现神经网络

这里实现的是多 GPU 训练，实现思想是把输入数据按照使用的 GPU 数量均分。

```python
inputs_batches = tf.split(self.inputs_, self.num_gpus, axis=0)
pi_batches = tf.split(self.pi_, self.num_gpus, axis=0)
z_batches = tf.split(self.z_, self.num_gpus, axis=0)

tower_grads = [None] * self.num_gpus

self.loss = 0
self.accuracy = 0
self.policy_head = []
self.value_head = []

with tf.variable_scope(tf.get_variable_scope()):
    # 定义每个GPU要做的事
    for i in range(self.num_gpus):
        with tf.device('/gpu:%d' % i):
            # 不同的GPU分别使用不同的名字域
            with tf.name_scope('TOWER_{}'.format(i)) as scope:
                # 将上面均分的输入数据输入给各自的GPU进行运算
                inputs_batch, pi_batch, z_batch = inputs_batches[i], pi_batches[i],
                                                    z_batches[i]
                # 重点：计算图的构建一定要单独写在新的函数中，这样运行才不会出
                # 错，否则TensorFlow会提示不能重复使用变量
                # 构建神经网络计算图的函数，一会详细说
                loss = self.tower_loss(inputs_batch, pi_batch, z_batch, i)
```

```
                    tf.get_variable_scope().reuse_variables()
                    grad = optimizer.compute_gradients(loss)
                    tower_grads[i] = grad     # 保存每一个GPU的梯度

self.loss /= self.num_gpus          # loss是多个GPU的损失总和，所以要取平均值
self.accuracy /= self.num_gpus      # acc也是同理
grads = self.average_gradients(tower_grads)      # 同理，对所有梯度取平均值
self.train_op = optimizer.apply_gradients(grads, global_step=global_step)
```

这里完全是按照论文中所述的神经网络结构来实现的，大家可以对照看前文给出的结构图，是一一对应的。稍有不同的是，滤波器个数被设为 128，没有使用 256。另外，残差块的数量默认使用了 7 层，没有使用 19 层或者 39 层，如果电脑"给力"的话，大家可以尝试修改。

```
def tower_loss(self, inputs_batch, pi_batch, z_batch, i):
    # 卷积块
    with tf.variable_scope('init'):
        layer=tf.layers.conv2d(inputs_batch, self.filters_size, 3, padding='SAME')

        layer = tf.contrib.layers.batch_norm(layer, center=False, epsilon=1e-5,
                                             fused=True, is_training=self.
                                             training,
                                             activation_fn=tf.nn.relu)

    # 残差块
    with tf.variable_scope("residual_block"):
        for _ in range(self.res_block_nums):
            layer = self.residual_block(layer)

    # 策略"头"
    with tf.variable_scope("policy_head"):
        policy_head = tf.layers.conv2d(layer, 2, 1, padding='SAME')
        policy_head = tf.contrib.layers.batch_norm(policy_head, center=False,
                                                   epsilon=1e-5, fused=True,
                                                   is_training=self.training,
                                                   activation_fn=tf.nn.relu)

        # print(self.policy_head.shape)    # (?, 9, 10, 2)
```

```python
            policy_head = tf.reshape(policy_head, [-1, 9 * 10 * 2])
            policy_head = tf.contrib.layers.fully_connected(policy_head,
                    self.prob_size, activation_fn=None)
            # 保存多个GPU的策略"头"结果（走子概率向量）
            self.policy_head.append(policy_head)

        # 价值"头"
        with tf.variable_scope("value_head"):
            value_head = tf.layers.conv2d(layer, 1, 1, padding='SAME')
            value_head = tf.contrib.layers.batch_norm(value_head, center=False,
                                                     epsilon=1e-5, fused=True,
                                                     is_training=self.training,
                                                     activation_fn=tf.nn.relu)
            # print(self.value_head.shape)  # (?, 9, 10, 1)
            value_head = tf.reshape(value_head, [-1, 9 * 10 * 1])
            value_head = tf.contrib.layers.fully_connected(value_head, 256,
                    activation_fn=tf.nn.relu)
            value_head = tf.contrib.layers.fully_connected(value_head, 1,
                    activation_fn=tf.nn.tanh)
            self.value_head.append(value_head)    # 保存多个GPU的价值"头"结果（胜率）

        # 损失
        with tf.variable_scope("loss"):
            # 走子概率的交叉熵损失
            policy_loss = tf.nn.softmax_cross_entropy_with_logits(labels=pi_batch,
                    logits=policy_head)
            policy_loss = tf.reduce_mean(policy_loss)

            # 胜率的MSE损失
            value_loss = tf.losses.mean_squared_error(labels=z_batch,
                    predictions=value_head)
            value_loss = tf.reduce_mean(value_loss)
            tf.summary.scalar('mse_tower_{}'.format(i), value_loss)

            regularizer = tf.contrib.layers.l2_regularizer(scale=self.c_l2)
            regular_variables = tf.trainable_variables()
```

```python
            l2_loss = tf.contrib.layers.apply_regularization(regularizer,
                regular_variables)

            # loss = value_loss - policy_loss + l2_loss
            loss = value_loss + policy_loss + l2_loss    # softmax交叉熵损失+MSE+l2损失
            self.loss += loss                            # 多个GPU的损失总和
            tf.summary.scalar('loss_tower_{}'.format(i), loss)

        with tf.variable_scope("accuracy"):
            # accuracy，这个准确率是预测概率向量和MCTS的概率向量的比较
            correct_prediction = tf.equal(tf.argmax(policy_head, 1),
                                tf.argmax(pi_batch, 1))
            correct_prediction = tf.cast(correct_prediction, tf.float32)
            accuracy = tf.reduce_mean(correct_prediction, name='accuracy')
            self.accuracy += accuracy
            tf.summary.scalar('move_accuracy_tower_{}'.format(i), accuracy)
        return loss

# 残差块生成函数，每调用一次生成一个残差块
def residual_block(self, in_layer):
    orig = tf.identity(in_layer)

    # filters 128(or 256)
    layer = tf.layers.conv2d(in_layer, self.filters_size, 3, padding='SAME')
    layer = tf.contrib.layers.batch_norm(layer, center=False, epsilon=1e-5,
                                fused=True, is_training=self.training,
                                activation_fn=tf.nn.relu)

    # filters 128(or 256)
    layer = tf.layers.conv2d(layer, self.filters_size, 3, padding='SAME')
    layer = tf.contrib.layers.batch_norm(layer, center=False, epsilon=1e-5,
            fused=True, is_training=self.training)
    out = tf.nn.relu(tf.add(orig, layer))

    return out
```

这样实现之后，基于多GPU的神经网络就构建出来了。

7.3.4 神经网络训练和预测

1. 训练网络

代码实现如下：

```
def train_step(self, positions, probs, winners, learning_rate):
    # 传入MCTS输出的棋谱、走子概率和胜者
    feed_dict = {
        self.inputs_: positions,
        self.training: True,
        self.learning_rate: learning_rate,
        self.pi_: probs,
        self.z_: winners
    }

    # 开始训练网络
    _, accuracy, loss, global_step, summary = self.sess.run([self.train_op,
        self.accuracy, self.loss, self.global_step, self.summaries_op],
        feed_dict=feed_dict)
    self.train_writer.add_summary(summary, global_step)

    return accuracy, loss, global_step
```

2. 使用神经网络预测

预测的代码稍微麻烦一些，因为用 MCTS 自我对弈时是多线程在执行的，传进来的输入数据可能并不能按照 GPU 数量均分，比如有 2 个 GPU，但是传进来的输入数据是 3 个，这样的话就有一个 GPU 处理 2 个数据，一个 GPU 处理 1 个数据。可实际上这样代码是运行不起来的，会报错。

笔者的解决方案是，先看看输入数据的长度能否被 GPU 数量整除，如果能，那么就一切正常，直接把输入数据传给网络就好，神经网络会将数据按照 GPU 数量均分。

一旦不能整除，就把输入数据分成两部分，一部分是能被 GPU 数量整除的数据，一部分是余下的数据。比如有 2 个 GPU，输入数据的长度是 5，就把这 5 个数据分成 4 个和 1 个。对 4 个数据的处理就是正常处理，直接把数据传给网络就好，神经网络会将数据按照 GPU 数量均分。

余下的那部分数据怎么处理呢？把余下的数据不断堆叠起来，直到数据能够按照 GPU 数量均分为止。假如余下 1 个数据，那就复制 1 份，变成 2 个相同的数据，这样正好被 2 个 GPU 均分。只不过这 2 个 GPU 处理后返回的数据，我们只要一个 GPU 的处理结果就行了，抛弃另外一个。

本代码在 AWS 的 2 个 GPU 的环境下运行正常，未尝试在使用更多 GPU 的环境下运行。

```
#@profile
def forward(self, positions):
    # print("positions.shape : ", positions.shape)
    positions = np.array(positions)
    batch_n = positions.shape[0] // self.num_gpus
    alone = positions.shape[0] % self.num_gpus

    # 判断是否不能被GPU均分
    if alone != 0:
        # 如果不止1个数据。因为有可能输入数据的长度是1，这样肯定不能被多个GPU均分了
        if(positions.shape[0] != 1):
            feed_dict = {
                # 先将能均分的这部分数据传入神经网络
                self.inputs_: positions[:positions.shape[0] - alone],
                self.training: False
            }
            # 经过前向传播推理得到策略"头"和价值"头"的结果
            action_probs, value = self.sess.run([self.policy_head,
                self.value_head], feed_dict=feed_dict)
            action_probs, value = np.vstack(action_probs), np.vstack(value)

        # 取余下的部分数据
        new_positions = positions[positions.shape[0] - alone:]
        pos_lst = []
        # 循环将余下的部分数据堆叠起来，直到数据的长度能被GPU数量整除
        while len(pos_lst) == 0 or (np.array(pos_lst).shape[0] * np.array(pos_lst).
            shape[1]) % self.num_gpus != 0:
            pos_lst.append(new_positions)

        if(len(pos_lst) != 0):
            shape = np.array(pos_lst).shape
            pos_lst = np.array(pos_lst).reshape([shape[0] * shape[1], 9, 10, 14])
```

```python
    # 将数据传入网络,得到不能被GPU均分的数据的计算结果
    feed_dict = {
        self.inputs_: pos_lst,
        self.training: False
    }
    # 经过前向传播推理得到余下的部分数据的策略"头"和价值"头"的结果
    action_probs_2, value_2 = self.sess.run([self.policy_head,
        self.value_head], feed_dict=feed_dict)
    action_probs_2, value_2 = action_probs_2[0], value_2[0]

    # 多个数据的计算结果
    if(positions.shape[0] != 1):
        # 如果有多个数据,则需要将被GPU均分的数据的网络输出结果和余下的部分
        # 数据的网络输出结果合并起来
        action_probs = np.concatenate((action_probs, action_probs_2),axis=0)
        value = np.concatenate((value, value_2),axis=0)

        # 返回所有数据的网络输出结果
        return action_probs, value
    else:
        # 只有1个数据的计算结果
        return action_probs_2, value_2
else:
    # 正常情况,数据能被GPU均分
    # 传入棋局数据
    feed_dict = {
        self.inputs_: positions,
        self.training: False
    }
    # 经过前向传播推理得到策略"头"和价值"头"的结果
    action_probs, value = self.sess.run([self.policy_head, self.value_head],
        feed_dict=feed_dict)

    # 将多个GPU的计算结果堆叠起来返回
    return np.vstack(action_probs), np.vstack(value)
```

7.3.5 通过自我对弈训练神经网络

本章有关自我对弈和 MCTS 的算法部分参考自围棋开源项目（https://github.com/yhyu13/AlphaGOZero-python-tensorflow）、五子棋开源项目（https://github.com/junxiaosong/AlphaZero_Gomoku）和黑白棋开源项目（https://github.com/mokemokechicken/reversi-alpha-zero）。

通过自我对弈训练神经网络的思想在上面分析论文时已经说过了，程序自己跟自己下棋，将每盘棋的数据保存起来，当数据量达到我们设置的批量大小时，将棋谱数据作为输入，开始训练神经网络。

```
def run(self):
    batch_iter = 0
    try:
        while(True):
            batch_iter += 1
            # 自我对弈，返回完整一盘棋的下棋数据（在下一节讲MCTS时讲解）
            play_data, episode_len = self.selfplay()
            # batch_iter表示第几盘棋，episode_len表示回合数
            print("batch i:{}, episode_len:{}".format(batch_iter, episode_len))
            extend_data = []
            # 返回的下棋数据包含所有回合的棋盘局面、走子概率和胜者(对于当前玩家胜
            # 是1，负是-1，平局是0)
            for state, mcts_prob, winner in play_data:
                # 为了将棋谱作为输入数据传给神经网络，需要将棋盘局面数据转换成
                # 我们最开始设计的14个特征平面。下文会具体说明转换方法
                states_data = self.mcts.state_to_positions(state)
                # 将棋盘特征平面、MCTS算出的概率向量、胜率保存起来
                extend_data.append((states_data, mcts_prob, winner))
            # data_buffer是一个超大的空间用来保存所有的对弈数据
            # 对这个buffer设置了大小，一旦数据量超出了该大小，新数据就会覆盖旧数据
            self.data_buffer.extend(extend_data)
            # 当保存的数据达到指定的批量数量时
            if len(self.data_buffer) > self.batch_size:
                # 开始训练
                self.policy_update()
    except KeyboardInterrupt:
        # 强制中断时保存网络参数
```

```
self.log_file.close()
self.policy_value_netowrk.save(self.global_step)
```

MCTS 自我对弈一局后返回的下棋数据包含所有回合的棋盘局面、走子概率和胜者（对于当前玩家胜是 1，负是 –1，平局是 0）。其中棋盘局面是之前说过的 FEN 格式串的第一部分：

"RNBAKABNR/9/1C5C1/P1P1P1P1P/9/9/p1p1p1p1p/1c5c1/9/rnbakabnr"

我们使用这种字符串来表示棋局，所以返回的下棋数据里面的棋盘局面就是这种字符串的集合。这样是不符合神经网络输入要求的，所以需要转换成 $9 \times 10 \times 14$ 的特征平面。

在"输入特征的设计"一节说过，中国象棋棋子分为：车、马、炮、象、士、将、兵，共 7 种，每个玩家 7 个特征平面，一共 14 个特征平面。比如，对于"车"，那就是在"车"所对应的特征平面上只记录有关"车"的位置，有"车"的地方就置 1，没有就是 0，其他棋子同理。

```
# 共14个棋子字母，14种棋子对应14个特征平面
pieces_order = 'KARBNPCkarbnpc'
# 定义棋子字母转特征平面下标的字典
ind = {pieces_order[i]: i for i in range(14)}


# 将棋局状态转换成神经网络的输入特征平面
def state_to_positions(self, state):
    # 先把FEN字符串中的数字(2~9)都转换成1，1就是1不用转换，并去掉/
    # 这样每行都是完整的9个字符了，比如/1C5C1转换成1C11111C1
    board_state = self.replace_board_tags(state)
    # 定义9×10×14的特征平面，默认值都是0
    pieces_plane = np.zeros(shape=(9, 10, 14), dtype=np.float32)
    # 循环遍历棋盘每一个位置
    for rank in range(9):      # 横线
        for file in range(10):      # 纵线
            # 得到每一个位置的状态，要么是数字1，要么是棋子字母
            v = board_state[rank * 9 + file]
            if v.isalpha():
                # 如果是棋子的话，将该棋子类型对应的特征平面的棋子位置置1
                pieces_plane[rank][file][ind[v]] = 1
    assert pieces_plane.shape == (9, 10, 14)
    return pieces_plane


def replace_board_tags(self, board):
```

```
        board = board.replace("2", "11")
        board = board.replace("3", "111")
        board = board.replace("4", "1111")
        board = board.replace("5", "11111")
        board = board.replace("6", "111111")
        board = board.replace("7", "1111111")
        board = board.replace("8", "11111111")
        board = board.replace("9", "111111111")
        return board.replace("/", "")
```

数据都准备好了,开始训练网络。

```
def policy_update(self):
    # 从数据中随机抽取一部分数据。注意,我们不会将棋局按照下棋的顺序传给网络,而是
    # 把每次的下棋数据都放在一个很大的buffer里,这个buffer保存了很多局数据,然后
    # 随机从里面抽取一部分互不相干的某一回合的牌面
    mini_batch = random.sample(self.data_buffer, self.batch_size)
    # 从抽取的数据中分离出特征平面、走子概率和胜者
    state_batch = [data[0] for data in mini_batch]
    mcts_probs_batch = [data[1] for data in mini_batch]
    winner_batch = [data[2] for data in mini_batch]
    winner_batch = np.expand_dims(winner_batch, 1)

    start_time = time.time()
    # 先通过正向传播预测网络输出结果,用于计算训练后的KL散度
    old_probs, old_v = self.mcts.forward(state_batch)
    # 一共训练5次
    for i in range(self.epochs):
        # 训练网络。这里的学习率需要特别注意
        # 笔者在AWS上用的是g2.2xlarge, 24小时只能下大约200盘棋,很慢
        # 所以需要动态调整学习率
        # 当然,也可以使用指数衰减学习率,那么在之前定义学习率的地方就需要修改成
        # self.learning_rate = tf.maximum(tf.train.exponential_decay(0.001, self.
        # global_step, 1e3, 0.66), 1e-5)
        # 然后,在这里训练网络的地方学习率就不用作为参数传递了,也可以在训练
        # 网络函数中不使用传递的学习率参数
        accuracy, loss, self.global_step = self.policy_value_netowrk.train_step(
```

```python
            state_batch, mcts_probs_batch, winner_batch,
            self.learning_rate * self.lr_multiplier)
    # 使用训练后的新网络预测结果，跟之前的结果计算KL散度
    new_probs, new_v = self.mcts.forward(state_batch)
    kl_tmp = old_probs * (np.log((old_probs + 1e-10) / (new_probs + 1e-10)))
    # print("kl_tmp.shape", kl_tmp.shape)
    kl_lst = []
    for line in kl_tmp:
        # 去掉inf值
        all_value = [x for x in line if str(x) != 'nan' and str(x)!= 'inf']
        kl_lst.append(np.sum(all_value))
    kl = np.mean(kl_lst)
    # kl = scipy.stats.entropy(old_probs, new_probs)
    # kl = np.mean(np.sum(old_probs * (np.log(old_probs + 1e-10) - np.log(
    # new_probs + 1e-10)), axis=1))

    # 如果KL散度变得糟糕，就从5次训练的循环中退出
    if kl > self.kl_targ * 4:
        break
self.policy_value_netowrk.save(self.global_step)
print("train using time {} s".format(time.time() - start_time))

# 通过计算调整学习率乘子
if kl > self.kl_targ * 2 and self.lr_multiplier > 0.1:
    self.lr_multiplier /= 1.5
elif kl < self.kl_targ / 2 and self.lr_multiplier < 10:
    self.lr_multiplier *= 1.5

explained_var_old = 1 - np.var(np.array(winner_batch) - old_v.flatten()) / np.
    var(np.array(winner_batch))
explained_var_new = 1 - np.var(np.array(winner_batch) - new_v.flatten()) / np.
    var(np.array(winner_batch))
print(
    "kl:{:.5f},lr_multiplier:{:.3f},loss:{},accuracy:{},explained_var_old:{:.3f
    },explained_var_new:{:.3f}".format(
        kl, self.lr_multiplier, loss, accuracy, explained_var_old,
```

```
            explained_var_new))
    self.log_file.write("kl:{:.5f},lr_multiplier:{:.3f},loss:{},accuracy:{},
    explained_var_old:{:.3f},explained_var_new:{:.3f}".format(
            kl, self.lr_multiplier, loss, accuracy, explained_var_old,
            explained_var_new) + '\n')
    self.log_file.flush()
```

总结起来,就是从数据 buffer 里取出随机的批量棋局作为输入数据,传给神经网络进行训练,并通过 KL 散度来衡量训练的效果,如果训练的效果较好,则同一批数据训练网络 5 次,否则就会提前停止训练。这个可以这样理解,比如我们训练的是分类神经网络,并通过损失来确认训练的效果,损失越小则训练的效果越好。如果用同一批数据进行训练,损失有可能是一次比一次小,因为每次对同一批数据进行学习,多次学习就会越来越"熟悉、认识"这些数据。但是有过拟合的风险。

7.3.6 自我对弈

自我对弈就是通过 MCTS 下每一步棋,直到分出胜负,并返回下棋数据。

```
# 取出给定棋盘状态下的棋子个数
def get_pieces_count(state):
    count = 0
    for s in state:
        if s.isalpha():
            count += 1
    return count


# 输入走子前和走子后的棋盘状态,判断是否是吃子的动作
def is_kill_move(state_prev, state_next):
    # 通过走子前和走子后的棋子个数的差,就知道是否有棋子被吃掉了
    return get_pieces_count(state_prev) - get_pieces_count(state_next)


def selfplay(self):
    # 初始化棋盘
    # 将棋局状态state初始化为"RNBAKABNR/9/1C5C1/P1P1P1P1P/9/9/p1p1p1p1p/1c5c1/9/
    # rnbakabnr"
    # 回合数round重置为1
```

```python
# 当前棋手current_player重置为w
# 没有吃子回合数restrict_round重置为0
# GameBoard是棋盘类,用来模拟下棋,保存棋局的各种状态
self.game_borad.reload()
states, mcts_probs, current_players = [], [], []
z = None
game_over = False
winnner = ""
start_time = time.time()
# 下棋循环,结束条件是分出胜负
while(not game_over):
    # 传入当前棋局状态,通过MCTS算出下哪一步棋
    # action是MCTS算出的在当前局面要下的一步棋
    # probs包含两项,probs[0][0]是走子动作的集合,即在当前局面所有合法的走子动作
    # probs[0][1]是这些走子动作的概率
    # win_rate是action所在节点的Q值(平均行动价值)
    action, probs, win_rate = self.get_action(self.game_borad.state,
        self.temperature)
    ####################################################
    # 这部分代码跟笔者的设计有关。因为在输入特征平面中没有使用颜色特征
    # 所以传给神经网络数据时,要确保当前棋手是红色,有可能需要把当前棋手转换成
    # 红色(先手),转换的其实是棋盘的棋子位置
    # 神经网络预测的始终是红色先手方向该如何下棋
    state, palyer = self.mcts.try_flip(self.game_borad.state, self.game_borad.
    current_player, self.mcts.is_black_turn(self.game_borad.current_player))
    # 保存棋局状态,在函数最后返回
    states.append(state)
    # labels_len是所有走子的数量,为2086,这里初始化一个所有走子数量的数组,在
    # 下面用来保存走子概率
    prob = np.zeros(labels_len)
    # 概率向量也可能需要转换,假如当前棋手是黑色,当棋盘转换成当前棋手是红色
    # 后,由于棋盘位置的变化,走子概率也要转换方向才行。比如原来是a0a9这步棋
    # 的概率,转换棋盘后,就是a9a0这步棋的概率了,因为棋盘位置变了
    if self.mcts.is_black_turn(self.game_borad.current_player):
        # probs[0][0][idx]是每一个走子动作,probs[0][1][idx]是每一个走子概率
```

```python
            # 它们是一一对应的
            for idx in range(len(probs[0][0])):
                # 如果当前棋手是黑色,则需要先转换probs中的动作,比如将a0a9
                # 转换成a9a0
                act = "".join((str(9 - int(a)) if a.isdigit() else a) for a in
                    probs[0][0][idx])
                # 然后把走子概率赋值给转换后的动作所在的位置
                # label2i是走子动作转下标的字典
                prob[label2i[act]] = probs[0][1][idx]
        else:
            # 当前棋手是红色,所以不用转换,直接将走子概率赋值给走子动作对应的位置
            for idx in range(len(probs[0][0])):
                prob[label2i[probs[0][0][idx]]] = probs[0][1][idx]
    # 保存走子概率,在函数最后返回
    mcts_probs.append(prob)

    # 上面的棋盘位置变化只发生在当前棋手是黑色时
    # 变化的目的是为了保存训练神经网络用的数据(棋局状态和走子概率)
    # 因为训练的神经网络预测的始终是红色先手方向该如何下棋
    ####################################################

    # 保存每一回合的当前棋手
    current_players.append(self.game_borad.current_player)

    last_state = self.game_borad.state
    # 在棋盘上下算出的action这步棋,得到新的棋盘状态
    self.game_borad.state = GameBoard.sim_do_action(action,
        self.game_borad.state)
    # 更新回合数
    self.game_borad.round += 1
    # 切换当前棋手
    self.game_borad.current_player = "w" if self.game_borad.current_player ==
        "b" else "b"
    # 判断刚刚下的棋是否吃子了
    if is_kill_move(last_state, self.game_borad.state) == 0:
        # 如果没有吃子发生,则更新没有进展回合数
```

```python
            self.game_borad.restrict_round += 1
        else:
            # 发生吃子，将没有进展回合数清零
            self.game_borad.restrict_round = 0

    # 判断在棋盘上能否找到将/帅，如果if判断条件为真，则说明将/帅被吃，游戏结束
    if (self.game_borad.state.find('K') == -1 or self.game_borad.state.find('k
') == -1):
        # z用来保存胜平负
        z = np.zeros(len(current_players))
        if (self.game_borad.state.find('K') == -1):
            winnner = "b"
        if (self.game_borad.state.find('k') == -1):
            winnner = "w"
        z[np.array(current_players) == winnner] = 1.0
        z[np.array(current_players) != winnner] = -1.0
        game_over = True
        print("Game end. Winner is player : ", winnner, " In {} steps".format(
            self.game_borad.round - 1))
    elif self.game_borad.restrict_round >= 60:
        # 60回合没有进展（吃子），平局
        z = np.zeros(len(current_players))
        game_over = True
        print("Game end. Tie in {} steps".format(self.game_borad.round - 1))
    # 认输的部分没有实现
    # elif(self.mcts.root.v < self.resign_threshold):
    #     pass
    # elif(self.mcts.root.Q < self.resign_threshold):
    #     pass
    if(game_over):
        # 游戏结束，重置棋盘
        self.mcts.reload()
print("Using time {} s".format(time.time() - start_time))
# 返回下棋数据
return zip(states, mcts_probs, z), len(z)
```

我们来看看棋盘转换的实现。

```python
# 判断当前棋手是不是黑色
def is_black_turn(self, current_player):
    return current_player == 'b'

# 尝试棋盘转换，通常flip传入的是is_black_turn的结果
def try_flip(self, state, current_player, flip=False):
    # 不需要转换，说明当前棋手是红色
    if not flip:
        return state, current_player

    # state是FEN格式字符串，先把棋盘状态一行分成多行
    rows = state.split('/')

    # 转换单个棋子，大写字母转小写，小写字母转大写，即红色转黑色，黑色转红色
    def swapcase(a):
        if a.isalpha():
            return a.lower() if a.isupper() else a.upper()
        return a

    # 按行批量转换棋子
    def swapall(aa):
        return "".join([swapcase(a) for a in aa])

    # 返回转换后的棋盘状态和棋手颜色
    # reversed(rows)，转换棋盘，通过swapall也转换每行的棋子
    # 最后用"/"连接，棋盘状态多行变成一行
    return "/".join([swapall(row) for row in reversed(rows)]), ('w' if
        current_player == 'b' else 'b')
```

接下来看看棋盘类的下棋实现。

```python
# 把棋盘状态(FEN格式字符串)转换成多行的完整棋盘表示，方便按照坐标方式操作棋盘
@staticmethod
def board_to_pos_name(board):
    board = board.replace("2", "11")
    board = board.replace("3", "111")
    board = board.replace("4", "1111")
```

```python
        board = board.replace("5", "11111")
        board = board.replace("6", "111111")
        board = board.replace("7", "1111111")
        board = board.replace("8", "11111111")
        board = board.replace("9", "111111111")
        # 棋盘状态一行分成多行
        return board.split("/")

# 传入的是要下的走子动作和棋盘状态
@staticmethod
def sim_do_action(in_action, in_state):
    x_trans = {'a':0, 'b':1, 'c':2, 'd':3, 'e':4, 'f':5, 'g':6, 'h':7, 'i':8}

    # 走子动作如a0a9,前两个是源src,后两个是目的dst,即从a0走到a9
    src = in_action[0:2]
    dst = in_action[2:4]

    # 将a0,a9转换成棋盘坐标x和y
    src_x = int(x_trans[src[0]])
    src_y = int(src[1])

    dst_x = int(x_trans[dst[0]])
    dst_y = int(dst[1])

    # 把棋盘状态(FEN格式字符串)转换成多行的完整棋盘表示,方便使用坐标操作棋盘
    board_positions = GameBoard.board_to_pos_name(in_state)
    line_lst = []
    # 把棋盘每一行放入列表中
    for line in board_positions:
        line_lst.append(list(line))
    lines = np.array(line_lst)

    # 将源棋子落入目的位置,并将源位置置1
    lines[dst_y][dst_x] = lines[src_y][src_x]
    lines[src_y][src_x] = '1'
```

```python
# 因为源和目的棋子所在的行都可能发生改变
# 所以将源行和目的行的状态保存到棋盘状态中
board_positions[dst_y] = ''.join(lines[dst_y])
board_positions[src_y] = ''.join(lines[src_y])

# 将多行的棋盘状态再变回FEN格式表示
board = "/".join(board_positions)
board = board.replace("111111111", "9")
board = board.replace("11111111", "8")
board = board.replace("1111111", "7")
board = board.replace("111111", "6")
board = board.replace("11111", "5")
board = board.replace("1111", "4")
board = board.replace("111", "3")
board = board.replace("11", "2")

# 返回新的棋盘状态
return board
```

7.3.7　实现蒙特卡罗树搜索：异步方式

关键的代码来了，函数通过蒙特卡罗树搜索（MCTS）进行若干次下棋模拟（论文中是1600次，笔者用了1200次），然后根据根节点的所有子节点的访问量决定要下哪一步棋，使用异步方式能够加快 MCTS 的速度。

```
#@profile
def get_action(self, state, temperature = 1e-3):
    # MCTS主函数，模拟下棋，次数是self.playout_counts
    self.mcts.main(state, self.game_borad.current_player, self.game_borad.
    restrict_round, self.playout_counts)
    # 取得当前局面（根节点）下所有子节点的合法走子和相应的访问量
    # 所有子节点可能并不会覆盖所有合法的走子，这个是由树搜索的质量决定的
    # 加大模拟次数就是加大思考的深度，考虑更多的局面，搜索更多不同的走法
    # 避免出现有些特别重要的棋步却没有考虑到的情况
    actions_visits = [(act, nod.N) for act, nod in self.mcts.root.child.items()]
    actions, visits = zip(*actions_visits)
```

```python
# 通过节点访问量计算走子概率
probs = softmax(1.0 / temperature * np.log(visits))    #+ 1e-10
move_probs = []
# 保存每一个走子动作和相应的走子概率
move_probs.append([actions, probs])

# 是否进行探索的标志
if(self.exploration):
    # 训练时，可以通过加入噪声来探索更多可能性的走子
    act = np.random.choice(
        actions, p=0.75 * probs + 0.25*np.random.dirichlet(0.3*np.ones(len(
        probs)))))
else:
    # 通过节点访问量的softmax选择最大可能性的走子
    # 走子的选择与节点的访问次数成正比
    act = np.random.choice(actions, p=probs)

# 将节点的Q值（平均行动价值）当作胜率
win_rate = self.mcts.Q(act)
# 更新搜索树，将算出的这步棋的局面作为树的根节点
self.mcts.update_tree(act)
# 返回所选择的走子、所有走子动作和概率、胜率
return act, move_probs, win_rate

# 返回指定动作下的胜率(走子节点对应的Q值)
def Q(self, move) -> float:
    ret = 0.0
    find = False
    for a, n in self.root.child.items():
        if move == a:
            ret = n.Q
            find = True
    if(find == False):
        print("{} not exist in the child".format(move))
    return ret
```

```python
# 更新搜索树, 将指定走子动作的节点更新为搜索树新的根节点
def update_tree(self, act):
    # 从已扩展的节点集合中删除根节点
    self.expanded.discard(self.root)
    # 将指定动作下的子节点更新为新的根节点
    self.root = self.root.child[act]
    # 删除根节点的父节点
    self.root.parent = None
```

论文中描述了在先验概率中添加 Dirichlet 噪声,以实现额外的探索。这种噪声确保所有的走子都可以被尝试,但是搜索仍然可能下臭棋。公式是 $P(s,a) = (1-\epsilon)p_a + \epsilon\eta_a$,其中 $\eta \sim \text{Dir}$ (0.03), ϵ 是 0.25。在笔者的实现中 ϵ 是 0.25, $\eta \sim \text{Dir}$ 是 0.3, 这个可以作为超参数多尝试。

我们来看看 MCTS 的类定义。

```python
from collections import deque, defaultdict, namedtuple
QueueItem = namedtuple("QueueItem", "feature future")
c_PUCT = 5
virtual_loss = 3
cut_off_depth = 30


class MCTS_tree(object):
    # 参数search_threads默认使用16个搜索线程
    def __init__(self, in_state, in_forward, search_threads):
        self.noise_eps = 0.25
        self.dirichlet_alpha = 0.3    # 0.03
        # 在根节点的先验概率中加入了噪声
        self.p_ = (1 - self.noise_eps) * 1 + self.noise_eps * np.random.dirichlet([
            self.dirichlet_alpha])
        # 定义根节点, 传入概率和棋盘状态 (棋子位置)
        # 因为是根节点, 所以参数父节点传入None
        self.root = leaf_node(None, self.p_, in_state)
        self.c_puct = 5    #1.5
        # 保存前向传播 (预测) 函数
        self.forward = in_forward
        self.node_lock = defaultdict(Lock)
        # 虚拟损失
```

```python
        self.virtual_loss = 3
        # 用来保存正在展开的节点
        self.now_expanding = set()
        # 保存被展开过的节点
        self.expanded = set()
        self.cut_off_depth = 30
        # search_threads是16，定义信号量用于异步运行
        self.sem = asyncio.Semaphore(search_threads)
        # 保存搜索线程的队列，用于异步预测，下文会说明
        self.queue = Queue(search_threads)
        self.loop = asyncio.get_event_loop()
        self.running_simulation_num = 0
```

叶子节点的类定义：

```python
class leaf_node(object):
    # 在定义节点时，传入父节点、概率和棋盘状态（棋子位置）
    def __init__(self, in_parent, in_prior_p, in_state):
        # 保存概率，其他值默认是0
        self.P = in_prior_p
        # 平均行动价值
        self.Q = 0
        # 访问计数
        self.N = 0
        # value值
        self.v = 0
        self.U = 0
        # 总行动价值
        self.W = 0
        # 保存父节点
        self.parent = in_parent
        # 子节点默认是空
        self.child = {}
        # 保存棋盘状态
        self.state = in_state
```

MCTS 主函数：

```python
def is_expanded(self, key) -> bool:
```

```python
        # 判断指定的节点是否被展开过
        return key in self.expanded

    # 根据传入的棋盘状态,生成要传给神经网络的特征平面。需要判断是否要转换棋盘,确保
    # 当前棋手是红色
    def generate_inputs(self, in_state, current_player):
        state, palyer = self.try_flip(in_state, current_player,
                                    self.is_black_turn(current_player))
        return self.state_to_positions(state)

    #@profile
    def main(self, state, current_player, restrict_round, playouts):
        node = self.root
        # 先通过神经网络展开根节点(如果根节点没有被展开过的话)
        if not self.is_expanded(node):
            # 将棋盘状态转换成特征平面
            positions = self.generate_inputs(node.state, current_player)
            positions = np.expand_dims(positions, 0)
            # 通过神经网络预测走子概率
            action_probs, value = self.forward(positions)
            # 判断走子概率是否需要根据先手/后手进行转换
            # 因为神经网络始终是以先手红色预测概率的
            # 如果当前棋手是黑色,则需要转换走子概率(棋盘转换)
            if self.is_black_turn(current_player):
                action_probs = cchess_main.flip_policy(action_probs)
            # 取得当前局面(棋盘状态)下所有合法的走子
            moves = GameBoard.get_legal_moves(node.state, current_player)

            # 展开节点
            node.expand(moves, action_probs)
            # 将当前节点加入已扩展的节点集合中
            self.expanded.add(node)

        coroutine_list = []
        # 模拟playouts(1200)次,以异步方式执行,一共使用16个线程
        for _ in range(playouts):
```

```
            coroutine_list.append(self.tree_search(node, current_player,
                restrict_round))
        # 运行异步预测函数，用于在MCTS中展开叶子节点，这个下文会说明
        coroutine_list.append(self.prediction_worker())
        # 运行所有的异步操作直到结束
        self.loop.run_until_complete(asyncio.gather(*coroutine_list))
```

走子概率的转换：

```
# 用来转换走子集合
def flipped_uci_labels(param):
    # 转换函数，主要是转换棋盘位置，比如a0a9转换成a9a0
    def repl(x):
        return "".join([(str(9 - int(a)) if a.isdigit() else a) for a in x])

    # 对走子集合中的每种着法进行转换
    return [repl(x) for x in param]


# 创建所有走子的集合，走子数量共2086个
labels_array = create_uci_labels()
# 传入走子集合，返回转换后的走子集合
flipped_labels = flipped_uci_labels(labels_array)
# 得到转换后的走子在原走子集合中的下标
unflipped_index = [labels_array.index(x) for x in flipped_labels]


# 转换走子概率，使用转换后的走子下标unflipped_index重新排列概率向量即可
@staticmethod
def flip_policy(prob):
    prob = prob.flatten()
    return np.asarray([prob[ind] for ind in unflipped_index])
```

接下来就是完整的蒙特卡罗树搜索算法了，通过16个线程以异步方式搜索1200次。

```
# 树搜索算法入口
async def tree_search(self, node, current_player, restrict_round) -> float:
    # running_simulation_num用来标记当前是否在执行树搜索，0表示没有执行，大于0表示
    # 正在进行树搜索的数量
    # 在异步预测函数prediction_worker中用来判断是否还有树搜索在执行
```

```python
        self.running_simulation_num += 1

        # 异步执行树搜索,共16个线程,信号量不足会阻塞,等待信号量被释放
        with await self.sem:
            # 开始树搜索
            value = await self.start_tree_search(node, current_player, restrict_round)
            # 树搜索结束,计数减1
            self.running_simulation_num -= 1

        return value

# 异步预测函数,管理队列数据,一旦队列中有数据,就统一传给神经网络,获得预测结果
async def prediction_worker(self):
    q = self.queue
    # 一个时间间隔,主要是刚开始时搜索线程还没有结束,此时队列中也没有数据,避
    # 免预测函数提前退出
    margin = 10
    while self.running_simulation_num > 0 or margin > 0:
        if q.empty():
            if margin > 0:
                margin -= 1
            await asyncio.sleep(1e-3)
            continue
        # 取出队列中的批量数据,是MCTS中放入队列的特征平面
        # type: list[QueueItem]
        item_list = [q.get_nowait() for _ in range(q.qsize())]
        features = np.asarray([item.feature for item in item_list])

        # 将特征平面作为输入数据传给神经网络,得到预测结果
        action_probs, value = self.forward(features)
        # 然后把神经网络的预测结果返回,此时在MCTS中等待的若干线程会被唤醒
        for p, v, item in zip(action_probs, value, item_list):
            item.future.set_result((p, v))

# 将特征平面放入队列中,返回等待的事件,事件发生会唤醒等待的线程
async def push_queue(self, features):
```

```python
        # 创建事件
        future = self.loop.create_future()
        # 将特征平面和事件绑定
        item = QueueItem(features, future)
        # 将特征平面放入队列中
        await self.queue.put(item)
        # 返回异步事件
        return future

    # ***树搜索函数***
    async def start_tree_search(self, node, current_player, restrict_round)->float:
        # 蒙特卡罗树搜索：选择、展开、评估、回传

        # 正在扩展的节点集合
        now_expanding = self.now_expanding

        # 如果当前节点正在被展开，就小睡一会儿
        while node in now_expanding:
            await asyncio.sleep(1e-4)

        # 如果节点没有被展开过（说明是叶子节点），要展开这个节点
        if not self.is_expanded(node):
            # 将节点加入正在扩展的节点集合中
            self.now_expanding.add(node)

            # 在前面的理论部分说过，叶子节点是使用神经网络的输出来展开的
            # 将棋盘状态转换为特征平面，作为神经网络的输入数据
            positions = self.generate_inputs(node.state, current_player)

            # 这里有一个trick，就是并不是逐个节点使用神经网络预测结果的，这样效率太低了
            # 而是放到队列中，通过prediction_worker函数统一管理队列
            # 将队列中的一组（16个）输入数据传给神经网络，得到预测结果，这一切都是异步
            # 进行的
            future = await self.push_queue(positions)  # type: Future
            # 等待预测结束，线程会被唤醒
            await future
```

```python
        # 从事件中获得神经网络预测的结果
        action_probs, value = future.result()

        # 根据当前棋手的颜色决定是否对走子概率进行转换
        if self.is_black_turn(current_player):
            action_probs = cchess_main.flip_policy(action_probs)

        # 取得当前局面（棋盘状态）下所有合法的走子
        moves = GameBoard.get_legal_moves(node.state, current_player)
        # 展开操作，使用神经网络预测的结果展开当前节点
        node.expand(moves, action_probs)
        # 将节点添加到已扩展的节点集合中
        self.expanded.add(node)
        # 并从正在扩展的节点集合中移除
        self.now_expanding.remove(node)

        # 返回神经网络预测的胜率，一定要取负，理由在分析论文时已经说过了
        return value[0] * -1
    else:
        # 如果节点被展开过，则执行选择操作
        # 选择操作会选出Q+U最大的节点
        last_state = node.state

        # 选择操作，根据Q+U最大选择节点，c_PUCT是一个决定探索水平的常数
        action, node = node.select_new(c_PUCT)
        current_player = "w" if current_player == "b" else "b"
        # 判断选择节点后的棋盘状态是否有棋子被吃了
        if is_kill_move(last_state, node.state) == 0:
            # 没有发生吃子，更新没有进展回合数
            restrict_round += 1
        else:
            # 发生吃子，将没有进展回合数清零
            restrict_round = 0
        last_state = node.state

        # 为所选择的节点添加虚拟损失，防止其他线程继续探索这个节点，增加探索多样性
```

```python
        node.N += virtual_loss
        node.W += -virtual_loss

        # 判断在这个节点状态下是否分出胜负
        if (node.state.find('K') == -1 or node.state.find('k') == -1):
            # 分出胜负了，设置胜率1或者-1
            if (node.state.find('K') == -1):
                value = 1.0 if current_player == "b" else -1.0
            if (node.state.find('k') == -1):
                value = -1.0 if current_player == "b" else 1.0
            # 一定要符号取反
            value = value * -1
        elif restrict_round >= 60:
            # 60回合无进展（吃子），平局
            value = 0.0
        else:
            # 没有分出胜负，在当前节点局面下继续树搜索
            value = await self.start_tree_search(node, current_player,
            restrict_round)

        # 当前节点搜索完毕，去掉虚拟损失，恢复节点状态
        node.N += -virtual_loss
        node.W += virtual_loss

        # 按照树搜索的路径更新各节点: N, W, Q, U
        # 执行节点的回传操作，更新节点的各类数值
        node.back_up_value(value)

        # 一定要符号取反
        return value * -1
```

最后介绍节点的选择、展开和回传的实现。

```python
# 选择，选出Q+U最大的节点
def select_new(self, c_puct):
    # 计算当前节点的所有子节点的Q+U，返回Q+U最大的节点
    return max(self.child.items(), key=lambda node: node[1].
```

```python
            get_Q_plus_U_new(c_puct))

# 返回节点的Q+U
def get_Q_plus_U_new(self, c_puct):
    # 按照公式计算即可
    U = c_puct * self.P * np.sqrt(self.parent.N) / ( 1 + self.N)
    return self.Q + U


# 参数是所有合法走子moves和神经网络预测的概率向量action_probs
#@profile
def expand(self, moves, action_probs):
    tot_p = 1e-8
    action_probs = action_probs.flatten()

    for action in moves:
        # 模拟执行每一个合法走子，得到走子后的棋盘状态
        in_state = GameBoard.sim_do_action(action, self.state)
        # 从概率向量中得到当前走子对应的概率，label2i是走子动作转下标的字典
        mov_p = action_probs[label2i[action]]
        # 创建新节点，传入父节点（因为是展开当前节点，所以当前节点是新节点的
        # 父节点）、概率、棋盘状态
        new_node = leaf_node(self, mov_p, in_state)
        # 将新节点添加到当前节点的子节点集合中
        self.child[action] = new_node
        tot_p += mov_p
    # 归一化所有子节点的概率
    for a, n in self.child.items():
        n.P /= tot_p


# 更新节点的各项参数
def back_up_value(self, value):
    self.N += 1      # 计数加1
    self.W += value      # 更新总行动价值
    self.v = value
    self.Q = self.W / self.N   # 更新平均行动价值
    self.U = c_PUCT * self.P * np.sqrt(self.parent.N) / ( 1 + self.N)   # 更新U
```

以上就是蒙特卡罗树搜索的全部内容，读者可以对照理论部分理解代码。

经过上述代码，神经网络和MCTS不断得到训练加强。当神经网络训练好后，如何利用神经网络或者MCTS进行下棋呢？来看下面的函数。

```python
# 利用MCTS或神经网络选择走子
def select_move(self, mcts_or_net):
    # 如果用MCTS，则直接传入当前棋盘状态即可，搜索树会返回走子动作、概率和胜率
    if mcts_or_net == "mcts":
        action, probs, win_rate = self.get_action(self.game_borad.state, self.
        temperature)
    elif mcts_or_net == "net":
        # 使用神经网络选择走子，需要先将棋盘状态转换成特征平面
        positions = self.mcts.generate_inputs(self.game_borad.state,
                                              self.game_borad.current_player)
        positions = np.expand_dims(positions, 0)
        # 将特征平面传给神经网络，得到预测的走子概率和胜率
        action_probs, value = self.mcts.forward(positions)
        win_rate = value[0, 0]
        # 因为是从神经网络得到预测结果的，所以需要判断是否翻转走子概率
        if self.mcts.is_black_turn(self.game_borad.current_player):
            action_probs = cchess_main.flip_policy(action_probs)
        # 取得当前棋盘状态下所有合法的走子
        # 这个很重要，比如将的上边和左边有士围着，那么向上走和向左走就是
        # 不合法的
        moves = GameBoard.get_legal_moves(self.game_borad.state,
                                          self.game_borad.current_player)

        tot_p = 1e-8
        action_probs = action_probs.flatten()
        act_prob_dict = defaultdict(float)

        for action in moves:
            # 取得每一个合法走子动作对应的概率
            # 毕竟当前状态下合法走子数量是少数，神经网络返回的是所有走子的概率
            mov_p = action_probs[label2i[action]]
            act_prob_dict[action] = mov_p
```

```python
            tot_p += mov_p

        # 归一化合法走子概率
        for a, _ in act_prob_dict.items():
            act_prob_dict[a] /= tot_p

        # 取得概率最大的走子动作
        action = max(act_prob_dict.items(), key=lambda node: node[1])[0]

print('Win rate for player {} is {:.4f}'.format(
    self.game_borad.current_player, win_rate))
last_state = self.game_borad.state
print(self.game_borad.current_player, " now take a action : ", action,
    "[Step {}]".format(self.game_borad.round))
# 在棋盘上按照选定的动作下棋,得到走子后的新棋盘状态
self.game_borad.state = GameBoard.sim_do_action(action,
    self.game_borad.state)
# 回合数加1
self.game_borad.round += 1
# 切换棋手
self.game_borad.current_player = "w" if self.game_borad.current_player ==
    "b" else "b"
# 判断是否是吃子动作
if is_kill_move(last_state, self.game_borad.state) == 0:
    # 没有发生吃子,更新没有进展回合数
    self.game_borad.restrict_round += 1
else:
    # 发生吃子,将没有进展回合数清零
    self.game_borad.restrict_round = 0

# 在终端打印棋盘状态
self.game_borad.print_borad(self.game_borad.state)

# 下面是将走子动作转换成坐标形式,以便在UI上显示
x_trans = {'a': 0, 'b': 1, 'c': 2, 'd': 3, 'e': 4, 'f': 5, 'g': 6, 'h': 7,
    'i': 8}
```

```
        if self.human_color == 'w':
            action = "".join(flipped_uci_labels(action))

        src = action[0:2]
        dst = action[2:4]

        src_x = int(x_trans[src[0]])
        src_y = int(src[1])

        dst_x = int(x_trans[dst[0]])
        dst_y = int(dst[1])

        return (src_x, src_y, dst_x - src_x, dst_y - src_y), win_rate
```

7.3.8 训练和运行

运行本项目除需要安装 TensorFlow 以外，还需要安装 uvloop，在终端输入如下命令安装：

`pip install uvloop`

安装好后就可以训练网络了，训练命令说明如下：

- `--mode`，指定是训练（train）还是下棋（play），默认是训练。
- `--train_playout`，指定 MCTS 的模拟次数，论文中是 1600 次，笔者做训练时模拟 1200 次。
- `--batch_size`，指定训练数据达到多少时开始训练，默认为 512。
- `--search_threads`，指定执行 MCTS 时的线程个数，默认为 16。
- `--processor`，指定是使用 CPU 还是 GPU，默认是 CPU。
- `--num_gpus`，指定 GPU 的个数，默认是 1。
- `--res_block_nums`，指定残差块的层数，论文中是 19 层或 39 层，默认是 7 层。

训练命令举例：

```
python main.py --mode train --train_playout 1200 --batch_size 512 --search_threads 16 --processor gpu --num_gpus 2 --res_block_nums 7
```

这里将笔者训练的模型文件也提供给大家，模型文件下载地址：https://pan.baidu.com/s/1dLvxFFpeWZK-aZ2Koewrvg，解压缩后放到项目根目录下即可，文件夹名为 gpu_models。

除训练以外，也可以尝试跟 AI 下棋，下棋命令说明如下：

- --ai_count，指定 AI 的个数，1 是人机对战，2 是两个 AI 下棋。
- --ai_function，指定 AI 的下棋方法，是思考（mcts，下棋慢）还是直觉（net，下棋快）。
- --play_playout，指定 AI 进行 MCTS 的模拟次数。
- --delay 和 --end_delay，保持默认就好，两个 AI 下棋太快，就不知道它们是怎么下的了。
- --human_color，指定人类棋手的颜色，w 是先手，b 是后手。

下棋命令举例：

```
python main.py --mode play --ai_count 1 --ai_function mcts --play_playout 1200 --human_color w
```

7.4 实现中国象棋：基于 TensorFlow 2.0，多 GPU 版

经过前几章介绍的 TensorFlow 2.0 的实现，相信读者已经很熟悉这种新的构建网络方式了，本章的技术点在于如何构建网络，以及如何使用多个 GPU 训练的实现方式。

主要的实现代码都在 `policy_value_network_gpus_tf2.py` 文件中，你需要安装 GPU 版的 TensorFlow。安装命令如下：

```
# GPU
pip install tensorflow-gpu==2.0.0-alpha0
```

运行的要求是，需要 GPU 算力值至少为 3.5，还要安装 CUDA 10，本节代码在 AWS p2.8xlarge 上进行了测试。

本节实现的网络结构稍稍有点区别，不再是一个输入、一个输出，或者多个输入、一个输出，而是一个输入、两个输出，即策略"头"结果（走子概率向量）和价值"头"结果（胜率）输出。

构建多 GPU 的训练方式，需要定义一个分布式策略，通常就是定义镜像策略，它支持在一台机器中的多个 GPU 上进行分布式训练（本示例代码引用自 https://github.com/tensorflow/docs/blob/master/site/en/r2/guide/distribute_strategy.ipynb）。

```
mirrored_strategy = tf.distribute.MirroredStrategy()
```

然后把需要多 GPU 分布训练的代码放在镜像策略的 scope 下即可，比如：

```python
with mirrored_strategy.scope():
    model = tf.keras.Sequential([tf.keras.layers.Dense(1, input_shape=(1,))])
    model.compile(loss='mse', optimizer='sgd')
```

接下来需要创建输入数据集，并调用 make_dataset_iterator 函数，根据策略向多个 GPU 分发数据集。

```python
with mirrored_strategy.scope():
    dataset = tf.data.Dataset.from_tensors(([1.], [1.])).repeat(1000).batch(
        global_batch_size)
    input_iterator = mirrored_strategy.make_dataset_iterator(dataset)
```

这样，网络和数据都准备好了。现在，我们定义单步训练函数 step_fn。为了分发这个训练步骤，让多个 GPU 都执行单步训练，需要调用策略的 strategy.experimental_run 函数，传入的参数是单步训练函数 step_fn 和分发的数据迭代器 input_iterator，如下所示。

```python
# 单步训练功能函数
@tf.function
def train_step():
    # 真正的单步训练函数，接收的参数是数据迭代器
    def step_fn(inputs):
        features, labels = inputs

        with tf.GradientTape() as tape:
            logits = model(features)
            cross_entropy = tf.nn.softmax_cross_entropy_with_logits(
                logits=logits, labels=labels)
            loss = tf.reduce_sum(cross_entropy) * (1.0 / global_batch_size)

        grads = tape.gradient(loss, model.trainable_variables)
        optimizer.apply_gradients(list(zip(grads, model.trainable_variables)))
        return loss
    # 为了调用单步训练函数，需要调用策略的 experimental_run 函数
    # 传入的参数是要调用的单步训练函数和数据迭代器
    per_replica_losses = mirrored_strategy.experimental_run(
        step_fn, input_iterator)
    # 计算多个GPU的平均损失
    # 注意，step_fn函数不能返回无法对多GPU的数据进行整合的张量，step_fn是分发在每个
```

```
    # GPU上执行的，会有多个结果，比如返回的loss，所以这里需要将多个GPU的loss进行整合
    mean_loss = mirrored_strategy.reduce(
        tf.distribute.ReduceOp.MEAN, per_replica_losses)
    return mean_loss
```

最后，在策略的 scope 下循环调用单步训练函数即可，在调用之前需要初始化数据迭代器。

```
with mirrored_strategy.scope():
    # 初始化数据迭代器
    input_iterator.initialize()
    # 循环调用单步训练函数
    for _ in range(10):
        print(train_step())
```

上面借用官方的例子解释了多 GPU 训练的方法，本次的实现稍微有些不同，没有直接调用 experimental_run 函数，而是通过使用 tf.function 函数封装的 distributed_train 函数间接调用了 experimental_run。另外，没有使用镜像策略的 reduce 方法计算多个 GPU 的平均损失，而是使用了平均指标 tf.keras.metrics.Mean 来计算平均损失，你可以修改本节代码使用官方例子中的实现来计算多个 GPU 的平均损失。现在我们来看看代码实现。

```
import tensorflow as tf
import numpy as np
from tensorflow.python.ops import summary_ops_v2
import os


# 定义网络类
class policy_value_network_gpus(object):
    # 参数分别是学习率回调函数（使学习率动态变化）、残差块层数（默认是7层）
    def __init__(self, learning_rate_fn, res_block_nums = 7):

        self.save_dir = "./models"

        # 如果模型目录不存在，则创建
        if tf.io.gfile.exists(self.save_dir):
            pass
        else:
            tf.io.gfile.makedirs(self.save_dir)
```

```python
train_dir = os.path.join(self.save_dir, 'summaries', 'train')
test_dir = os.path.join(self.save_dir, 'summaries', 'eval')

# 定义日志记录器，可以不使用
self.train_summary_writer = summary_ops_v2.create_file_writer(train_dir,
                                                              flush_millis
                                                              =10000)
self.test_summary_writer = summary_ops_v2.create_file_writer(test_dir,
                                                             flush_millis
                                                             =10000,
                                                             name='test')
# 定义镜像策略
self.strategy = tf.distribute.MirroredStrategy()
# 打印设备（GPU）数量
print ('Number of devices: {}'.format(self.strategy.num_replicas_in_sync))

# 定义了一个lambda函数，接收的参数it是数据迭代器，这样我们可以间接调用函数
# experimental_run
# experimental_run函数的两个参数分别是单步训练函数和数据迭代器
# 这里你也可以修改成使用官方例子中的方式完成训练
self.distributed_train = lambda it: self.strategy.experimental_run(
    self.train_step, it)
# 调用tf.function函数将lambda函数封装成TensorFlow函数
# 在训练时调用distributed_train函数进行分布式训练
self.distributed_train = tf.function(self.distributed_train)

# global_step这个变量在本次实现中并没有使用，已经不需要global_step了
# 不能在GPU上定义该变量
with tf.device('/cpu:0'):
    self.global_step = tf.Variable(0, name="global_step", trainable=False)

# 在策略的scope下构建网络模型
with self.strategy.scope():
    # 定义若干超参数
    self.filters_size = 128    # 或256
    self.prob_size = 2086
```

```python
# 定义输入占位符(棋盘状态)，每个玩家7个特征平面，一共14个特征平面，棋盘
# 大小是9×10
self.inputs_ = tf.keras.layers.Input([9, 10, 14], dtype='float32',
                                     name='inputs')
# L2正则的参数
self.c_l2 = 0.0001
# Momentum优化器的参数
self.momentum = 0.9

# 卷积块
self.layer = tf.keras.layers.Conv2D(kernel_size=3,
                                    filters=self.filters_size,
                                    padding='same')(self.inputs_)
self.layer = tf.keras.layers.BatchNormalization(epsilon=1e-5,
                                                fused=True)(self.layer)
self.layer = tf.keras.layers.ReLU()(self.layer)

# 残差块
with tf.name_scope("residual_block"):
    for _ in range(res_block_nums):
        self.layer = self.residual_block(self.layer)

# 策略"头"
with tf.name_scope("policy_head"):
    self.policy_head = tf.keras.layers.Conv2D(filters=2, kernel_size=1,
                                              padding='same')(self.
                                              layer)
    self.policy_head = tf.keras.layers.BatchNormalization(epsilon=1e-5,
        fused=True)(self.policy_head)
    self.policy_head = tf.keras.layers.ReLU()(self.policy_head)

    self.policy_head = tf.keras.layers.Reshape([9 * 10 * 2])(
        self.policy_head)
    self.policy_head = tf.keras.layers.Dense(self.prob_size)(
        self.policy_head)
```

```python
# 价值"头"
with tf.name_scope("value_head"):
    self.value_head=tf.keras.layers.Conv2D(filters=1, kernel_size=1,
                                           padding='same')(self.layer)
    self.value_head = tf.keras.layers.BatchNormalization(epsilon=1e-5,
        fused=True)(self.value_head)
    self.value_head = tf.keras.layers.ReLU()(self.value_head)

    self.value_head = tf.keras.layers.Reshape([9 * 10 * 1])(
        self.value_head)
    self.value_head = tf.keras.layers.Dense(256,
                                            activation='relu')(self.
                                            value_head)
    self.value_head = tf.keras.layers.Dense(1,
                                            activation='tanh')(self.
                                            value_head)

# 定义模型，一个输入、两个输出
self.model = tf.keras.Model(
    inputs=[self.inputs_],
    outputs=[self.policy_head, self.value_head])

self.model.summary()

# Momentum优化器，传入的学习率是一个回调函数，可以动态修改学习率
self.optimizer = tf.compat.v1.train.MomentumOptimizer(
    learning_rate=learning_rate_fn, momentum=self.momentum,
    use_nesterov=True)

# 定义分类准确率指标
self.ComputeMetrics = tf.keras.metrics.CategoricalAccuracy()
# 定义平均指标，用于计算平均损失
self.avg_loss = tf.keras.metrics.Mean('loss', dtype=tf.float32)

# 定义保存模型参数检查点
```

```python
        self.checkpoint_dir = os.path.join(self.save_dir, 'checkpoints')
        self.checkpoint_prefix = os.path.join(self.checkpoint_dir, 'ckpt')
        self.checkpoint = tf.train.Checkpoint(model=self.model,
                                              optimizer=self.optimizer)

        # 加载存在的参数
        self.checkpoint.restore(tf.train.latest_checkpoint(
                                self.checkpoint_dir))

# 残差块生成函数，每调用一次生成一个残差块
def residual_block(self, in_layer):
    orig = tf.convert_to_tensor(in_layer)
    layer = tf.keras.layers.Conv2D(kernel_size=3, filters=self.filters_size,
        padding='same')(in_layer)
    layer = tf.keras.layers.BatchNormalization(epsilon=1e-5, fused=True)(layer)
    layer = tf.keras.layers.ReLU()(layer)

    layer = tf.keras.layers.Conv2D(kernel_size=3, filters=self.filters_size,
        padding='same')(layer)
    layer = tf.keras.layers.BatchNormalization(epsilon=1e-5, fused=True)(layer)
    add_layer = tf.keras.layers.add([orig, layer])
    out = tf.keras.layers.ReLU()(add_layer)

    return out

# 保存模型，注意也要放在策略的scope下
def save(self, in_global_step):
    with self.strategy.scope():
        self.checkpoint.save(self.checkpoint_prefix)
    # print("Model saved in file: {}".format(save_path))

# 本函数引用自官方实现：tensorflow/tensorflow/contrib/layers/python/layers/
# regularizers.py
# 用于计算本节的L2正则
def apply_regularization(self, regularizer, weights_list=None):
    # if not weights_list:
```

```python
#         weights_list = ops.get_collection(ops.GraphKeys.WEIGHTS)
    if not weights_list:
        raise ValueError('No weights to regularize.')
    with tf.name_scope('get_regularization_penalty',
                       values=weights_list) as scope:
        penalties = [regularizer(w) for w in weights_list]
        penalties = [
            p if p is not None else tf.constant(0.0) for p in penalties
        ]
        for p in penalties:
            if p.get_shape().ndims != 0:
                raise ValueError('regularizer must return a scalar Tensor'
                                 'instead of a '
                                 'Tensor with rank %d.' % p.get_shape().ndims)

        summed_penalty = tf.add_n(penalties, name=scope)
        # ops.add_to_collection(ops.GraphKeys.REGULARIZATION_LOSSES,
        #                       summed_penalty)
        return summed_penalty

# 计算联合损失：交叉熵损失 + MSE + L2损失
def compute_loss(self, pi_, z_, policy_head, value_head):

    # loss
    with tf.name_scope("loss"):
        # 分类交叉熵损失，用于计算走子概率的损失
        policy_loss = tf.keras.losses.categorical_crossentropy(y_true=pi_,
                                                               y_pred=
                                                               policy_head,
                                                               from_logits=True
                                                               )
        policy_loss = tf.reduce_mean(policy_loss)

        # 胜率的MSE损失
        value_loss = tf.keras.losses.mean_squared_error(z_, value_head)
        value_loss = tf.reduce_mean(value_loss)
```

```python
        # summary_ops_v2.scalar('mse_loss', value_loss)

        # 计算L2正则
        regularizer = tf.keras.regularizers.l2(self.c_l2)
        regular_variables = self.model.trainable_variables
        l2_loss = self.apply_regularization(regularizer, regular_variables)

        # 得到联合损失
        # self.loss = value_loss - policy_loss + l2_loss
        self.loss = value_loss + policy_loss + l2_loss
        # summary_ops_v2.scalar('loss', self.loss)

        return self.loss

# 单步训练函数，传入的it是数据迭代器，参数learning_rate已经没有用了，因为
# 是动态改变学习率的
# TODO(yashkatariya): Add tf.function when b/123315763 is resolved
# @tf.function
def train_step(self, it, learning_rate=0):
    positions = it[0]
    pi = it[1]
    z = it[2]

    if True:
        # 使用磁带记录计算损失的操作
        with tf.GradientTape() as tape:
            # 调用模型得到模型输出，参数training是True
            policy_head, value_head = self.model(positions, training=True)
            # 计算联合损失
            loss = self.compute_loss(pi, z, policy_head, value_head)

            # 计算预测概率向量和MCTS概率向量的分类准确率
            self.ComputeMetrics(pi, policy_head)
            # 计算平均损失
            self.avg_loss(loss)
```

```python
        # 使用磁带计算模型参数的梯度
        grads = tape.gradient(loss, self.model.trainable_variables)
        self.optimizer.apply_gradients(zip(grads, self.model.
                                      trainable_variables))
    return self.ComputeMetrics.result(), self.avg_loss.result(),
        self.global_step

#@profile
def forward(self, positions):
    # 前向传播，调用模型得到模型输出
    with self.strategy.scope():
        positions=np.array(positions)
        if len(positions.shape) == 3:
            sp = positions.shape
            positions=np.reshape(positions, [1, sp[0], sp[1], sp[2]])
        action_probs, value = self.model(positions, training=False)

    return action_probs, value
```

以上就是模型结构的定义，变化并不大，只要按照模型的定义去构建网络即可。接下来看看如何进行多 GPU 训练，代码在 `main_tf2.py` 中，修改的代码并不多，主要改动是将 flatten() 的函数调用改成 tf.squeeze。另外，定义了学习率回调函数，在定义网络类实例时传入即可。

```python
# 可以动态改变学习率
def lr_callback(self):
    return self.learning_rate * self.lr_multiplier
```

最后的改动就是 policy_update 函数，这里是训练网络的地方，主要就是以下几句代码：

```python
# 训练的代码要放在策略的scope下
with self.policy_value_netowrk.strategy.scope():
    # 将MCTS输出的棋谱、走子概率和胜者组织成数据集
    train_dataset = tf.data.Dataset.from_tensor_slices((state_batch,
        mcts_probs_batch, winner_batch)).batch(len(winner_batch))
    # 得到数据迭代器
    train_iterator = self.policy_value_netowrk.strategy.make_dataset_iterator(
        train_dataset)
    # 初始化数据迭代器
```

```
train_iterator.initialize()
# 开始多GPU训练
accuracy, loss, self.global_step = self.policy_value_netowrk.distributed_train(
    train_iterator)
```

7.5　本章小结

　　这里说说训练情况，因为是从"白板"开始训练的，刚开始都是乱下，从乱下的经验中学会下棋是需要大量对弈才行的。论文中训练了 700 000 次的 mini-batch，国际象棋开源项目 chess-alpha-zero 也训练了 10 000 次。笔者只训练了不到 4 000 次，模型刚刚学会用象和士防守，总之还没有学会下棋。本章开始也介绍过，现在已经有人在用分布式的方式进行训练，如果有条件的话你可以再多训练试试。

　　本章对 AlphaGo Zero 和 AlphaZero 的论文进行了讨论，并讲解了各部分的理论知识，最后通过完成中国象棋的实践复现了 AlphaZero，相信读者通过代码能够更直观地理解论文中所讲的理论。

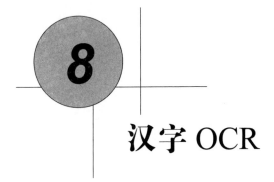

第 8 章 汉字 OCR

8.1 概述

本章要讨论的是 OCR（Optical Character Recognition，光学字符识别），OCR 技术可以使计算机能够理解图片中的字符信息，可以说它是人们生活当中使用非常普遍的一项技术。比如手机上的"Google 翻译"App，就提供了图片翻译的功能，可以选择图片给 App，然后 App 识别出指定的字符进行翻译。再比如金山的 WPS 软件，还有汉王的 OCR 软件等，都提供了 OCR 的功能，可以帮助用户将图片中的文字提取出来。除了这类通用的图片转文字的需求，还有一些应用场景，比如车牌识别，现在已经有很多停车场等车辆通行场所，在出入口处架设了摄像头，然后利用 OCR 技术实现了自动识别车牌的功能，利用该技术可以做到闸机无人值守。再比如基于特定情景下的 OCR，如身份证识别、驾驶证识别、发票识别、化验单识别等，这些应用场景都离不开 OCR 技术，可以说它的用处是非常广泛的。

本章将讨论实现 OCR 的两种思路，其中一种是将汉字 OCR 看成分类问题；另一种是使用端到端的方式实现汉字 OCR。

8.2 分类网络实现汉字 OCR

如果你只想识别数字和英文字母的话，则可以考虑使用 Google 维护的开源项目 Tesseract，虽然它也支持汉字识别，但是想提高精度的话，还需要多训练。

在第 1 章中，已经介绍过 Yann LeCun 用于手写数字识别的 LeNet-5 网络，那时是 1998 年，已经可以使用卷积神经网络进行手写数字的识别了。训练用的手写数字数据集是 MNIST，可以说 MNIST 手写数字识别是深度学习的入门级项目，相信学习过第 2 章中介绍的分类网络实践以后，大家已经可以解决很多现实中的分类问题了。在第 2 章中举的例子是 CIFAR-100 图像分类识别，我们也可以做动物识别、交通灯识别，甚至是物体检测识别（比如用来检测物体是人、自行车、汽车、花、某种动物等），可以说应用领域很广泛，只要是可以分类的问题，就可以使用分类网络来解决。

相信读者也想到了，本节要介绍的汉字 OCR，也可以理解成汉字的分类问题，那么读者实际上已经具备了解决汉字 OCR 的能力。所以，本节不会再一次实现分类网络，而是讲解使用分类网络实现汉字 OCR 的相关步骤和一些思路，读者可以通过互联网查找相关思路的文章，具体实现就由感兴趣的读者自行实践了。

8.2.1 图片矫正

假设我们要对手机拍摄的图片或者扫描仪扫描的图片进行汉字识别，通常手机拍摄的图片都会有些变形，呈现出某种梯形，不利于识别，如图 8.1 所示。

图 8.1 手机拍摄的图片

而扫描仪扫描的图片，如果摆放不好，即使人眼看是对齐的，但是使用标尺衡量的话，发现纸张多少也会偏一些。如图 8.2 所示，这张图片不是扫描的，而且特意偏得夸张了一些，好让读者能够理解笔者想要表达的意思。想象这是扫描的图片，即使摆放得足够对齐，多少也会偏几度。

针对第一种情况，我们需要使用透视变换的方法，摆正图片的位置。思路是利用 OpenCV 的 canny 函数找到纸张的边缘，并利用 findContours 函数查找纸张的轮廓，然后针对找到的轮廓，使用 approxPolyDP 函数拟合多边形曲线，使用 contourArea 函数计算该曲线的面积（需要定义一个阈值，大于阈值认为是有效的轮廓），使用 isContourConvex 函数判断曲线的凸性（轮廓必须是凸的）。通过这几个函数可以找到轮廓的四个顶点，当然，你也可以使用其他方法找到轮廓的四个顶点。当找到四个顶点之后，就可以调用 warpPerspective 函数做透视变换，将变形的图片摆正了。

图 8.2　假设是扫描仪扫描的图片

针对第二种情况，只要知道图片相对于水平方向的夹角，然后做旋转即可。找角度的方法有很多种，这里提供一个思路，同样是调用 canny 函数做边缘检测，然后使用 fitLine 函数对感兴趣的区域拟合一条直线，假如这个区域是纸张的边线部分，则会得到 $[v_x, v_y, x_0, y_0]$ 这样一组数据，利用如下公式可以计算出 φ（phi）和 ρ（rho）：

```
phi = math.atan2(vy, vx) + np.pi / 2.0
rho = y0 * vx - x0 * vy
```

通过 φ 和 ρ 可以计算出拟合直线的两个端点，格式是 $(x_1, y_1), (x_2, y_2)$：

```
(rho / math.cos(phi), 0), ((rho - rows * math.sin(phi)) / math.cos(phi), rows)
```

或者

```
(0, rho / math.sin(phi)), (cols, (rho - cols * math.cos(phi)) / math.sin(phi))
```

有了直线的两个端点，就可以计算出这条直线相对于水平线的夹角了：

```
angle = math.atan2(y2 - y1, x2 - x1) / np.pi *180
```

最后调用 getRotationMatrix2D 和 warpAffine 函数做旋转变换即可将图片矫正。

8.2.2 文本切割

现在图片已经矫正过了，接下来需要对图片上的文本进行切割。文本切割可以说极其重要，一旦字切错了，肯定会影响模型的识别。文本切割分为行切割和列切割。行切割就是把一大段文字一行一行地切分出来；列切割就是把每行的文字一个字一个字地切分出。比如要切割如图 8.3 所示的文本。

> **话说天下大势，分久必合，合久必分。周末七国分争，并入于秦。及秦灭之后，楚、汉分争，又并入于汉。汉朝自高祖斩白蛇而起义，一统天下，后来光武中兴，传至献帝，遂分为三国。推其致乱之由，殆始于桓、灵二帝。桓帝禁锢善类，崇信宦官。及桓帝崩，灵帝即位，大将军窦武、太傅陈蕃共相辅佐。时有宦官曹节等弄权，窦武、陈蕃谋诛之，机事不密，反为所害，中涓自此愈横。**

图 8.3　待切割文本

切割以后的文本如图 8.4 所示。

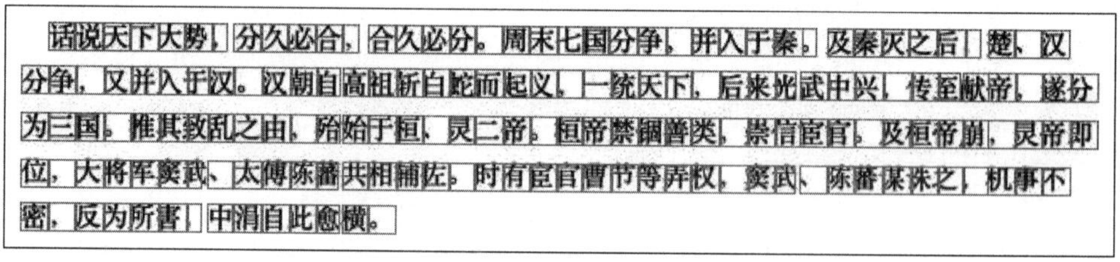

图 8.4　切割以后的文本

这样就把文本中的每一个字都切分出来，然后把每一个字的图片输入给神经网络，由神经网络识别每张图片都是什么字。这样就可以把汉字 OCR 问题看成汉字的分类问题，每一个字都是一个分类，通过训练一个汉字分类网络就可以进行汉字的识别。这里一共有两个重点，一是汉字分类网络的设计和训练；二是文本切割。

汉字是很复杂的，除汉字种类很多以外，还有各种各样的字体，且每种字体的写法都不一样。除不同的印刷体之外，还有手写体，而每个人的写法又各不相同。除这些差异之外，还有输入图片的质量，比如分辨率的大小等都会影响汉字的识别。汉字本身的特点是具有上下结构、左右结构，甚至左中右、上中下结构。尤其是手写体，有些时候字与字之间是连起来的，并没有明显的分割点。

这些都给文字的切割带来了挑战，所以有时只能接受切割算法 99% 的成功率。也就是说，允许部分文字切割错误的情况存在，这样的话，就需要在切割程序之后再进行容错处理。比如对切割后的文字做一个宽度比较，如果某个字的宽度很夸张，假如是单个字宽度的两倍，那么可以认为这是两个字连在一起了，可以再切割变成两个字；再比如某个字的宽度很短，则有可能是偏旁部首被单独切割下来，可以考虑跟下一个字连接起来合并成一个字。

还可以将错就错，将这些切割好的文字包含切错的都输入给神经网络，然后根据神经网络的识别结果做修正处理。对于切错的字，神经网络识别的置信度可能不是很高，一旦发现这样的字存在，就尝试将相邻的两个字合并在一起再次输入给神经网络识别；或者再训练一个词嵌入矩阵，将神经网络识别出的句子交给词嵌入矩阵做修正处理。比如某个字因为切错了，导致识别出的字在整个句子当中很奇怪，这样可以利用词嵌入矩阵得到字与字之间的距离，一旦发现某个词语跟词嵌入矩阵中的距离相差很多，就可以认为这个字不是识别错了就是切错了，甚至可以根据词嵌入矩阵得到最有可能的词组结果对错误的结果做修正。

当然，思路不限于上面提到的，读者可以尽情发挥。

8.2.3 汉字分类网络

接下来讲述另一个重点，训练一个汉字分类网络。OCR 就是一个看图识字的过程，对于神经网络来说，CNN 正适合这类任务。从 MNIST 手写数字识别的 LeNet 网络开始构建汉字分类网络是一个不错的开始，当然，你也可以尝试其他有效的分类网络模型，比如 AlexNet、GoogLeNet、VGG 等网络都可以尝试。当确定好要用的网络结构之后，还要设计好输入数据的尺寸，以便设计每一个卷积层的参数。对于网络的每一层的形状，你一定要心中有数。

当有了输入数据的尺寸，设计好了网络的每一层，确定了要用的网络结构之后，还要决定到底分多少类。根据 GB2312 国标码中国标一、二级字库的定义，分为一级字库（常用汉字，共 3755 个汉字）和二级字库（非常用汉字，共 3008 个汉字），还包含数字、一般符号、拉丁字母、日本假名、希腊字母、俄文字母、拼音符号、注音字母等 682 个字符。所以想识别出多少个字就看自己的需求了，比如想训练一个识别一级汉字的分类网络，那么一共是 3755 个分类。关于一级汉字和二级汉字都有哪些字，大家在网络上搜索就能够找到相关的资料。

这样关于网络结构的部分就完成了，接下来需要生成训练集来训练网络。汉字训练集的生成，就是不断地生成每个汉字的图片，这个可以通过字体文件（.ttf 文件）来完成。为了提高识别汉字的精度，建议使用的字体越丰富越好，可以多找一些字体文件，使得相同的文字有不同的字体。

通过字体文件生成文字图片可以使用 pillow 库提供的方法来实现，示例代码如下：

```
# 引入pillow库
from PIL import Image
```

```python
from PIL.ImageDraw import Draw
from PIL import ImageFont
import cv2

# 创建一个64×64的彩色（RGB）画布(就是我们要生成的文字图片)，(0, 0, 0)是背景色黑色
img = Image.new("RGB", (64, 64), (0, 0, 0))
draw = Draw(img)
# 创建字体，64是字体大小
font = ImageFont.truetype("fonts/fangsong.ttf", 64,)
# 在画布上画出文字，(0, 0)是左上角的位置，(255, 255, 255)是字体颜色白色
draw.text((0, 0), "哈", (255, 255, 255), font=font)
cv2.imwrite("./ha.png", np.asarray(img, dtype='uint8'))
```

这样保存的 ha.png 文件如图 8.5 所示。

图 8.5　ha.png 文件

有一点要注意，一定要生成黑底白字的图片进行训练，这样会加速收敛。对于极端情况，当一张图片只有黑色（值是 0）和白色（值是 255）时，神经网络在做各种乘法计算时，黑色的 0 都不会被激活，只有白色的 255 对应的参数值参与了计算，所以学习到的都是白色文字的特征，没有多余的计算，并且会加快计算的速度，也就加快了学习的速度。对于白底黑字的图片，因为底色的特征没有任何意义，然后又要参与各种乘法计算，这类无用的计算变得更多了，而真正要学习的特征又都是 0，在进行训练（乘法计算）时无法有效学习到文字的特征，反而不容易收敛。

现在相关的工具已经有了，可以通过多种字体文件来生成训练数据了。为了提高网络识别的精度，还可以对生成的训练集做适当的数据增强处理，比如旋转、缩放、扭曲、仿射变换等。你也可以在开源社区查找使用开源的数据增强库，这些开源的数据增强库功能已经实现好了，你可以根据自己的需求自行决定使用哪些数据增强方法。

剩下的就是分类网络的训练了，这部分就不多说了，相信读者已经掌握了分类网络的训练方法。一旦网络训练好之后，你会发现汉字的分类识别精度还是很高的。网络的训练相对容易一些，真正的难点在于文本切割。接下来给大家介绍一种不用切割文本，即可实现汉字识别的端到端的 OCR 方法。

8.3　端到端的汉字 OCR：基于 TensorFlow 1.x

端到端的识别就很方便了，不用一个字一个字地切割，但是仍然要切分成行，然后将一行文字的图片作为输入数据传给神经网络，无论里面有多少个字都可以识别出来。

网络的开始仍然是 CNN，看图识字，CNN 就是我们的眼睛。CNN 提取出汉字特征之后，传给 BiLSTM（Bidirectional LSTM，双向 LSTM）得到基于时间步的特征。然后将 BiLSTM 的输出传入全连接层，得到每个时间步的分类结果，就是每个时间步预测的汉字。最后跟标签（真实汉字）计算 CTC 损失，CTC（Connectionist Temporal Classification，连接时序分类）可以解决基于时序（时间步）的分类问题。网络结构如图 8.6 所示。

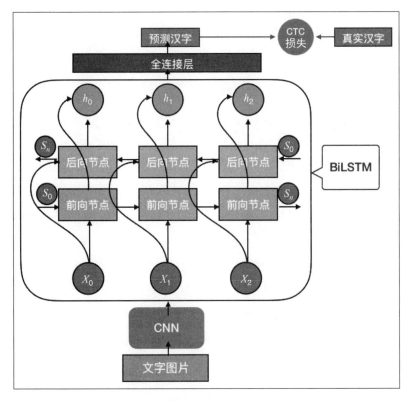

图 8.6　网络结构

限于篇幅，本章重要代码放在书里，完整代码可以在 GitHub 上下载，地址：https://github.com/chengstone/end2end-chinese-ocr，TensorFlow 1.x 的代码文件是 ocr.py，TensorFlow 2.0 的代码文件是 ocr_tf2.py。

8.3.1 CNN 设计

这里介绍笔者设计的 CNN 结构,仅供参考。

首先设计输入数据的尺寸。对于要识别的文字图片,宽度是不固定的,因为并不能确定输入图片中有多少个汉字。最终定义一个汉字的高度和宽度都是 64,输入数据的高度也是 64。

```
class ocr_network(object):
    def __init__(self, birnn_type=2, lstm_num_layers=1, batch_size=32, max_timestep
    =None):
        # 批量大小
        self.batch_size = batch_size
        # 图像通道数(1,表示是灰度图)
        self.channels = 1
        # 定义一张文字图片的高度和宽度
        self.img_height = 64
        self.img_width = self.img_height
```

定义输入数据的维度是 [批量大小(batch_size),高度(64),宽度(不固定),通道数(1)],上面说过,因为要识别的文字图片中的汉字个数是不固定的,所以输入图片的宽度未知。而且,我们不需要图片的彩色信息,只传入灰度图即可,所以通道数是 1。

```
# Batch, Height, Width, Channel
self.inputs_ = tf.placeholder(tf.float32, [self.batch_size, self.img_height, None,
self.channels], name='inputs')
```

接下来定义 CNN 结构,这里定义的 CNN 一共有 4 层卷积层,每一层滤波器(Filter)的个数分别是 32、64、128、128,激活函数使用 ELU。每层卷积层的最后是最大池化层,缩小数据的尺寸。

```
self.filters_size = [32, 64, 128, 128]
self.cnn_layer_num = 4
```

CNN 结构如图 8.7 所示。

8 汉字 OCR

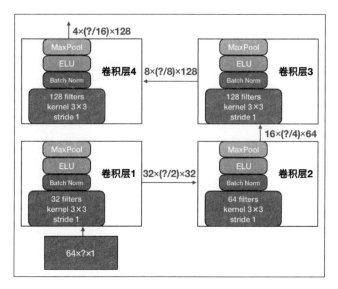

图 8.7　CNN 结构

输入图片是一个（高度 64× 宽度未知）的灰度图，经过第一层卷积层之后，因为最大池化层会将数据尺寸缩小一半，所以得到的输出尺寸是 $(64/2) \times (?/2) \times 32$(滤波器1的个数) $= 32 \times (?/2) \times 32$。同理，经过第二层卷积层后，得到的输出尺寸是 $(64/4) \times (?/4) \times 64$(滤波器2的个数) $= (32/2) \times (?/4) \times 64 = 16 \times (?/4) \times 64$；经过第三层卷积层后，得到的输出尺寸是 $(16/2) \times (?/8) \times 128 = 8 \times (?/8) \times 128$；经过最后一层卷积层后，得到最终的 CNN 输出尺寸是 $(8/2) \times (?/16) \times 128 = 4 \times (?/16) \times 128$。加上批量数据之后，CNN 最后输出的维度是：[批量大小（batch_size），特征高度（feature_h），特征宽度（feature_w），CNN 最终滤波器个数（cnn_out_channels）]。这里之所以把每一层的输出都写出来，是因为这部分非常重要，在设计 CNN 时应清楚地知道数据经过每一层的维度。

虽然输入图片的宽度未知，但是我们知道 CNN 最后一层的输出宽度是（图片原宽度 / 16），记住这个值，在后面会用到。

CNN 的代码实现如下：

```
with tf.name_scope('cnn'):
    self.layer = self.inputs_
    # 4层卷积层，循环4次
    for i in range(self.cnn_layer_num):
        with tf.name_scope('cnn_layer-%d' % i):
            # 定义卷积层
            self.layer = self.cnn_layer(self.layer, self.filters_size[i])
```

373

```
            print(self.layer.get_shape())
        # 得到CNN最后输出的维度
        _, feature_h, feature_w, cnn_out_channels=self.layer.get_shape().as_list()

    # 卷积层定义函数
    def cnn_layer(self, layer, filter_size):
        # 卷积层,3×3的卷积核,步幅是1
        layer = tf.layers.conv2d(layer, filter_size, 3, padding='SAME')
        # batch_norm层,激活函数是ELU
        layer = tf.contrib.layers.batch_norm(layer, center=False, epsilon=1e-5,
                                             fused=True, activation_fn=tf.nn.elu)
        # 最大池化层,缩小数据的尺寸
        layer = tf.nn.max_pool(layer, ksize=[1, 2, 2, 1], strides=[1, 2, 2, 1],
                                padding='SAME')
        return layer
```

8.3.2 双向 LSTM 设计

在第 3 章中介绍了 RNN 和 LSTM,当时使用的是单向 LSTM,也就是说,RNN 沿着时间顺序从左向右移动。这样的结构是否合理呢?如果某事物不仅取决于时间序列前面的特征,时间序列后面的特征也对当前时间的特征起到作用,那么这时就需要一种结构,同时计算前向时间和后向时间的特征,这就是双向 LSTM 的作用。以本章的汉字 OCR 为例,把一个汉字按照时间序列拆分的话,如图 8.8 所示,汉字的宽度就是时间步的集合。假如宽度是 64,那么就是 64 个时间步(timestep)。对于每一个时间步,汉字的高度就是该时间步的文字特征。

图 8.8　按照时间序列拆分汉字

因为汉字的组成是按照笔画写出来的,一横一竖、一撇一捺都是连续的,可以认为在写字的过程中,每一个时间点的特征都与它之前写过的笔画以及之后继续写的笔画有关联性,正适合双向 LSTM 提取前后时间序列的特征。比如"哈"字,口字旁的汉字有很多,除按照时间序列从前往后计算汉字的特征以外，从后往前计算汉字的特征对识别出"哈"字是有帮助的。同理，对于"刚"字，刂部的汉字也有很多，从两个方向去计算汉字的特征要比单独一个方向更有助于识别出某个汉字。

但是我们并不会把这个文字图片传给 LSTM，而是把 CNN 的输出作为 LSTM 的输入数据，回忆一下 CNN 最后一层输出的维度是 [特征高度（feature_h），特征宽度（feature_w），CNN 最终滤波器个数（cnn_out_channels）]，这里不考虑批量大小。其中 feature_w 的值是（图片原宽度 / 16），这样时间步也就确定了。因为有 cnn_out_channels 个滤波器，而每个滤波器提取的特征高度是 feature_h，所以对于每一个时间步，一共有 cnn_out_channels × feature_h 个特征。在传给 LSTM 之前，先把数据的维度调整成 [batch_size, max_timestep, feature_h * cnn_out_channels]。

```
# 接着CNN的输出，self.layer的维度是[batch_size, feature_h, feature_w,
# cnn_out_channels]
# 现在进行位置调整，转成：[batch_size, feature_w, feature_h, cnn_out_channels]
self.layer = tf.transpose(self.layer, [0, 2, 1, 3])
# 在LSTM中，feature_w是时间步(max_timestep)，接下来要调整成
# [batch_size, max_timestep, feature_h * cnn_out_channels]
self.layer = tf.reshape(self.layer, [self.batch_size, -1,
                                     feature_h * cnn_out_channels])
```

是时候说明双向 LSTM（BiLSTM）了。BiLSTM 包含两个方向的 LSTM 序列，一个是前向的，状态和特征按照时间步从前往后传递；一个是后向的，状态和特征按照时间步从后往前传递。

创建双向 LSTM 也很简单，TensorFlow 提供了多个 API 供我们选择，分别是：

```
tf.nn.static_bidirectional_rnn
tf.nn.bidirectional_dynamic_rnn
tf.contrib.rnn.stack_bidirectional_dynamic_rnn
tf.contrib.rnn.stack_bidirectional_rnn
```

1. 栈式 BiLSTM 和非栈式 BiLSTM

从网络结构的不同来区分的话，可以分成两类，一类是栈式 BiLSTM（tf.contrib.rnn.stack_bidirectional_dynamic_rnn、tf.contrib.rnn.stack_bidirectional_rnn）；一类是非栈式 BiLSTM（tf.nn.static_bidirectional_rnn、tf.nn.bidirectional_dynamic_rnn）。

非栈式 BiLSTM 是将 LSTM 分成两组，一组前向，一组后向，每组都可以是多层的 LSTM。输入数据同时传给前向 LSTM 和后向 LSTM，经过 LSTM 的传递计算之后，将两组 LSTM 的输出组合在一起作为整个 BiLSTM 的输出。大家回忆一下第 3 章中介绍的 LSTM，在构建 LSTM 时

要定义 LSTM 隐节点的个数，也就是 LSTM 输出的维度。而 BiLSTM 的输出是前向和后向两组 LSTM 输出的组合，也就是 LSTM 输出维度的两倍。非栈式 BiLSTM 结构如图 8.9 所示，这里假设前向 LSTM 和后向 LSTM 的层数是两层。

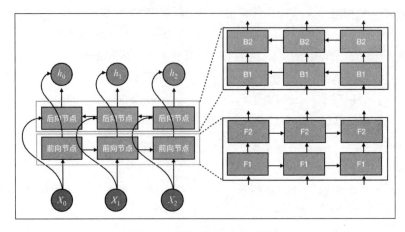

图 8.9　非栈式 BiLSTM 结构

栈式 BiLSTM 是将多个双向 LSTM 层堆叠在一起，每一个双向 LSTM 层都由一个前向 LSTM 和一个后向 LSTM 组成，结构如图 8.10 所示，这里假设栈式 BiLSTM 由两层双向 LSTM 层组成。

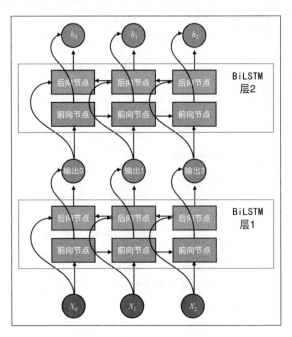

图 8.10　栈式 BiLSTM 结构

如果栈式 BiLSTM 只有一层，而且非栈式 BiLSTM 的前向 LSTM 和后向 LSTM 都是一层的话，栈式 BiLSTM 和非栈式 BiLSTM 是等价的。

2. 静态 BiLSTM 和动态 BiLSTM

除栈式和非栈式的区别之外，BiLSTM 还可以分成静态 BiLSTM（tf.nn.static_bidirectional_rnn、tf.con trib.rnn.stack_bidirectional_rnn）和动态 BiLSTM（tf.nn.bidirectional_dynamic_rnn、tf.contrib.rnn.stack _bidirectional_dynamic_rnn）。

静态 BiLSTM 和动态 BiLSTM 的区别在于，在创建 BiLSTM 时，对于静态 BiLSTM 来说，输入数据的时间步必须是确定的；而对于动态 BiLSTM 来说，输入数据的时间步可以不确定。上面说过，时间步实际上是数据的宽度，如果要创建静态 BiLSTM，那么数据的宽度必须是固定的。比如之前定义的输入数据的高度是 64，但是宽度未知，这样不行，宽度必须要定下来，比如可以定成 128，即输入数据的宽度是两个汉字宽（64 的 2 倍）。无论定成多少，必须要定下来，不能宽度未知。

而动态 BiLSTM 要友好一些，允许在创建时数据宽度不固定，这样我们就可以输入包含不同文字个数的图片了。

思路回到在将 CNN 的输出传给 LSTM 之前，要先把数据的维度调整成 [batch_size, max_timestep, feature_h * cnn_out_channels]。之前已经给出了实现代码：

```
# [batch_size, max_timestep, feature_h * cnn_out_channels]
self.layer = tf.reshape(self.layer, [self.batch_size, -1,
                                     feature_h * cnn_out_channels])
```

这里有一点要注意，在 reshape 时，max_timestep 的位置传入的是 −1，也就是说，时间步不确定。那么这段代码是给动态 BiLSTM 用的，如果用在静态 BiLSTM 中，一定要确定 max_timestep，并且用 max_timestep 替换 −1。

```
self.layer = tf.reshape(self.layer, [self.batch_size, self.max_timestep,
                                     feature_h * cnn_out_channels])
```

还记得如何确定 max_timestep 吗？首先要确定输入数据（就是图片）的宽度，如果宽度确定了，那么此时的时间步 max_timestep=(图片原宽度 / 16)。

无论是静态 BiLSTM 还是动态 BiLSTM，栈式 BiLSTM 还是非栈式 BiLSTM，最终 BiLSTM 的输出都是前向 LSTM 和后向 LSTM 输出的组合，也就是 LSTM 输出维度的两倍。我们需要把 BiLSTM 的输出再交给全连接层，得到分类个数的输出，也就是网络最终的输出。为了跟标签计算 CTC 损失，需要把网络最终的输出（logits）维度调整成 [时间步（max_timestep），批量大小（batch_size），分类个数（num_classes）]。这叫作 Time Major，表示时间步排列优先。

下面是创建各类 BiLSTM 的代码实现。

```python
# LSTM的层数，默认是1
self.lstm_num_layers = lstm_num_layers
# 定义LSTM的隐节点数，也就是LSTM的输出维度
self.num_hidden = 128
# seq_len实际就是max_timestep，因为后面会创建动态BiLSTM，所以要定义一个不确定的时间
# 步，就是seq_len
self.seq_len = tf.placeholder(tf.int32)
# charset是字符集，就不在这里放3000多个汉字了，改成用数字代替，训练时需要替换成这里
# 要识别的汉字字符集
self.charset="1234567890"
# 分类个数，字符集长度 + 2(blank字符和space字符)
self.num_classes=len(self.charset)+2


with tf.name_scope('lstm'):
    # 第一种情况，创建一个普通的两层RNN
    if self.birnn_type == 0:  # dynamic_rnn
        # 将CNN的输出维度调整成[batch_size, feature_w, feature_h, cnn_out_channels]
        self.layer = tf.transpose(self.layer, [0, 2, 1, 3])
        # 在LSTM中，feature_w是时间步(max_timestep)，接下来要调整成
        # [batch_size, max_timestep, feature_h * cnn_out_channels]
        self.layer = tf.reshape(self.layer, [self.batch_size, -1,
                                             feature_h * cnn_out_channels])
        print('lstm input shape: {}'.format(self.layer.get_shape().as_list()))

        # 创建两个LSTM单元，也可以尝试RNNCell或GRUCell
        self.cell1 = tf.nn.rnn_cell.LSTMCell(self.num_hidden, state_is_tuple=True)
        self.cell2 = tf.nn.rnn_cell.LSTMCell(self.num_hidden, state_is_tuple=True)
        # 将两个LSTM单元组合起来
        self.stack = tf.nn.rnn_cell.MultiRNNCell([self.cell1, self.cell2],
                                                  state_is_tuple=True)
        initial_state = self.stack.zero_state(self.batch_size, dtype=tf.float32)
        # 将组合的两个LSTM单元传给dynamic_rnn创建RNN网络
        # 输出outputs的维度是[batch_size, max_timestep, self.num_hidden]
```

```python
        # 不是两倍的num_hidden的原因是，这是一个普通的两层单向LSTM网络
        # 不是双向LSTM，所以输出维度是num_hidden
        # 传入的sequence_length是self.seq_len，时间步未知，只有在得到数据的那一刻
        # 才能知道(宽度/16)
        outputs, _ = tf.nn.dynamic_rnn(
            cell=self.stack,
            inputs=self.layer,
            sequence_length=self.seq_len,
            initial_state=initial_state,
            dtype=tf.float32,
            time_major=False
        )
        # 调整输出维度：[batch_size * max_timestep, self.num_hidden]
        outputs = tf.reshape(outputs, [-1, self.num_hidden])

        # 全连接层，做从num_hidden（LSTM隐节点数）到num_classes（分类个数）的映射
        W = tf.get_variable(name='W_out',
                            shape=[self.num_hidden, self.num_classes],
                            dtype=tf.float32,
                            initializer=tf.glorot_uniform_initializer())
        b = tf.get_variable(name='b_out',
                            shape=[self.num_classes],
                            dtype=tf.float32,
                            initializer=tf.constant_initializer())

        # 全连接层的输出，维度是[batch_size * max_timestep, self.num_classes]
        self.logits = tf.matmul(outputs, W) + b

        # 调整logits的维度：[batch_size, max_timestep, num_classes]
        self.logits = tf.reshape(self.logits, [self.batch_size, -1, self.num_classes])
        # 最终调整成Time Major的形式：[max_timestep, batch_size, num_classes]
        self.logits = tf.transpose(self.logits, (1, 0, 2))

        # 第二种情况，创建静态非栈式BiLSTM，static_bidirectional_rnn，要求max_timestep
        # 必须是确定的
```

```python
        elif self.birnn_type == 1 and max_timestep is not None:
            # 将CNN的输出维度调整成[batch_size, feature_w, feature_h, cnn_out_channels]
            self.layer = tf.transpose(self.layer, [0, 2, 1, 3])
            # 在LSTM中，feature_w是时间步(max_timestep)，接下来要调整成
            # [batch_size, max_timestep, feature_h * cnn_out_channels]
            # 此时传入的max_timestep是确定的
            self.layer = tf.reshape(self.layer, [self.batch_size, self.max_timestep,
                                                feature_h * cnn_out_channels])
            print('lstm input shape: {}'.format(self.layer.get_shape().as_list()))

            # 创建前向LSTM，隐节点数是num_hidden，层数lstm_num_layers默认是1
            self.forward_cell = [tf.nn.rnn_cell.LSTMCell(self.num_hidden) for _ in range(self.lstm_num_layers)]
            # 将前向LSTM组合起来(如果有多层的话)
            self.forward_stack = tf.nn.rnn_cell.MultiRNNCell(self.forward_cell,
                                                            state_is_tuple=True)
            # 创建后向LSTM，隐节点数是num_hidden，层数lstm_num_layers默认是1
            self.backward_cell = [tf.nn.rnn_cell.LSTMCell(self.num_hidden) for _ in range(self.lstm_num_layers)]
            # 将后向LSTM组合起来(如果有多层的话)
            self.backward_stack = tf.nn.rnn_cell.MultiRNNCell(self.backward_cell,
                                                             state_is_tuple=True)

            # 再次调整CNN的输出维度，Time Major的形式，将时间步排在前面
            # layer的维度：[max_timestep, batch_size, feature_h * cnn_out_channels]
            layer = tf.unstack(self.layer, self.max_timestep, axis=1)

            # 创建静态非栈式BiLSTM，输出outputs的维度是 [max_timestep, batch_size,
            # 2 * num_hidden]
            # 因为是BiLSTM，输出的是两个LSTM输出组合在一起的结果，所以是2 * num_hidden
            outputs, _, _ = tf.nn.static_bidirectional_rnn(self.forward_stack,
                self.backward_stack, layer, dtype=tf.float32)
            # 调整输出outputs的维度：[max_timestep, batch_size, 2, num_hidden]
            # 把2 * num_hidden拆开成[2, num_hidden]，这样要做两次全连接。当然，此处也
            # 可以不用拆开，只进行一次全连接即可，就像self.birnn_type == 2时做的那样
            # 这里提供两种方法，选择哪种方法都可以，算法不是固定的
```

```python
            outputs = [tf.reshape(t, [self.batch_size, 2, self.num_hidden]) for t in
outputs]

            W1 = tf.Variable(tf.truncated_normal([2, self.num_hidden], stddev=np.sqrt
(2.0 / (2 * self.num_hidden))))
            b1 = tf.Variable(tf.zeros([self.num_hidden]))
            W2 = tf.Variable(tf.truncated_normal([self.num_hidden, self.num_classes],
                                                stddev=np.sqrt(2.0/self.num_hidden)))
            b2 = tf.Variable(tf.zeros([self.num_classes]))

            # 第一次全连接,得到的形状是:[max_timestep, batch_size, num_hidden]
            outputs = [tf.reduce_sum(tf.multiply(t, W1), reduction_indices=1) + b1 for
t in outputs]
            # 第二次全连接,Time Major的形式,[max_timestep, batch_size, num_classes]
            self.logits = [tf.matmul(t, W2) + b2 for t in outputs]
            self.logits = tf.stack(self.logits)

        # 第三种情况(默认),创建动态非栈式BiLSTM,bidirectional_dynamic_rnn,
        # max_timestep可以不确定
        elif self.birnn_type == 2:
            # 将CNN的输出维度调整成[batch_size, feature_w, feature_h, cnn_out_channels]
            self.layer = tf.transpose(self.layer, [0, 2, 1, 3])
            # 在LSTM中,feature_w是时间步(max_timestep),接下来要调整成
            # [batch_size, max_timestep, feature_h * cnn_out_channels]
            # max_timestep不确定,所以传入-1
            self.layer = tf.reshape(self.layer, [self.batch_size, -1,
                                                feature_h * cnn_out_channels])
            print('lstm input shape: {}'.format(self.layer.get_shape().as_list()))

            # 创建前向LSTM,隐节点数是num_hidden,层数lstm_num_layers默认是1
            self.forward_cell = [tf.nn.rnn_cell.LSTMCell(self.num_hidden) for _ in
range(self.lstm_num_layers)]
            # 将前向LSTM组合起来(如果有多层的话)
            self.forward_stack = tf.nn.rnn_cell.MultiRNNCell(self.forward_cell,
                                                            state_is_tuple=True)
```

```python
        # 创建后向LSTM, 隐节点数是num_hidden, 层数lstm_num_layers默认是1
        self.backward_cell = [tf.nn.rnn_cell.LSTMCell(self.num_hidden) for _ in
range(self.lstm_num_layers)]
        # 将后向LSTM组合起来(如果有多层的话)
        self.backward_stack = tf.nn.rnn_cell.MultiRNNCell(self.backward_cell,
                                                          state_is_tuple=True)

        # 创建动态非栈式BiLSTM
        outputs, _ = tf.nn.bidirectional_dynamic_rnn(self.forward_stack,
                                                     self.backward_stack,
                                                     self.layer, dtype=tf.float32)
        # 输出outputs的维度是: [batch_size, max_timestep, 2 * num_hidden]
        # 输出维度同样是2 * num_hidden, 但这次就不拆分成[2, num_hidden]了
        # 不拆分的话只需做一次全连接即可, 下面看看不拆分的做法
        outputs = tf.concat(outputs, axis=2)
        # 调整outputs的维度: [batch_size * max_timestep, 2 * num_hidden]
        outputs = tf.reshape(outputs, [-1, 2 * self.num_hidden])

        W = tf.get_variable(name='W_out',
                            shape=[2 * self.num_hidden, self.num_classes],
                            dtype=tf.float32,
                            initializer=tf.glorot_uniform_initializer())
        b = tf.get_variable(name='b_out',
                            shape=[self.num_classes],
                            dtype=tf.float32,
                            initializer=tf.constant_initializer())
        # 全连接层, 做从2 * num_hidden到num_classes (分类个数) 的映射
        # 维度是[batch_size * max_timestep, num_classes]
        self.logits = tf.matmul(outputs, W) + b

        # 调整全连接层的输出维度: [batch_size, max_timestep, num_classes]
        self.logits = tf.reshape(self.logits, [self.batch_size, -1,
                                               self.num_classes])
        # 再次调整logits的维度为Time Major的形式: [max_timestep, batch_size,
        # num_classes]
        self.logits = tf.transpose(self.logits, (1, 0, 2))
```

```python
        # 第四种情况，创建动态栈式BiLSTM, stack_bidirectional_dynamic_rnn, max_timestep
        # 可以不确定
        elif self.birnn_type == 3:
            # 将CNN的输出维度调整成[batch_size, feature_w, feature_h, cnn_out_channels]
            self.layer = tf.transpose(self.layer, [0, 2, 1, 3])
            # 在LSTM中, feature_w是时间步(max_timestep), 接下来要调整成
            # [batch_size, max_timestep, feature_h * cnn_out_channels]
            # max_timestep不确定, 所以传入-1
            self.layer = tf.reshape(self.layer, [self.batch_size, -1,
                                                feature_h * cnn_out_channels])
            print('lstm input shape: {}'.format(self.layer.get_shape().as_list()))

            # 创建前向LSTM, 隐节点数是num_hidden, 层数lstm_num_layers默认是1
            self.forward_cell = [tf.nn.rnn_cell.LSTMCell(self.num_hidden) for _ in
range(self.lstm_num_layers)]

            # 创建后向LSTM, 隐节点数是num_hidden, 层数lstm_num_layers默认是1
            self.backward_cell = [tf.nn.rnn_cell.LSTMCell(self.num_hidden) for _ in
range(self.lstm_num_layers)]

            # 创建栈式BiLSTM不用组合LSTM, 将上面创建的前向LSTM和后向LSTM直接传入即可
            # 得到的输出维度是: [batch_size, max_timestep, 2 * num_hidden]
            outputs, _, _ = tf.contrib.rnn.stack_bidirectional_dynamic_rnn(
self.forward_cell, self.backward_cell, self.layer, dtype=tf.float32)
            # 调整输出维度: [batch_size * max_timestep, 2 * num_hidden]
            outputs = tf.reshape(outputs, [-1, 2 * self.num_hidden])

            W = tf.get_variable(name='W_out',
                                shape=[2 * self.num_hidden, self.num_classes],
                                dtype=tf.float32,
                                initializer=tf.glorot_uniform_initializer())
            b = tf.get_variable(name='b_out',
                                shape=[self.num_classes],
                                dtype=tf.float32,
                                initializer=tf.constant_initializer())
```

```python
        # 进行一次全连接，做从2 * num_hidden到num_classes（分类个数）的映射
        # 维度是[batch_size * max_timestep, num_classes]
        self.logits = tf.matmul(outputs, W) + b

        # 调整全连接层的输出维度：[batch_size, max_timestep, num_classes]
        self.logits = tf.reshape(self.logits, [self.batch_size, -1,
                                               self.num_classes])
        # 调整logits的维度为Time Major的形式：[max_timestep, batch_size,
        # num_classes]
        self.logits = tf.transpose(self.logits, (1, 0, 2))

# 第五种情况，创建静态栈式BiLSTM, stack_bidirectional_rnn, 要求max_timestep
# 必须是确定的
        elif self.birnn_type == 4 and max_timestep is not None:
            # 将CNN的输出维度调整成[batch_size, feature_w, feature_h, cnn_out_channels]
            self.layer = tf.transpose(self.layer, [0, 2, 1, 3])
            # 在LSTM中，feature_w是时间步(max_timestep)，接下来要调整成
            # [batch_size, max_timestep, feature_h * cnn_out_channels]
            # 此时传入的max_timestep是确定的
            self.layer = tf.reshape(self.layer, [self.batch_size, self.max_timestep,
                                                  feature_h * cnn_out_channels])
            print('lstm input shape: {}'.format(self.layer.get_shape().as_list()))

            # 创建前向LSTM，隐节点数是num_hidden，层数lstm_num_layers默认是1
            self.forward_cell = [tf.nn.rnn_cell.LSTMCell(self.num_hidden) for _ in
range(self.lstm_num_layers)]

            # 创建后向LSTM，隐节点数是num_hidden，层数lstm_num_layers默认是1
            self.backward_cell = [tf.nn.rnn_cell.LSTMCell(self.num_hidden) for _ in
range(self.lstm_num_layers)]

            # 再次调整CNN的输出维度，Time Major的形式，将时间步排在前面
            # layer的维度：[max_timestep, batch_size, feature_h * cnn_out_channels]
            layer = tf.unstack(self.layer, self.max_timestep, axis=1)

            # 创建静态栈式BiLSTM，输出outputs的维度是[max_timestep, batch_size,
```

```
                        # 2 * num_hidden]
            outputs, _, _ = tf.contrib.rnn.stack_bidirectional_rnn(self.forward_cell,
self.backward_cell, layer, dtype=tf.float32)
            # 将输出outputs的维度2 * num_hidden拆开: [max_timestep, batch_size, 2,
            # num_hidden]
            outputs = [tf.reshape(t, [self.batch_size, 2, self.num_hidden]) for t in
outputs]
            # 做两次全连接
            W1 = tf.Variable(tf.truncated_normal([2, self.num_hidden],
                            stddev=np.sqrt(2.0 / (2 * self.num_hidden))))
            b1 = tf.Variable(tf.zeros([self.num_hidden]))
            W2 = tf.Variable(tf.truncated_normal([self.num_hidden, self.num_classes],
                                        stddev=np.sqrt(2.0 / self.
                                        num_hidden)))
            b2 = tf.Variable(tf.zeros([self.num_classes]))

            # 第一次全连接，得到维度: [max_timestep, batch_size, num_hidden]
            outputs = [tf.reduce_sum(tf.multiply(t, W1), reduction_indices=1) + b1 for
t in outputs]
            # 第二次全连接，输出维度是Time Major的形式: [max_timestep, batch_size,
            # num_classes]
            self.logits = [tf.matmul(t, W2) + b2 for t in outputs]
            self.logits = tf.stack(self.logits)
```

8.3.3 CTC 损失

此时我们的网络已经得到最终的 logits 输出了，现在来分析最终调整后的 logits 的维度：[max_timestep, batch_size, num_classes]。这里先假设 batch_size 是 1，这样我们得到的是每个时间步对应的分类结果。

还是以 "哈" 字为例，如图 8.11 所示。

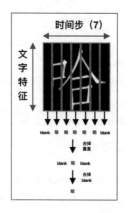

图 8.11 "哈"字例子

假设"哈"字的时间步是 7，logits 输出的维度是 [max_timestep, batch_size, num_classes] = [7, 1, 分类个数]，那么 logits 输出就是每一列汉字特征（就是每一个时间步的特征）对应的分类结果。我们看"哈"字的例子，一共 7 个时间步，第一个和最后一个时间步的分类结果是空（blank），中间的 5 个时间步得到的分类结果都是"哈"，通过跟标签计算 CTC 损失，会去掉重复的和空字符，最终得到了预测结果是"哈"字。当然，这是直观来看的结果，实际上我们使用的时间步的特征不是汉字的图片，而是神经网络的 logits 输出。

CTC 用来解决标签和神经网络的输出不对齐的问题，比如"哈"字的例子，时间步的个数总是大于标签个数的，按照通常的损失计算方法都是一对一的，这种多对一的映射只能靠 CTC 来计算。这样通过计算多对一的 CTC 损失，使得预测结果跟真实标签越来越接近。注意，在计算 CTC 损失时，标签个数不能大于时间步的个数。

通过调用 tf.nn.ctc_loss 函数计算 CTC 损失。需要传入三个参数：标签（labels）、输入（inputs）和序列长度（sequence_length）。

标签要求输入的是稀疏张量（Sparse Tensor），需要将原标签（汉字类别对应的下标，如果是多个汉字就是下标序列）转换成稀疏张量。定义标签稀疏张量如下：

```
self.labels_ = tf.sparse_placeholder(tf.int32, name='labels')
```

稀疏张量是由（indices, values, shape）组成的，indices 表示值所在的下标，比如：indices = np.array([[3, 2], [4, 5]], dtype = np.int64)，表示有两个值，一个在下标 [3, 2]（3 行 2 列）处，一个在 [4, 5]（4 行 5 列）处。values 是具体的值，比如：values = np.array([1.0, 2.0], dtype = np.float32)，说明有两个值分别是 1.0 和 2.0。shape 就是数据本来的形状，比如：shape = np.array([7, 9], dtype = np.int64)，说明数据的形状是 7 行 9 列的。

我们来看完整的代码。

```python
indices = np.array([[3, 2], [4, 5]], dtype=np.int64)
values = np.array([1.0, 2.0], dtype=np.float32)
shape = np.array([7, 9], dtype=np.int64)

# 这样将三个值合在一起就表示一个稀疏张量了
# 这个稀疏张量用来说明有一个7行9列的数据，分别在[3, 2]和[4, 5]的位置上有值1.0和2.0
(indices, values, shape)
```

接下来是 tf.nn.ctc_loss 函数的第二个参数 inputs，这个只要传入神经网络的输出 logits 即可。

第三个参数 sequence_length 需要传入时间步 self.seq_len。注意，这是一个长度为 batch_size 的一维数组，数组内的值是每一个 logits 的时间步，也就是（原图片宽度 / 16）的值。

这样传入三个参数之后，就完成了计算 CTC 损失的任务。

最后来看损失和优化相关的代码实现。

```python
with tf.name_scope("loss"):
    self.global_step = tf.train.get_or_create_global_step()
    # 传入三个参数计算CTC损失
    self.loss = tf.nn.ctc_loss(labels=self.labels_,
                               inputs=self.logits,
                               sequence_length=self.seq_len)
    self.loss = tf.reduce_mean(self.loss)
    tf.summary.scalar('loss', self.loss)
    # 定义指数衰减的学习率
    self.learning_rate = tf.train.exponential_decay(1e-3,
                                                    self.global_step,
                                                    10000,
                                                    0.98,
                                                    staircase=True)
    tf.summary.scalar('learning_rate', self.learning_rate)

    # 使用Adam优化器
    self.optimizer = tf.train.AdamOptimizer(learning_rate=self.learning_rate,
                                            beta1=0.9, beta2=0.999).\
                                            minimize(self.loss,
                                                     global_step=self.global_step)

    # 调用ctc_beam_search_decoder函数得到最有可能的预测结果
```

```
        self.decoded, self.log_prob = tf.nn.ctc_beam_search_decoder(self.logits,
                                        self.seq_len, merge_repeated=False)
        # 将self.decoded的稀疏值转换成非稀疏值，在self.dense_decoded中保存的是预测的
        # 每一个汉字的分类ID
        # 默认值是-1，说明不是任何汉字
        self.dense_decoded = tf.sparse_tensor_to_dense(self.decoded[0],
                                        default_value=-1)
        # 根据预测结果和标签真实结果计算错误率
        self.label_error_rate = tf.reduce_mean(tf.edit_distance(tf.cast(self.decoded
[0], tf.int32), self.labels_))
```

8.3.4 端到端汉字 OCR 的网络训练

经过上一节的代码实现，我们的网络结构就定义完成了，接下来需要生成训练集训练网络。跟 8.2 节介绍的一样，可以使用 pillow 库提供的方法实现使用字体文件生成文字图片。不过，这回我们不再一次生成一个字了，而是每张图片生成多个汉字。之前我们定义了 5 种 LSTM 结构，程序默认使用的是第三种，创建动态非栈式 BiLSTM，bidirectional_dynamic_rnn。所以，理论上汉字个数可以是任意的。为了简化代码，在实现中没有特意做到每张图片的宽度都不一样，而是每个批次（batch）的输入数据的宽度都是一样的，不同批次的图片宽度不一样，并且定义了汉字个数的上下限。

```
# 定义最少汉字个数，4个
self.MIN_LEN = 4
# 定义最多汉字个数，6个。这两个值可以任意调整
self.MAX_LEN = 6
```

1. 生成批量训练数据

我们来看看生成批量训练数据的代码实现。

```
# 定义汉字转ID的字典
self.encode_maps = {}
# 定义ID转汉字的字典
self.decode_maps = {}
# 字符集的循环，创建汉字和ID之间转换的字典
for i, char in enumerate(self.charset, 1):
```

```python
            self.encode_maps[char] = i
            self.decode_maps[i] = char
        # 定义space字符，ID是0，字符是''，加入字典中
        SPACE_INDEX = 0
        SPACE_TOKEN = ''
        self.encode_maps[SPACE_TOKEN] = SPACE_INDEX
        self.decode_maps[SPACE_INDEX] = SPACE_TOKEN

    # 参数imgaug_process用来指定是否做数据增强处理
    def get_batches(self, imgaug_process=False):
        images = []
        labels = []
        # 每一个批次随机选择生成一张图片中汉字的个数
        self.char_nums = random.randint(self.MIN_LEN, self.MAX_LEN)

        while True:
            # 调用generateImg函数，根据字体文件生成一张汉字图片（黑底白字）和汉字序列
            # 比如生成的label是"你好"，im就是"你好"的图片
            im, label = self.generateImg()
            # 判断是否做数据增强处理
            if imgaug_process:
                im = self.imgaug_process(im)
            # 如果图片不是灰度图，则转换成灰度图
            if len(im.shape) > 2:
                im = Image.fromarray(im).convert('L')
                im = np.array(im, dtype=np.uint8)

            # 如果处理过的图片高度不是规定的高度，则裁剪图片使之符合规定的图片高度
            # 这里我们定义的高度是64
            if im.shape[0] > self.img_height:
                im = im[int((im.shape[0] - self.img_height) / 2):int(
                    (im.shape[0] - self.img_height) / 2) + self.img_height, :]
            elif im.shape[0] < self.img_height:
                dst = np.reshape(np.array([0] * (im.shape[1] * int(self.img_height)),
                                          dtype=np.uint8),
                                 [int(self.img_height), im.shape[1]]).astype(np.uint8)
```

```python
            dst[int((self.img_height - im.shape[0]) / 2):int((self.img_height -
                im.shape[0]) / 2) + im.shape[0], :] = im
            im = dst
        # 同上面一样，这里判断处理过的图片宽度是否是固定的宽度，因为每个批次内的
        # 图片宽度都必须一致
        # 图片宽度是汉字个数 × 汉字宽度，因为宽度定义的也是64，所以图片宽度是
        # 汉字个数 × 64。
        # 如果处理过的图片宽度不符合要求，则裁剪图片使之符合规定的图片宽度
        if im.shape[1] > self.char_nums * self.img_width:
            im=im[:, int((im.shape[1] - self.char_nums * self.img_width) / 2):int(
                (im.shape[1] - self.char_nums * self.img_width) / 2) + self.
                char_nums * self.img_width]
        elif im.shape[1] < self.char_nums * self.img_width:
            dst = np.reshape(
                np.array([0] * (self.char_nums * self.img_width * int(self.
                img_height)), dtype=np.uint8),
                [int(self.img_height), self.char_nums * self.img_width]).astype(np.
                uint8)

            dst[:, int((self.char_nums * self.img_width - im.shape[1]) / 2):int(
                (self.char_nums * self.img_width - im.shape[1]) / 2) + im.shape[1]]
                 = im
            im = dst
        # 现在，汉字图片的高度和宽度都符合输入数据的要求了
        # 灰度图只有高度和宽度两个维度，这里再扩展一维
        # 使图片的维度是(高度，宽度，1)
        im=np.expand_dims(im, 2)
        # print(im.shape)
        # cv2.imwrite("./data/train/{}.png".format(label), im)

        # 将汉字图片保存到数组里，这个就是网络的输入[batch_size, height, width, 1]
        images.append(im)
        # 将汉字序列label转换成汉字对应的ID序列，这样label就都是数字了
        # 比如label是"你好"，通过字典得到"你好"两个字对应的ID
        # code的值可能是[22('你'的ID), 45('好'的ID)]，这里的22和45是随便说的，用来
        # 表示两个汉字的ID
```

```
code = [self.encode_maps[c] for c in list(label)]
# 将ID序列保存到数组中作为标签，后面还要把这个标签转换成稀疏张量，以便
# 传给self.labels_
labels.append(code)

# 现在，输入数据和标签都有了，如果数组长度没有达到一个批次，则继续生成
# 图片和标签；如果数组长度达到一个批次，就可以返回这个批次的数据了
if len(images) == self.batch_size:
    # batch_seq_len保存一个批次内每一张图片的时间步，因为一个批次内的图片宽
    # 度都一样，图片宽度是汉字个数 × 汉字宽度64，而时间步是（图片宽度/16）
    # 所以batch_seq_len的值是batch_size个值为（图片宽度/16）的数组
    batch_seq_len = np.asarray([(self.char_nums * self.img_width) // 16] *
        self.batch_size, dtype=np.int32)
    # 图片数组，作为网络的输入数据
    batch_inputs = np.array(images)
    # 将标签转换成稀疏张量，用来传给self.labels_
    batch_labels = self.sparse_tuple_from_label(labels)
    # 返回一个批次内的输入图片、时间步、原始标签（数值）、标签的稀疏张量
    yield batch_inputs, batch_seq_len, np.array(labels), batch_labels
    # 重新随机生成下一个批次的汉字个数
    self.char_nums = random.randint(self.MIN_LEN, self.MAX_LEN)
    # 清空输入图片数组和标签数组，开始下一个批次的生成循环
    images = []
    labels = []
```

生成的训练图片如图8.12所示。

图8.12　训练图片

2. OCR网络的训练

生成的批量数据包含输入图片、时间步、原始标签（数值）和标签的稀疏张量，然后将这些数据传给网络开始训练。

```python
# 参数分别是输入图片、标签的稀疏张量、时间步、原始标签（数值）
def train_step(self, batch_inputs, batch_labels, batch_seq_len, orig_labels):
    # 将输入图片、标签的稀疏张量和时间步传给网络
    feed_dict = {
        self.inputs_: batch_inputs,
        self.labels_: batch_labels,
        self.seq_len: batch_seq_len
    }

    start_time = time.time()
    # 开始训练，得到标签错误率（label_error_rate）、损失（loss）和预测的
    # 汉字ID序列dense_decoded
    _, label_error_rate, loss, global_step, summary, dense_decoded = self.sess.run(
        [self.optimizer, self.label_error_rate, self.loss, self.global_step,
        self.summaries_op, self.dense_decoded], feed_dict=feed_dict)
    self.train_writer.add_summary(summary, global_step)
    # 根据原始标签和预测的汉字序列计算预测准确率
    accuracy = self.accuracy_calculation(orig_labels, dense_decoded,
                                        ignore_value=-1, isPrint=True)

    now = datetime.datetime.now()
    log = "{}/{} {}:{}:{} global_step {}, " \
          "accuracy = {:.3f},train_loss = {:.3f}, " \
          "label_error_rate = {:.3f}, train using time = {:.3f}"
    print(log.format(now.month, now.day, now.hour, now.minute, now.second,
                     global_step, accuracy, loss,
                     label_error_rate, time.time() - start_time))
    if global_step % 5 == 0:
        self.save(self.global_step)

def train():
    # 定义网络模型对象
    model = ocr_network()
    # 获得批量数据生成器
    train_batches = model.get_batches()
    try:
```

```python
        # 训练的循环
        while True:
            # 获得一个批次数据
            batch_inputs, batch_seq_len, orig_labels, batch_labels = next(
                train_batches)
            # 传给网络，执行训练
            model.train_step(batch_inputs,batch_labels,batch_seq_len,orig_labels)
    except KeyboardInterrupt:
        model.save(model.global_step)

if __name__ == '__main__':
    train()
```

训练 10 160 次的打印结果如下：

```
1/10 21:8:36 global_step 10160, accuracy = 0.938,train_loss = 0.479,
label_error_rate = 0.013, train using time = 3.876
Model saved in file: ./models/best_model.ckpt-10160
```

3. 汉字 OCR

当网络训练好之后，就可以使用该网络来识别汉字了，而且不再需要对图片切割每一个汉字，直接将图片作为输入数据传给网络即可。

```python
    # 传入的参数是灰度图
    def forward(self, img):
        # 对灰度图扩展一维，变成[高度，宽度，1]
        im = np.expand_dims(img, 2)
        # 再对图片扩展一维，变成[1,高度,宽度,1]，以满足输入数据的维度[batch_size,
        # height, width, 1]
        # 这样的话，batch_size就是1
        im = np.expand_dims(im, 0)
        # 将图片im传给输入inputs_，并且计算该图片的时间步（图片宽度/16）传给seq_len
        feed_dict = {
            self.inputs_: im,
            self.seq_len: np.asarray([(img.shape[1]) // 16], dtype=np.int32)
        }
        # 运行网络得到预测的汉字ID序列
```

```python
        dense_decoded = self.sess.run([self.dense_decoded], feed_dict=feed_dict)
        # 根据字典，将汉字ID序列转换成汉字字符串，默认值-1表示不是汉字
        decoded_label = [self.decode_maps[j] for j in dense_decoded[0][0] if j !=
            -1]
        # 打印汉字字符串，看看预测的结果
        print("-------- prediction : {} --------".format(decoded_label))


def inference():
    # 定义网络模型对象，batch_size设成1，一次只识别一张图片
    model = ocr_network(batch_size=1)
    # 读取要识别的图片
    im = cv2.imread("./newimg.png", cv2.IMREAD_GRAYSCALE)
    # 如果图片是白底黑字，则可以通过下面代码转换成黑底白字
    # im = ~im
    # 如果图片的高度不符合要求，则可以缩放图片的大小使之符合要求
    # ratio = 64 / im.shape[0]
    # im = cv2.resize(im, None, fx=ratio, fy=ratio, interpolation=cv2.INTER_LINEAR)

    # 如果图片的高度不符合要求，则裁剪图片使之符合要求
    if im.shape[0] > model.img_height:
        im = im[int((im.shape[0] - model.img_height) / 2):int(
            (im.shape[0] - model.img_height) / 2) + model.img_height, :]
    elif im.shape[0] < model.img_height:
        dst = np.reshape(np.array([0] * (im.shape[1] * int(model.img_height)),
                         dtype=np.uint8),
                         [int(model.img_height), im.shape[1]]).astype(np.uint8)
        dst[int((model.img_height - im.shape[0]) / 2):int((model.img_height - im.
            shape[0]) / 2) + im.shape[0], :] = im
        im = dst
    # 将图片传给网络，得到预测的结果
    model.forward(im)


if __name__ == '__main__':
    inference()
```

待识别的图片如图 8.13 所示。

图 8.13 待识别的图片

识别结果如下：

```
(1, 32, ?, 32)
(1, 16, ?, 64)
(1, 8, ?, 128)
(1, 4, ?, 128)
lstm input shape: [1, None, 512]
Successfully loaded: ./models/best_model.ckpt-10160
-------- prediction : ['滚', '滚', '长', '江', '东', '逝', '水', '浪', '花', '淘',
'尽', '英', '雄'] --------
```

8.4 汉字 OCR：基于 TensorFlow 2.0

现在，我们按照上一节的网络结构图，使用 TensorFlow 2.0 来实现。

8.4.1 CNN 的实现

首先是 CNN 的实现，此处只使用 keras 来定义 CNN 的各层。

```
# 定义输入占位符，时间步未知
self.inputs_ = tf.keras.layers.Input(shape=[self.img_height, None, self.channels],
                                     dtype="float32", name='inputs')

    # 卷积层定义函数
    def cnn_layer(self, layer, filter_size):
        # 卷积层，3×3的卷积核，步幅是1
        layer = tf.keras.layers.Conv2D(kernel_size=3, filters=filter_size,
                                       padding='same')(layer)
        # batch_norm层，激活函数是ELU
        layer = tf.keras.layers.BatchNormalization(epsilon=1e-5, fused=True)(layer)
        layer = tf.keras.layers.ELU()(layer)
        # 最大池化层，缩小数据的尺寸
```

```
layer = tf.keras.layers.MaxPool2D(strides=2, padding="same")(layer)

return layer
```

8.4.2 双向 LSTM 的实现

基于 TensorFlow 2.0 构建双向 LSTM 非常简单，使用 `tf.keras.layers.Bidirectional` 函数即可，没有上一节基于 TensorFlow 1.x 实现那么复杂。

Bidirectional 只接收两个参数，第一个参数是 RNN 的实例；第二个参数是网络返回值的组合模式，取值范围是 'sum'、'mul'、'concat'、'ave'、None，默认值是 'concat'，也就是说，前向网络和后向网络的返回值连接在一起作为双向网络的返回值。

举个例子：

`tf.keras.layers.Bidirectional(tf.keras.layers.LSTM(10))`

这样就定义了一个双向 LSTM 层，参数是一个 LSTM（输出维度是 10），Bidirectional 会使用这个 LSTM 作为前向 LSTM 和后向 LSTM 来构建 BiLSTM 层，没有动态、静态、栈式、非栈式这些概念。

那么上一节中介绍的多种 BiLSTM 结构如何实现呢？对于动态 BiLSTM 和静态 BiLSTM 无非就是时间步是否确定这个区别，这不影响 Bidirectional 的构建方式，所以本节主要介绍如何实现栈式 BiLSTM 和非栈式 BiLSTM 结构。

在 TensorFlow 2.0 中，栈式和非栈式的概念跟上一节不同，上一节中栈的概念是用于描述 BiLSTM 的，而本节中栈的概念是用于 RNN 层的，堆叠的目标不一样。

使用 `tf.keras.layers.StackedRNNCells` 函数来堆叠 RNN 单元，举个例子：

```
# 输入占位符
layer = tf.keras.layers.Input((timesteps, input_dim))
# 定义两个LSTM单元
cells = [
        tf.keras.layers.LSTMCell(output_dim),
        tf.keras.layers.LSTMCell(output_dim),
    ]
# 将两个LSTM单元堆叠在一起，作为一个RNN单元（两层LSTM的RNN）
cells_stack = tf.keras.layers.StackedRNNCells(cells)
# 将上面的RNN单元封装成RNN实例，现在就可以把这个RNN实例作为参数传给Bidirectional了
```

```python
rnn_cells_stack = tf.keras.layers.RNN(cells_stack)
# 这样创建的BiLSTM,就是每个前向LSTM和后向LSTM都是由两层LSTM构成的
# 如果要按照上一节中的概念来说的话,这是一个非栈式BiLSTM
outputs = tf.keras.layers.Bidirectional(rnn_cells_stack)(layer)
```

如果要构建上一节中所说的栈式 BiLSTM 该怎么做呢?也很容易,无非就是将上一层的 BiLSTM 的输出作为下一层的 BiLSTM 的输入,循环创建即可,这样就是 BiLSTM 的堆叠了。示例代码如下:

```python
# 输入占位符
layer = tf.keras.layers.Input((timesteps, input_dim))
# 循环次数是要堆叠的个数,表示要构建几层BiLSTM
for _ in range(lstm_num_layers):
    # 定义前向和后向要用的LSTM
    lstm_cell = tf.keras.layers.LSTM(output_dim)
    # 创建BiLSTM,上一层的输出作为这一层的输入,循环堆叠BiLSTM
    layer = tf.keras.layers.Bidirectional(lstm_cell)(layer)
```

要说明的概念就是这些,下面是创建各类 BiLSTM 的代码实现。

```python
        # CNN输出的形状
        _, feature_h, feature_w, cnn_out_channels=self.layer.get_shape().as_list()
        with tf.name_scope('lstm'):
            # 第一种情况,创建一个普通的两层RNN
            if self.birnn_type == 0:
                # 将CNN的输出维度调整成[batch_size, feature_w, feature_h,
                # cnn_out_channels]
                # 注意,因为使用的tf.transpose不是keras的层,所以要用tf.keras.
                # layers.Lambda封装
                self.layer = tf.keras.layers.Lambda(lambda layer: tf.transpose(
                    layer, [0, 2, 1, 3]))(self.layer)
                
                # 在LSTM中,feature_w是时间步(max_timestep),接下来要调整成
                # [batch_size, max_timestep, feature_h * cnn_out_channels],
                # keras的形状可以不明确写出batch_size的维度
                self.layer = tf.keras.layers.Reshape([-1, feature_h *
                    cnn_out_channels])(self.layer)
```

```python
    print('lstm input shape: {}'.format(self.layer.get_shape().
        as_list()))

    # 创建两个LSTM单元，也可以尝试RNNCell或GRUCell
    self.cell1 = tf.keras.layers.LSTMCell(self.num_hidden)
    self.cell2 = tf.keras.layers.LSTMCell(self.num_hidden)

    # 创建RNN网络，输出outputs的维度是[batch_size, max_timestep,
    # self.num_hidden]
    outputs = tf.keras.layers.RNN([self.cell1, self.cell2],
        return_sequences=True)(self.layer)

    # 全连接层的输出，维度是[batch_size, max_timestep, num_classes]
    self.logits = tf.keras.layers.Dense(self.num_classes)(outputs)

    # 最终调整成Time Major的形式：[max_timestep, batch_size,
    # num_classes]
    self.logits = tf.keras.layers.Lambda(lambda layer: tf.transpose(
        layer, [1, 0, 2]))(self.logits)
    # 构建模型
    self.model = tf.keras.Model(inputs=[self.inputs_],
                                outputs=[self.logits])

    self.model.summary()
# 第二种情况，创建静态非栈式BiLSTM，要求max_timestep必须是确定的
elif self.birnn_type == 1 and max_timestep is not None:
    # 将CNN的输出维度调整成[batch_size, feature_w, feature_h,
    # cnn_out_channels]
    self.layer = tf.keras.layers.Lambda(lambda layer: tf.transpose(
        layer, [0, 2, 1, 3]))(self.layer)

    # 在LSTM中，feature_w是时间步(max_timestep)，接下来要调整成
    # [batch_size, max_timestep, feature_h * cnn_out_channels]
    # 此时传入的max_timestep是确定的
    self.layer = tf.keras.layers.Reshape([self.max_timestep,
        feature_h * cnn_out_channels])(self.layer)
```

```python
            print('lstm input shape: {}'.format(self.layer.get_shape().
                as_list()))

            # 创建多个LSTM单元,输出维度是num_hidden,层数lstm_num_layers默认是1
            self.cells = [tf.keras.layers.LSTMCell(self.num_hidden) for _ in
                range(self.lstm_num_layers)]
            # 将多个LSTM单元堆叠起来
            self.cells_stack = tf.keras.layers.StackedRNNCells(self.cells)
            # 将堆叠的多个LSTM单元封装成一个RNN层
            self.rnn_cells_stack = tf.keras.layers.RNN(self.cells_stack,
                                                    return_sequences=True)

            # 调整成Time Major的形式
            # [max_timestep, batch_size, feature_h * cnn_out_channels]
            self.layer = tf.keras.layers.Lambda(lambda layer: tf.transpose(
                layer, [1, 0, 2]))(self.layer)

            # 创建BiLSTM,前向LSTM和后向LSTM都是rnn_cells_stack的结构
            # 输出outputs的维度是: [max_timestep, batch_size, 2 * num_hidden]
            outputs = tf.keras.layers.Bidirectional(self.rnn_cells_stack)
                (self.layer)

            # 全连接层,输出的形状是: [max_timestep, batch_size, num_classes]
            self.logits = tf.keras.layers.Dense(self.num_classes)(outputs)

            # 构建模型
            self.model = tf.keras.Model(inputs=[self.inputs_], outputs=[self.
                logits])

            self.model.summary()
        # 第三种情况(默认),创建动态非栈式BiLSTM, max_timestep可以不确定
        elif self.birnn_type == 2:
            # 将CNN的输出维度调整成[batch_size, feature_w, feature_h,
            # cnn_out_channels]
            self.layer = tf.keras.layers.Lambda(lambda layer: tf.transpose(
```

```python
            layer, [0, 2, 1, 3]))(self.layer)

        # 在LSTM中，feature_w是时间步(max_timestep)，接下来要调整成
        # [batch_size, max_timestep, feature_h * cnn_out_channels]
        # max_timestep不确定，所以传入-1
        self.layer = tf.keras.layers.Reshape([-1,
            feature_h * cnn_out_channels])(self.layer)

        print('lstm input shape: {}'.format(self.layer.get_shape().
            as_list()))
        # 这里跟第二种情况的实现一样，区别只是时间步max_timestep是否确定
        self.cells = [tf.keras.layers.LSTMCell(self.num_hidden) for _ in
range(self.lstm_num_layers)]
        self.cells_stack = tf.keras.layers.StackedRNNCells(self.cells)
        self.rnn_cells_stack = tf.keras.layers.RNN(self.cells_stack,
                                        return_sequences=True)

        # 创建BiLSTM，输出维度是：[batch_size,max_timestep, 2 * num_hidden]
        outputs = tf.keras.layers.Bidirectional(self.rnn_cells_stack)
            (self.layer)

        # 全连接层，输出维度是：[batch_size, max_timestep, num_classes]
        self.logits = tf.keras.layers.Dense(self.num_classes)(outputs)

        # 调整logits维度为Time Major的形式：[max_timestep, batch_size,
        # num_classes]
        self.logits = tf.keras.layers.Lambda(lambda layer: tf.transpose(
            layer, [1, 0, 2]))(self.logits)

        # 构建模型
        self.model = tf.keras.Model(inputs=[self.inputs_],
                                outputs=[self.logits])

        self.model.summary()
        # 第四种情况，创建动态栈式BiLSTM，max_timestep可以不确定
    elif self.birnn_type == 3:
```

```python
# 将CNN的输出维度调整成[batch_size, feature_w, feature_h,
# cnn_out_channels]
self.layer = tf.keras.layers.Lambda(lambda layer: tf.transpose(
    layer, [0, 2, 1, 3]))(self.layer)

# 在LSTM中,feature_w是时间步(max_timestep),接下来要调整成
# [batch_size, max_timestep, feature_h * cnn_out_channels]
# max_timestep不确定,所以传入-1
self.layer = tf.keras.layers.Reshape([-1,
    feature_h * cnn_out_channels])(self.layer)

print('lstm input shape: {}'.format(self.layer.get_shape().
    as_list()))

# 循环堆叠每一个BiLSTM,输出的形状是:[batch_size, max_timestep,
# 2 * num_hidden]
for _ in range(self.lstm_num_layers):
    # 定义前向和后向使用的LSTM
    self.lstm_cell = tf.keras.layers.LSTM(self.num_hidden,
                                          return_sequences=True)
    # 构建BiLSTM,上一层的输出作为这一层的输入,循环堆叠
    self.layer = tf.keras.layers.Bidirectional(self.lstm_cell)
        (self.layer)

# 全连接层,输出的形状是:[batch_size, max_timestep, num_classes]
self.logits = tf.keras.layers.Dense(self.num_classes)(self.layer)

# 调整logits维度为Time Major的形式:[max_timestep, batch_size,
# num_classes]
self.logits = tf.keras.layers.Lambda(lambda layer: tf.transpose(
    layer, [1, 0, 2]))(self.logits)

# 构建模型
self.model = tf.keras.Model(inputs=[self.inputs_],
                            outputs=[self.logits])
```

```python
        self.model.summary()
    # 第五种情况, 创建静态栈式BiLSTM, 跟上一种情况的实现一样, 区别是
    # max_timestep必须是确定的
    elif self.birnn_type == 4 and max_timestep is not None:
        # [batch_size, feature_w, feature_h, cnn_out_channels]
        self.layer = tf.keras.layers.Lambda(lambda layer: tf.transpose(
            layer, [0, 2, 1, 3]))(self.layer)

        # [batch_size, max_timestep, feature_h * cnn_out_channels]
        self.layer = tf.keras.layers.Reshape([self.max_timestep,
            feature_h * cnn_out_channels])(self.layer)

        print('lstm input shape: {}'.format(self.layer.get_shape().
            as_list()))

        # [max_timestep, batch_size, feature_h * cnn_out_channels]
        self.layer = tf.keras.layers.Lambda(lambda layer: tf.transpose(
            layer, [1, 0, 2]))(self.layer)

        # [max_timestep, batch_size, 2 * num_hidden]
        for _ in range(self.lstm_num_layers):
            self.lstm_cell = tf.keras.layers.LSTM(self.num_hidden,
                                                  return_sequences=True)
            self.layer = tf.keras.layers.Bidirectional(self.lstm_cell)\
                (self.layer)

        # [max_timestep, batch_size, num_classes]
        self.logits = tf.keras.layers.Dense(self.num_classes)(self.layer)

        # 构建模型
        self.model = tf.keras.Model(inputs=[self.inputs_],
                                    outputs=[self.logits])

        self.model.summary()
```

8.4.3　OCR 网络的训练

最后是网络的训练。

```python
# 使用Adam优化器
self.optimizer = tf.keras.optimizers.Adam(1e-3)


# 计算CTC损失
def compute_loss(self, labels, logits, seq_len):
    with tf.name_scope("loss"):
        # 注意，TensorFlow 2.0对ctc_loss函数做了改变
        # 如果要计算TensorFlow 1.x形式的CTC损失
        # 需要把参数label_length设成None，把参数blank_index设成-1
        loss = tf.nn.ctc_loss(labels=labels,
                              logits=logits,
                              label_length=None, logit_length=seq_len,
                              blank_index=-1)
        loss = tf.reduce_mean(loss)
    return loss


def compute_metrics(self, labels, logits, seq_len):

    # 调用ctc_beam_search_decoder函数得到最有可能的预测结果
    # 当然，考虑到速度，训练时可以不使用这个函数
    # decoded, log_prob = \
    #     tf.nn.ctc_beam_search_decoder(inputs=logits,
    #                                   sequence_length=seq_len)

    # 考虑到训练速度，使用ctc_greedy_decoder来代替ctc_beam_search_decoder
    # 这个查找的精度高低不会影响训练的效果，所以精度低些也没关系
    decoded, log_prob = tf.nn.ctc_greedy_decoder(logits, seq_len,
                                                 merge_repeated=False)
    # 将decoded的稀疏值转换成非稀疏值，在dense_decoded中保存的是预测的每一个
    # 汉字的分类ID
    # 默认值是-1，说明不是任何汉字
    dense_decoded = tf.sparse.to_dense(decoded[0], default_value=-1)
```

```python
        # 根据预测结果和标签真实结果计算错误率
        label_error_rate = tf.reduce_mean(
            input_tensor=tf.edit_distance(tf.cast(decoded[0], tf.int32), labels))

        return dense_decoded, label_error_rate

    # 单步训练函数
    @tf.function
    def train_step(self, batch_inputs, batch_labels, batch_seq_len, orig_labels):
        # 使用磁带记录计算损失的操作
        with tf.GradientTape() as tape:
            # 调用模型得到模型输出
            logits = self.model(batch_inputs, training=True)
            # 计算CTC损失
            loss = self.compute_loss(batch_labels, logits, batch_seq_len)
        # 使用磁带计算模型参数的梯度
        grads = tape.gradient(loss, self.model.trainable_variables)
        self.optimizer.apply_gradients(zip(grads, self.model.trainable_variables))

        return loss, logits

    # 训练函数，参数分别是输入图片、标签的稀疏张量、时间步、原始标签（数值）
    def training(self, batch_inputs, batch_labels, batch_seq_len,
                 orig_labels):

        start_time = time.time()
        # 单步训练
        loss, logits = self.train_step(batch_inputs, batch_labels, batch_seq_len,
                                        orig_labels)
        # 计算预测的错误率
        dense_decoded, label_error_rate = self.compute_metrics(batch_labels,
            logits, batch_seq_len)
        # 根据原始标签和预测的汉字序列计算预测准确率
        accuracy = self.accuracy_calculation(orig_labels, dense_decoded,
                                              ignore_value=-1, isPrint=True)
```

```python
            # summary_ops_v2.scalar('loss', loss)
            # summary_ops_v2.scalar('label_error_rate', label_error_rate)

            now = datetime.datetime.now()
            log = "{}/{} {}:{}:{} global_step {}, " \
                  "accuracy = {:.3f},train_loss = {:.3f}, " \
                  "label_error_rate = {:.3f}, train using time = {:.3f}"

            print(log.format(now.month, now.day, now.hour, now.minute, now.second,
                             self.optimizer.iterations.numpy(), accuracy, loss,
                             label_error_rate, time.time() - start_time))
            if self.optimizer.iterations.numpy() % 10 == 0:
                # 保存模型
                self.checkpoint.save(self.checkpoint_prefix)


def train():
    # 定义网络模型对象
    model = ocr_network()
    # 获得批量数据生成器
    train_batches = model.get_batches()
    try:
        # 训练的循环
        while True:
            # 获得一个批次数据
            batch_inputs, batch_seq_len, orig_labels, batch_labels = next(
                train_batches)
            # 传给网络,执行训练
            model.training(batch_inputs, batch_labels, batch_seq_len, orig_labels)
    except KeyboardInterrupt:
        model.save()


if __name__ == '__main__':
    train()
```

8.5 本章小结

本章首先讨论了 OCR 的分类网络的实现思路，然后详细介绍了如何构建一个端到端的汉字 OCR 网络，接着介绍了 TensorFlow 1.x 提供的创建 BiLSTM 的 4 个 API 的作用和区别，也介绍了基于 TensorFlow 2.0 创建 BiLSTM 跟 TensorFlow 1.x 的区别，最后通过代码讲解了端到端汉字 OCR 的具体实现。

当然，这只是一个 Demo，实际效果并不完美，留给读者要做的事情还有很多。比如本章的训练只使用了一种字体，并且使用的是一级汉字字符集。如果你想用在自己的项目中，则可以加入更多的字体（在代码中找 font_list），并且扩展字符集（在代码中找 charset）重新训练，比如加入二级汉字、数字、字母和标点符号等。

本次只训练了 10 000 次，训练并不充分，你也可以使用栈式的网络结构训练试试。对于网络结构也有很多地方可以微调，比如 CNN 的结构、CNN 的层数、滤波器的个数、卷积核的参数、隐节点的个数、输入数据的尺寸、批次大小、LSTM 的种类、LSTM 的层数等。

为了得到更好的识别效果，建议做适当的数据增强，包括适当缩小汉字的尺寸，以适应分辨率低的文字图片。关于数据增强的入口，在代码中已经预留了，你只要在 imgaug_process 函数中加入相应的实现即可。

本章实现的汉字识别只是最基本的 OCR 应用场景，还有一些情况是对自然场景下的文字做识别，比如拍摄的街道图片、路牌、商店广告牌等情况。不像文本的扫描图片背景很干净，这类图片背景更加复杂，文字可能五颜六色，甚至形状是扭曲的，这些都给文字识别带来了更大的挑战。如果读者对这类复杂情况下的文字识别感兴趣，则可以继续研究诸如 CTPN、SegLink、EAST 等算法的理论和实现。

9 强化学习：玩转 Flappy Bird 和超级马里奥

9.1 概述

强化学习是一项很有意思的技术，可以通过它来进行训练，比如训练一架飞机学会自动飞行和降落，训练机器人走路、奔跑、跳跃、游泳、驾驶甚至飞翔，训练出下棋的 AI，训练玩各种游戏，等等。DeepMind 发表过一篇论文 *Playing Atari with Deep Reinforcement Learning*（论文地址：https://www.cs.toronto.edu/~vmnih/docs/dqn.pdf），讲述他们使用 DQN（Deep Q-Learning Network）来玩各种 Atari 游戏。OpenAI 的 Gym 提供了各种游戏的环境，用来评估强化学习算法。本章将讲解 DQN 算法，以及训练 DQN 玩 *Flappy Bird*；还会讲解如何使用 OpenAI 的强化学习 Baselines 玩《超级马里奥》，这一切都是通过 OpenAI 的 Baselines 和 Gym 来完成的。

9.2 DQN 算法

Q-Learning 的目的在于维护一张 Q 表（Q-Table），表中记录的是在每一个状态（state）下进行的每一个动作（action）所对应的奖励（reward）。

如表 9.1 所示，Q 表包含若干状态，并且只有两个动作。假如当前在 s_1（state 1）状态下，a_2（action 2）动作的奖励 $Q(s_1, a_2) = 5$，比 a_1 的奖励 $Q(s_1, a_1) = 0.4$ 高，所以应该选择 a_2 动作进

入下一个状态 s_2，假定状态是顺序变化的。然后以同样的方法继续判断，在状态 s_2 下选择奖励较高的动作，一直持续下去。这一切都依赖于 Q 表是准确的。

表 9.1 Q 表示例

	action 1	action 2
state 1	0.4	5
state 2	−3	1.2
state 3	0.9	3
...		
state N		

而在 Q-Learning 一开始，Q 表内的值都是随机的，我们需要不断地迭代尝试，根据环境给出的反馈不断修正 Q 表，使得 Q 表不断逼近真实的 Q 函数。Q 函数就是我们要训练的 $Q(s,a)$。

这里有一个问题，对于离散的状态而言，个数是有限的，上面的 Q 表还可用。一旦状态空间是连续值，就有无穷多个状态，这样上面的 Q 表就不适用了。这时可以引入神经网络来完成 Q 表的任务，那么训练 Q 函数就变成了训练神经网络，使得神经网络不断逼近真实的 Q 函数，如图 9.1 所示。

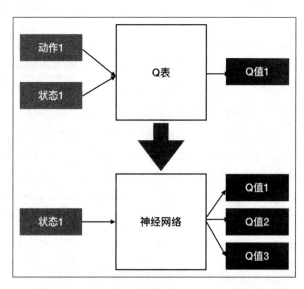

图 9.1 引入神经网络来完成 Q 表的任务

这样只要输入状态给神经网络，网络就会输出在该状态下每一个动作所对应的 Q 值。

我们来看看 DQN 算法，如图 9.2 所示（图片出自 DeepMind 的论文 *Playing Atari with Deep Reinforcement Learning*）。

```
Algorithm 1 Deep Q-learning with Experience Replay
    Initialize replay memory D to capacity N
    Initialize action-value function Q with random weights
    for episode = 1, M do
        Initialise sequence s₁ = {x₁} and preprocessed sequenced φ₁ = φ(s₁)
        for t = 1, T do
            With probability ε select a random action aₜ
            otherwise select aₜ = maxₐ Q*(φ(sₜ), a; θ)
            Execute action aₜ in emulator and observe reward rₜ and image xₜ₊₁
            Set sₜ₊₁ = sₜ, aₜ, xₜ₊₁ and preprocess φₜ₊₁ = φ(sₜ₊₁)
            Store transition (φₜ, aₜ, rₜ, φₜ₊₁) in D
            Sample random minibatch of transitions (φⱼ, aⱼ, rⱼ, φⱼ₊₁) from D
            Set yⱼ = { rⱼ                              for terminal φⱼ₊₁
                      { rⱼ + γ maxₐ' Q(φⱼ₊₁, a'; θ)   for non-terminal φⱼ₊₁
            Perform a gradient descent step on (yⱼ − Q(φⱼ, aⱼ; θ))² according to equation 3
        end for
    end for
```

图 9.2　DQN 算法

由于每个状态按照事件发展的顺序之间的相关性会造成采样分布的不均衡，给强化学习算法带来稳定性问题。需要将这些状态以及在该状态下所做的动作、下一个状态和得到的奖励都保存起来，这个保存区域就是回放内存（Replay Memory）D。在训练时，从内存 D 中随机选取一部分数据进行训练，这样做是为了减少样本之间的连续相关性所带来的分布影响，这就是经验回放（Experience Replay）。

算法解释如下：

（1）初始化回放内存 D，大小是 N。

（2）神经网络作为 Q 函数，初始化网络权重参数。

（3）设定 episode 次数为 M，这个可以理解成 epochs，就是玩多少次游戏。

- 状态 s_1 是特征 x 的集合（就是图像），如果有对图像的预处理操作就做预处理，得到 φ_1，这个其实就是状态 s_1。

- 本次游戏的时间步循环：

    - 此处有两个分支，即探索和利用。探索是以 ϵ 为概率选择一个随机的动作 a_t 的；利用是从神经网络输出的 Q 值选择动作 a_t 的。

    - 执行所选择的动作 a_t，得到执行动作后的奖励 r_t 和下一个状态的画面 x_{t+1}，相当于状态 s_{t+1}。

    - 预处理画面，得到下一个状态 φ_{t+1}，也就是 s_{t+1}。当然，这个是根据网络情况而定的，如果网络输入不需要对画面做任何处理，那么画面就是状态。

- 保存当前时间步 t 的 4 个参数（当前状态 φ_t、所做的动作 a_t、得到的奖励 r_t、下一个状态 φ_{t+1}）到内存 D 中。
- 随机从 D 中采样出小批量样本（$\varphi_j, a_j, r_j, \varphi_{j+1}$）。
- 根据 Bellman 方程计算每一个状态的 Q 值作为目标值 y，用于下一步训练网络。这一步计算有两个分支，如果在状态 φ_{j+1} 游戏结束了，则使用当时的奖励 r_j 作为目标值；如果游戏没有结束，则使用公式计算目标值，$y_j = r_j + \gamma *$ 状态 φ_{j+1} 下动作 a 中最大的 Q 值，也就是说，当前状态的 Q 值 = 奖励 +γ 倍的下一个状态的最大 Q 值，γ 是衰减系数。
- 使用梯度下降更新网络参数，损失是 $(y_j - Q(s_j, a_j))^2$。其中 y_j 是利用公式计算出的真实 Q 值，$Q(s_j, a_j)$ 是网络计算出的 Q 值，优化的目标是让两者尽量接近，这种优化方法是我们经常使用的。

简单点说：

（1）初始化内存和网络参数。

（2）游戏循环：

- 根据探索或利用选择动作 a_t。
- 执行动作 a_t。
- 保存（s_t, a_t, r_t, s_{t+1}）。
- 当保存的样本达到一定批量以后，随机采样数据。
- 根据 Bellman 方程计算 Q 值。
- 使用梯度下降更新网络参数。

1. 关于探索和利用

在选择动作 a_t 的步骤中，分别有探索和利用，那么程序什么时候做探索、什么时候做利用呢？我们知道 DQN 是使用神经网络近似 Q 表的，那么在网络还没有训练好时，网络的参数都是随机的，或者可以理解成 Q 表中的 Q 值都是 0 或都是随机无效的，那么在一开始如何选择动作 a_t 就很关键了。

算法使用 ϵ（epsilon）作为探索的概率，在最开始使概率 ϵ 最大，这样一开始就通过随机选择动作进行了大量的探索。随着网络的逐步训练，逐渐减小概率 ϵ，因为网络估算出的 Q 值越来越准确。

这样做的意义是什么呢？比如在训练的最开始，网络的预测很不可靠，在某个状态下可以选择的动作有很多，那么哪个动作带来的奖励最高还不清楚，所以需要使用探索机制尽量保证在每个状态下每个动作都尝试一下，这样能够覆盖多种组合。当各种尝试都做了以后，因为有了经验，对每种尝试的奖励都已经清楚明白，这时就不再需要探索了，使用过往的经验就能够知道在哪种场合使用哪个动作了。

如果只做了很少的几次探索，没有覆盖所有的可能性会发生什么？比如在状态 1 下有 5 个可能的动作，如果只尝试了其中一个动作，如动作 1，然后更新了相对应的 Q 值，那么对于状态 1 的 Q 值，显示如表 9.2 所示。

表 9.2 探索举例

Q 值	a_1	a_2	a_3	a_4	a_5
s_1	4	0	0	0	0

如果到此为止不再进行探索的话，那么以后只要在状态 1 下查找动作，返回的就永远是动作 1，因为动作 1 的 Q 值最大。但是其他 4 个动作都没有尝试过，怎么知道 a_1 就是最优的动作呢？这就是早期做大量探索的重要性。

当神经网络学到更多的知识后，就可以根据所学知识选择动作了，这就是利用。

2. **关于 Bellman 方程**

在计算目标值 y 时，使用了 Bellman 方程：

$$y = \hat{Q}(s, a) = r + \gamma \max Q(s', a')$$

其中，r 是在当前状态 s 下采取动作 a 的奖励，$\max Q(s', a')$ 是下一个状态 s' 的最大 Q 值，即未来的奖励。γ 是衰减系数，γ 值越小说明越重视眼前的奖励 r，γ 值越大说明越重视未来的奖励。这个公式表示，在每个状态 s 下执行了所选动作 a 后，观察新的状态和奖励，然后根据奖励和新状态的最大 Q 值来更新前一个状态的 Q 值。

3. **关于神经网络**

神经网络的定义，根据要解决的事情来做有针对性的设计，可以是任意网络，比如一个经典的 CNN 网络。网络的输出是多个动作的 Q 值，写作 $Q(s, a)$。要训练的目标就是让网络输出的 Q 值尽量与 Bellman 方程计算出的 Q 值相接近，所以要优化的目标就是 $(\hat{Q}(s, a) - Q(s, a))^2$。

至此，算法部分就解释完了，接下来看看如何实现。

9.3 实现 DQN 玩 Flappy Bird：基于 TensorFlow 1.x

本次的实现来自 GitHub 上的开源代码，是由 yenchenlin 实现的 DeepLearningFlappyBird，源码地址是：https://github.com/yenchenlin/DeepLearningFlappyBird。

我们主要看的是 deep_q_network.py 文件，这里只摘取部分主要代码，完整的代码请读者自行阅读。

首先是构建网络函数 createNetwork，这里构建了一个典型的 CNN 网络，因为使用 CNN 网络观察游戏的画面很合适。关于网络的结构并不建议大家特别去观察，关于设计几个卷积层，以及卷积核的大小、输出的维度等内容还是根据具体的项目情况来定。这里需要重点关注的是网络的输入、输出和优化部分。

```
def createNetwork():
    # 网络参数
    W_conv1 = weight_variable([8, 8, 4, 32])
    b_conv1 = bias_variable([32])

    W_conv2 = weight_variable([4, 4, 32, 64])
    b_conv2 = bias_variable([64])

    W_conv3 = weight_variable([3, 3, 64, 64])
    b_conv3 = bias_variable([64])

    W_fc1 = weight_variable([1600, 512])
    b_fc1 = bias_variable([512])

    W_fc2 = weight_variable([512, ACTIONS])
    b_fc2 = bias_variable([ACTIONS])

    # s是网络输入，就是状态。这是一个80×80×4的输入，为什么要×4呢？看到后面的代码
    # 大家就清楚了。实际上是由4个80×80的画面组成的，传入的是连续4个状态，就是4个
    # 画面，这个特点就根据项目情况大家自行设计
    s = tf.placeholder("float", [None, 80, 80, 4])

    # 一系列卷积层
    h_conv1 = tf.nn.relu(conv2d(s, W_conv1, 4) + b_conv1)
```

```python
        h_pool1 = max_pool_2×2(h_conv1)

        h_conv2 = tf.nn.relu(conv2d(h_pool1, W_conv2, 2) + b_conv2)
        #h_pool2 = max_pool_2×2(h_conv2)

        h_conv3 = tf.nn.relu(conv2d(h_conv2, W_conv3, 1) + b_conv3)
        #h_pool3 = max_pool_2×2(h_conv3)

        #h_pool3_flat = tf.reshape(h_pool3, [-1, 256])
        h_conv3_flat = tf.reshape(h_conv3, [-1, 1600])

        h_fc1 = tf.nn.relu(tf.matmul(h_conv3_flat, W_fc1) + b_fc1)

        # readout就是网络输出的Q值向量
        readout = tf.matmul(h_fc1, W_fc2) + b_fc2

        return s, readout, h_fc1
```

接下来就是最主要的函数trainNetwork。

```python
def trainNetwork(s, readout, h_fc1, sess):
    # a是动作向量，one-hot编码
    a = tf.placeholder("float", [None, ACTIONS])
    # 目标值y是Bellman方程计算出的真实Q值
    y = tf.placeholder("float", [None])
    # readout是上面网络定义的网络输出的Q值向量，跟动作向量a做乘法得到对应动作的Q值
    # 比如，Q值向量是[0.2, 0.4, 0.1]，动作向量是[0, 1, 0]，那么得到的就是0.4
    readout_action = tf.reduce_sum(tf.multiply(readout, a), reduction_indices=1)
    # 让网络输出的Q值与计算出的Q值尽量接近，得到MSE损失
    cost = tf.reduce_mean(tf.square(y - readout_action))
    train_step = tf.train.AdamOptimizer(1e-6).minimize(cost)

    # 先得到游戏的状态对象，用于后面执行游戏动作获取游戏状态
    game_state = game.GameState()

    # 定义回放内存
    D = deque()
```

```python
# log文件
a_file = open("logs_" + GAME + "/readout.txt", 'w')
h_file = open("logs_" + GAME + "/hidden.txt", 'w')

# 设定初始动作
do_nothing = np.zeros(ACTIONS)
do_nothing[0] = 1
# 执行动作得到游戏画面x_t
x_t, r_0, terminal = game_state.frame_step(do_nothing)
# 对画面做预处理
x_t = cv2.cvtColor(cv2.resize(x_t, (80, 80)), cv2.COLOR_BGR2GRAY)
ret, x_t = cv2.threshold(x_t,1,255,cv2.THRESH_BINARY)
# 将画面堆叠成4个，得到游戏状态s_t，大小是80×80×4
# 因为作者将游戏状态设定为连续4个画面，此时是第一个画面，所以就是连续
# 4个同一个画面堆叠在一起
s_t = np.stack((x_t, x_t, x_t, x_t), axis=2)

# 加载网络模型（如果有的话）
saver = tf.train.Saver()
sess.run(tf.initialize_all_variables())
checkpoint = tf.train.get_checkpoint_state("saved_networks")
if checkpoint and checkpoint.model_checkpoint_path:
    saver.restore(sess, checkpoint.model_checkpoint_path)
    print("Successfully loaded:", checkpoint.model_checkpoint_path)
else:
    print("Could not find old network weights")

# 开始训练，游戏开始
# epsilon是探索的概率
epsilon = INITIAL_EPSILON
# t是时间步step
t = 0
while "flappy bird" != "angry bird":
    # 先从网络中得到初始状态下的Q值向量
    readout_t = readout.eval(feed_dict={s : [s_t]})[0]
```

```python
# 定义并初始化动作向量
a_t = np.zeros([ACTIONS])
action_index = 0
# 因为FRAME_PER_ACTION的定义是1，所以此条件永远成立
if t % FRAME_PER_ACTION == 0:
    # 根据概率决定是探索还是利用
    if random.random() <= epsilon:
        # 探索模式，随机选择一个动作
        print("----------Random Action----------")
        action_index = random.randrange(ACTIONS)
        # 设定动作向量，one-hot编码，所以对应的动作下标值是1
        # 其实这有一个问题，上面得到了随机的下标值action_index，但是下面
        # 的赋值却没有使用，而是又随机选择了一次，不过后面没有再使用
        # action_index，所以也无所谓，没有什么影响
        a_t[random.randrange(ACTIONS)] = 1
    else:
        # 利用模式，根据网络输出的最大Q值决定动作
        action_index = np.argmax(readout_t)
        a_t[action_index] = 1
else:
    a_t[0] = 1

# 根据游戏的step次数不断减小随机探索的概率
if epsilon > FINAL_EPSILON and t > OBSERVE:
    epsilon -= (INITIAL_EPSILON - FINAL_EPSILON) / EXPLORE

# 执行选择的动作，得到游戏画面和奖励
x_t1_colored, r_t, terminal = game_state.frame_step(a_t)
# 仍然是预处理游戏画面
x_t1 = cv2.cvtColor(cv2.resize(x_t1_colored, (80, 80)), cv2.COLOR_BGR2GRAY)
ret, x_t1 = cv2.threshold(x_t1, 1, 255, cv2.THRESH_BINARY)
x_t1 = np.reshape(x_t1, (80, 80, 1))
#s_t1 = np.append(x_t1, s_t[:,:,1:], axis = 2)
# 将游戏画面放进状态中
s_t1 = np.append(x_t1, s_t[:, :, :3], axis=2)
```

```python
# 将状态、动作、奖励和下一个状态保存到内存D中
D.append((s_t, a_t, r_t, s_t1, terminal))
# 如果内存D超过了规定的大小，则从D中弹出一组数据
if len(D) > REPLAY_MEMORY:
    D.popleft()

# 当游戏玩了一定时间步之后，开始训练。这个OBSERVE相当于观察次数，为了多
# 尝试，多采样数据
if t > OBSERVE:
    # 从D中随机采样批量数据
    minibatch = random.sample(D, BATCH)

    # 将数据分成4组，分别是状态、动作、奖励和下一个状态
    s_j_batch = [d[0] for d in minibatch]
    a_batch = [d[1] for d in minibatch]
    r_batch = [d[2] for d in minibatch]
    s_j1_batch = [d[3] for d in minibatch]

    y_batch = []
    # 传给网络的下一个状态，得到下一个状态的Q值向量，用在下面的Bellman
    # 方程计算中，公式是y = r + γ * MaxQ(s', a')，MaxQ(s', a')就是下一个
    # 状态的最大Q值
    readout_j1_batch = readout.eval(feed_dict = {s : s_j1_batch})
    for i in range(0, len(minibatch)):
        terminal = minibatch[i][4]
        # 判断游戏是否结束，如果游戏结束，那么只使用奖励作为y值
        if terminal:
            y_batch.append(r_batch[i])
        else:
            # 用Bellman方程计算目标值y
            y_batch.append(r_batch[i]+GAMMA * np.max(readout_j1_batch[i]))

    # 将目标值y、动作a和状态s作为输入数据训练网络
    train_step.run(feed_dict = {
        y : y_batch,
        a : a_batch,
```

```
                s : s_j_batch}
        )

        # 更新状态
        s_t = s_t1
        # 增加时间步
        t += 1

        # 每10000次迭代保存一次网络参数
        if t % 10000 == 0:
            saver.save(sess, 'saved_networks/' + GAME + '-dqn', global_step = t)

        # 打印日志
        state = ""
        if t <= OBSERVE:
            state = "observe"
        elif t > OBSERVE and t <= OBSERVE + EXPLORE:
            state = "explore"
        else:
            state = "train"

        print("TIMESTEP", t, "/ STATE", state, \
            "/ EPSILON", epsilon, "/ ACTION", action_index, "/ REWARD", r_t, \
            "/ Q_MAX %e" % np.max(readout_t))
```

以上就是 DQN 的实现,只要按照算法描述中的步骤实现即可。

读者可以在自己的电脑上运行这个程序,看看实际训练的效果,只要在命令行中输入以下命令即可:

```
python deep_q_network.py
```

有关该项目的更多信息请参考原作者的项目链接。

9.4 实现 DQN 玩 Flappy Bird:基于 TensorFlow 2.0

本节的代码地址是:https://github.com/chengstone/DeepLearningFlappyBird,文件名是 deep_q_network_tf2.py。

首先实现构建网络的函数。

```python
def createNetwork():

    # s是网络输入,就是状态。这是一个80×80×4的输入,由4个80×80的画面组成
    # 传入的是连续4个状态,就是4个画面,这个特点就根据项目情况大家自行设计了
    s = tf.keras.layers.Input(shape=(80, 80, 4), dtype='float32')

    # 构建网络结构
    h_conv1 = tf.keras.layers.Conv2D(filters=32, kernel_size=8, strides=4,
                                     padding='same', activation="relu")(s)
    h_pool1 = tf.keras.layers.MaxPool2D(strides=2, padding='same')(h_conv1)

    h_conv2 = tf.keras.layers.Conv2D(filters=64, kernel_size=4, strides=2,
                                     padding='same', activation="relu")(h_pool1)

    h_conv3 = tf.keras.layers.Conv2D(filters=64, kernel_size=3, strides=1,
                                     padding='same', activation="relu")(h_conv2)

    h_conv3_flat = tf.keras.layers.Flatten()(h_conv3)

    h_fc1 = tf.keras.layers.Dense(units=512, activation='relu')(h_conv3_flat)

    # readout就是网络输出的Q值向量
    readout = tf.keras.layers.Dense(units=ACTIONS)(h_fc1)

    # 构建模型
    model = tf.keras.Model(
        inputs=[s],
        outputs=[readout])

    model.summary()

    # 返回模型
    return model
```

现在开始训练智能体。

```python
# 计算损失，这里跟原代码是一样的
def compute_loss(readout, a, y):
    readout_action = tf.reduce_sum(tf.multiply(readout, a), axis=1)
    cost = tf.reduce_mean(tf.square(tf.convert_to_tensor(y, dtype="float32") -
    readout_action))

    return cost

# 单步训练函数
@tf.function
def train_step(model, minibatch, optimizer):
    # 使用磁带记录计算损失的操作
    with tf.GradientTape() as tape:
        # 将数据分成4组，分别是状态、动作、奖励和下一个状态
        s_j_batch = [d[0] for d in minibatch]
        a_batch = [d[1] for d in minibatch]
        r_batch = [d[2] for d in minibatch]
        s_j1_batch = [d[3] for d in minibatch]

        y_batch = []
        s_j1_batch = tf.stack(s_j1_batch)
        # 传给网络的下一个状态，得到下一个状态的Q值向量，用在下面的Bellman方程计算
        # 中。公式是y = r + γ * MaxQ(s', a')，MaxQ(s', a')就是下一个状态的最大Q值
        readout_j1_batch = model(s_j1_batch, training=True)
        for i in range(0, len(minibatch)):
            terminal = minibatch[i][4]
            # 判断游戏是否结束，如果游戏结束，那么只使用奖励作为y值
            if terminal:
                y_batch.append(r_batch[i])
            else:
                # 用Bellman方程计算目标值y
                y_batch.append(r_batch[i] +
                            GAMMA * tf.keras.backend.max(readout_j1_batch[i]))

        s_j_batch = tf.stack(s_j_batch)
        # 调用模型，得到输出的Q值向量
```

```python
        readout = model(s_j_batch, training=True)
        # 计算损失
        loss = compute_loss(readout, a_batch, y_batch)
    # 使用磁带计算模型参数的梯度
    grads = tape.gradient(loss, model.trainable_variables)
    optimizer.apply_gradients(zip(grads, model.trainable_variables))
    return loss

# 网络训练函数
def trainNetwork(model):
    # Adam优化器
    optimizer = tf.keras.optimizers.Adam(1e-6)
    MODEL_DIR = "./saved_networks"
    # 如果模型目录不存在，则创建
    if tf.io.gfile.exists(MODEL_DIR):
        pass
    else:
        tf.io.gfile.makedirs(MODEL_DIR)

    train_dir = os.path.join(MODEL_DIR, 'summaries', 'train')
    test_dir = os.path.join(MODEL_DIR, 'summaries', 'eval')

    # 定义日志记录器，可以不使用
    train_summary_writer = summary_ops_v2.create_file_writer(train_dir,
flush_millis=10000)
    test_summary_writer = summary_ops_v2.create_file_writer(test_dir,
flush_millis=10000, name='test')

    # 定义保存模型参数检查点
    checkpoint_dir = os.path.join(MODEL_DIR, 'checkpoints')
    checkpoint_prefix = os.path.join(checkpoint_dir, 'ckpt')

    checkpoint = tf.train.Checkpoint(model=model, optimizer=optimizer)

    # 加载存在的参数
    checkpoint.restore(tf.train.latest_checkpoint(checkpoint_dir))
```

```python
# 先得到游戏的状态对象，用于后面执行游戏动作获取游戏状态
game_state = game.GameState()

# 定义replay memory
D = deque()

# log文件
a_file = open("logs_" + GAME + "/readout.txt", 'w')
h_file = open("logs_" + GAME + "/hidden.txt", 'w')

# 设定初始动作
do_nothing = np.zeros(ACTIONS)
do_nothing[0] = 1
# 执行动作得到游戏画面x_t
x_t, r_0, terminal = game_state.frame_step(do_nothing)
# 对画面做预处理
x_t = cv2.cvtColor(cv2.resize(x_t, (80, 80)), cv2.COLOR_BGR2GRAY)

ret, x_t = cv2.threshold(x_t, 1, 255, cv2.THRESH_BINARY)
# 将画面堆叠成4个，得到游戏状态s_t，大小是80×80×4
# 因为作者将游戏状态设定为连续4个画面，此时是第一个画面，所以就是连续4个同
# 一个画面堆叠在一起
s_t = np.stack((x_t, x_t, x_t, x_t), axis=2)
s_t = np.expand_dims(s_t, 0)

# 开始训练，游戏开始
# epsilon是探索的概率
epsilon = INITIAL_EPSILON
t = 0
while "flappy bird" != "angry bird":
    # 先从网络中得到初始状态下的Q值向量
    readout_t = model(s_t.astype(np.float32), training=True)
    # 定义并初始化动作向量
    a_t = np.zeros([ACTIONS])
    action_index = 0
```

```python
# 因为FRAME_PER_ACTION的定义是1，所以此条件永远成立
if t % FRAME_PER_ACTION == 0:
    # 根据概率决定是探索还是利用
    if random.random() <= epsilon:
        # 探索模式，随机选择一个动作
        print("----------Random Action----------")
        action_index = random.randrange(ACTIONS)
        # 设定动作向量，one-hot编码，所以对应的动作下标值是1
        a_t[random.randrange(ACTIONS)] = 1
    else:
        # 利用模式，根据网络输出的最大Q值决定动作
        action_index = np.argmax(readout_t)
        a_t[action_index] = 1
else:
    a_t[0] = 1

# 根据游戏的step次数不断减小随机探索的概率
if epsilon > FINAL_EPSILON and t > OBSERVE:
    epsilon -= (INITIAL_EPSILON - FINAL_EPSILON) / EXPLORE

# 执行选择的动作，得到游戏画面和奖励
x_t1_colored, r_t, terminal = game_state.frame_step(a_t)
# 仍然是预处理游戏画面
x_t1 = cv2.cvtColor(cv2.resize(x_t1_colored, (80, 80)), cv2.COLOR_BGR2GRAY)
ret, x_t1 = cv2.threshold(x_t1, 1, 255, cv2.THRESH_BINARY)
x_t1 = np.reshape(x_t1, (1, 80, 80, 1))
# 将游戏画面放进状态中
s_t1 = np.append(x_t1, s_t[:, :, :, :3], axis=3)
# 将状态、动作、奖励和下一个状态保存到内存D中
D.append((s_t[0].astype(np.float32), a_t, r_t, s_t1[0].astype(np.float32),
    terminal))
# 如果内存D超过了规定的大小，则从D中弹出一组数据
if len(D) > REPLAY_MEMORY:
    D.popleft()

# 当游戏玩了一定时间步之后，开始训练。这个OBSERVE相当于观察次数，为了
```

```python
            # 多尝试，多采样数据
            if t > OBSERVE:
                # 从D中随机采样批量数据
                minibatch = random.sample(D, BATCH)
                # 单步训练
                train_step(model, minibatch, optimizer)

            # 更新状态
            s_t = s_t1
            # 增加时间步
            t += 1

            # 每10000次迭代保存一次网络参数
            if t % 10000 == 0:
                checkpoint.save(checkpoint_prefix)

            # 打印日志
            state = ""
            if t <= OBSERVE:
                state = "observe"
            elif t > OBSERVE and t <= OBSERVE + EXPLORE:
                state = "explore"
            else:
                state = "train"

            print("TIMESTEP", t, "/ STATE", state, \
                "/ EPSILON", epsilon, "/ ACTION", action_index, "/ REWARD", r_t, \
                "/ Q_MAX %e" % np.max(readout_t))

def playGame():
    # 构建网络模型
    model = createNetwork()
    # 开始训练
    trainNetwork(model)

def main():
```

```
    playGame()

if __name__ == "__main__":
    main()
```

9.5 使用 OpenAI Baselines 玩超级马里奥

OpenAI 开源的 Baselines 项目包含了很多强化学习算法的实现，其中也包含了 DQN 算法的实现。本节以训练智能体玩《超级马里奥》为例，介绍如何使用 OpenAI Baselines 实现的强化学习算法 deepq（DQN）。

9.5.1 Gym

在使用 Baselines 之前，先介绍 OpenAI Gym。强化学习的训练包含两个要素，一个是环境（即外部世界，在本例中就是指游戏世界）；另一个就是智能体 Agent（也就是我们编写的强化学习算法，是要训练的对象）。智能体可以向环境发送动作，环境执行动作并返回观察（observation）和奖励（reward）。看到这些大家应该不会陌生吧，在上一节中训练 DQN 智能体玩 *Flappy Bird* 就是这个过程，其中 DQN 算法就是智能体，game_state 就是环境，game_state 环境返回的游戏画面就是观察。OpenAI 开源的 Gym 即是一个环境系统，可以用来模拟各种游戏的环境。只要遵循 Gym 的协议，我们也可以实现自定义游戏的 Gym 环境，这样就可以使用强化学习算法来训练自定义游戏的智能体了。当然，游戏只是一个例子，我们可以将强化学习算法用于游戏之外的其他场景中。

接下来介绍 Gym 的安装和使用。

Gym 的安装很简单，在命令行输入以下命令即可：

```
pip install gym
```

或者执行下面命令从源码构建安装：

```
git clone https://github.com/openai/gym
cd gym
pip install -e .
```

先来运行一个简单的环境，体验 Gym 的使用。

新建一个 Python 文件，输入如下代码，对代码的解释请看注释。

```python
# 首先引入gym包
import gym
# 创建Gym环境，其中'CartPole-v0'是环境id。CartPole是环境名，v0是版本号
# 当我们自定义环境时也要定义相应的环境id，然后同样使用gym.make创建环境
# 可以通过改变环境id来使用不同的环境(游戏)
env = gym.make('CartPole-v0')
# 重置环境的状态，相当于环境初始化
env.reset()
# 游戏的时间步循环
for _ in range(1000):
    # 绘制一帧环境（游戏）画面
    env.render()
    # 通过环境执行一个动作，这里是在动作空间中随机选择一个动作的
    env.step(env.action_space.sample())
```

至于 'CartPole-v0' 所对应的是什么游戏并不重要，大家运行以后能够看到游戏环境是如何运行的。

在上面的示例代码中，环境的重置函数 env.reset() 会返回初始观察，也就是画面；env.step（action）用来通过环境执行一个动作，该动作执行后会返回环境对游戏的观察、奖励、游戏结束标志（done）和调试信息（info）。如果 done 的值是 True，则说明游戏结束了。智能体和环境的游戏循环如图 9.3 所示（该图出自 OpenAI 的 Gym 说明文档，地址：https://gym.openai.com/docs/）。

图 9.3　智能体和环境的游戏循环

更正规一点的代码写法如下，同样来自 OpenAI 的 Gym 说明文档。

```python
import gym
env = gym.make('CartPole-v0')
# 20次游戏循环
for i_episode in range(20):
    # 通过重置环境得到初始观察
```

```python
observation = env.reset()
# 每次游戏执行100个时间步
for t in range(100):
    env.render()
    print(observation)
    action = env.action_space.sample()
    # 执行随机动作，得到环境返回的状态
    observation, reward, done, info = env.step(action)
    # 如果游戏结束，则退出本次游戏
    if done:
        print("Episode finished after {} timesteps".format(t+1))
        break
```

Gym 的用法大体就是这样的，所以本节要训练 DQN 智能体，首先要有 Gym 环境。

9.5.2 自定义 Gym 环境

除 Gym 预先提供的环境之外，我们可以根据自己的任务自定义 Gym 环境，用来训练自定义任务的智能体。接下来利用 Gym 提供的环境来介绍如何自定义 Gym 环境。

在命令行输入以下命令：

```
python -c "import gym; print(gym)"
```

会打印出类似于下面的内容：

```
<module 'gym' from '/Applications/anaconda/envs/deeplearning/lib/python3.5/site-packages/gym/__init__.py'>
```

这样就能够找到安装的 gym 所在的文件路径了，在 gym 路径下能够看到一个 env 文件夹，这里面就是安装在 gym 下的所有环境了。本节我们以上一节中使用的 CartPole 环境为例进行讲解。

首先打开 gym/env/ 文件夹下的 __init__.py 文件，能够看到里面都是由 register 括起来的代码，这些代码的作用是将环境注册到 gym 中。找到 CartPole 注册的地方：

```
# Classic
# ----------------------------------------

register(
    id='CartPole-v0',
```

```
    entry_point='gym.envs.classic_control:CartPoleEnv',
    max_episode_steps=200,
    reward_threshold=195.0,
)

register(
    id='CartPole-v1',
    entry_point='gym.envs.classic_control:CartPoleEnv',
    max_episode_steps=500,
    reward_threshold=475.0,
)
```

对于注册的每一个环境，id 用来标识唯一的环境，在训练当中通过 id 找到要使用的环境。在训练代码中，使用 gym.make(环境id) 来创建一个环境实例，例如：gym.make('CartPole-v0')。

entry_point 给出环境类所在的位置，例如：'gym.envs.classic_control:CartPoleEnv'，说明环境类 CartPoleEnv 的位置在 gym/envs/classic_control 文件夹下。大家可以找找看，在 gym/envs/classic_control 文件夹下有一个名为 cartpole.py 的文件，里面定义了类 CartPoleEnv。

max_episode_steps 是可选参数。当训练智能体的循环始终没有结束时，也就是游戏还没有结束，当循环迭代次数达到 max_episode_steps 时，结束本次游戏。每次游戏从开始到结束就是一个 episode。max_episode_steps 就是定义每个 episode 的最大 step 数。

reward_threshold 是奖励门槛，可以理解成在训练智能体时，每次游戏从开始到结束会有一个累计的奖励值，如果奖励值小于奖励门槛，则说明这次训练得并不好。也就是说，人为指定一个数值，当奖励值大于这个数值时才认为智能体是合格的。如果把每次游戏的累计奖励值看作是考试的分数的话，那么奖励门槛就是及格分数线。这是一个可选参数。

以上是在 __init__.py 文件中注册一个定义好的环境，接下来看看环境类是如何实现的. 打开 gym/envs/classic_control/cartpole.py 文件。

还记得上一节中我们使用 CartPole 环境的几个方法吗？有 env.reset()、env.render() 和 env.step，这几个函数是我们自定义环境时必须要实现的。

```
class CartPoleEnv(gym.Env):

    metadata = {
        'render.modes': ['human', 'rgb_array'],
        'video.frames_per_second' : 50
```

}

首先看到在 CartPoleEnv 类中定义了 metadata，其中 render.modes 定义的是给 render() 函数使用的 mode 种类，因为 render() 函数是用来渲染一帧画面的，通常人类观察画面时就使用 human 模式，因为人要一帧一帧画面慢速描画才能看清动画，如果速度太快，就什么也看不清了。在调用 render() 时就可以传入 mode 参数：env.render(mode="human")。

接下来是初始化函数。

```python
def __init__(self):
    self.gravity = 9.8
    self.masscart = 1.0
    self.masspole = 0.1
    self.total_mass = (self.masspole + self.masscart)
    self.length = 0.5
    self.polemass_length = (self.masspole * self.length)
    self.force_mag = 10.0
    self.tau = 0.02
    self.kinematics_integrator = 'euler'

    self.theta_threshold_radians = 12 * 2 * math.pi / 360
    self.x_threshold = 2.4

    high = np.array([
        self.x_threshold * 2,
        np.finfo(np.float32).max,
        self.theta_threshold_radians * 2,
        np.finfo(np.float32).max])

    self.action_space = spaces.Discrete(2)
    self.observation_space = spaces.Box(-high, high, dtype=np.float32)

    self.seed()
    self.viewer = None
    self.state = None

    self.steps_beyond_done = None
```

除开始定义的一些跟要模拟的游戏算法相关的变量之外，还有一些重要的变量是每个环境都需要的。

`self.action_space = spaces.Discrete(2)` 定义本次游戏的动作空间，包含可用的动作个数，在这里定义了两个离散的动作，在代码中也有注释说明了 CartPole 的两个动作分别是向左推（0）和向右推（1）。

`observation_space` 定义环境状态的观察空间，这个观察空间的定义根据特定的任务来决定取值范围和形状。大家可以回忆一下前面章节中的代码，环境每次调用 step 执行一个动作之后，都会返回 observation 作为环境的状态。所以这里定义的观察空间可以理解成定义状态。比如在 CartPole 中定义的是 `spaces.Box(-high, high, dtype=np.float32)`，这样定义的 observation 最小值就是 –high，最大值是 high。high 在上面已经定义了，形状是 4 个元素的数组 (4,)。这样就定义了状态空间的形状和取值范围。

假如我们要定义的状态是画面，可以这样定义：`self.observation_space = spaces.Box(low=0, high=255, shape=(640, 720, 3), dtype=np.uint8)`。这样就定义了一个形状是 640×720×3（高×宽、3 通道）的画面。

`self.viewer = None` 定义观察器对象，在后面渲染画面时要用到。

`self.state = None` 定义状态变量，用来保存环境的状态。

其他变量就不解释了，根据要实现的算法自行定义即可。

`self.seed()` 调用 seed 函数，初始化随机种子。固定写法，在你自己定义的环境类实现中，可以直接复制下面的 seed 函数。

```
def seed(self, seed=None):
    self.np_random, seed = seeding.np_random(seed)
    return [seed]
```

接着是 reset 函数，其作用是重置环境的状态，这个可以根据自己的具体情况来实现。对于 CartPole 来说，设置状态为初始的状态：`self.np_random.uniform(low=-0.05, high=0.05, size=(4,))`。

```
def reset(self):
    self.state = self.np_random.uniform(low=-0.05, high=0.05, size=(4,))
    self.steps_beyond_done = None
    return np.array(self.state)
```

最重要的就是 step 和 render 函数。

对于 step 函数，我们要实现执行一个动作，并且返回执行该动作后的状态、奖励和游戏是否结束。这段代码 CartPole 的实现这里就不贴出来了，它怎么实现的并不重要，这个需要根据实际情况来实现。

render 函数通过定义 rendering.Viewer 对象来显示画面。同样，这个函数的实现也不是固定的，下面给出另一种实现的例子，大家可以参照多种 render 函数的实现来体会如何在自己的特定函数中使用。当然，你还可以查看 rendering 提供的其他函数，或者看看其他环境类是如何实现 render 的。

```python
# 本函数跟CartPole无关，给大家提供另一种render函数的实现思路
# 函数默认的模式是人类观察模式
def render(self, mode='human'):
    # 这里的state状态是画面
    img = self.state
    if mode == 'rgb_array':
        return img
    elif mode == 'human':
        from gym.envs.classic_control import rendering
        if self.viewer is None:
            # 对于人类观察模式，如果观察器对象是None，则定义一个
            # SimpleImageViewer对象
            self.viewer = rendering.SimpleImageViewer()
        # 使用观察器显示画面
        self.viewer.imshow(img)
        return self.viewer.isopen
```

最后是关闭函数，只关闭观察器即可。

```python
def close(self):
    if self.viewer:
        self.viewer.close()
        self.viewer = None
```

代码到这里，一个环境的实现就完成了。总结，如果读者想自己实现某个场景（或者游戏）的环境类，那么要实现的代码格式如下：

```python
class CustomizeEnv(gym.Env):
    metadata = {
        'render.modes': ['human', 'rgb_array'],
```

```python
        'video.frames_per_second': 2
    }

    def __init__(self):
        self.action_space = None
        self.observation_space = None

    def step(self, action):
        return self.state, reward, done, {}

    def reset(self):
        return self.state

    def render(self, self, mode='human'):
        return None

    def close(self):
        return None
```

9.5.3 使用 Baselines 训练

现在环境有了，我们来看看如何使用 Baselines 提供的强化学习算法训练智能体。在 `baselines/deepq/experiments` 文件夹下，有事先提供的如何使用 Baseline 训练的代码，其中 `train_cartpole.py` 是使用 deepq 训练 CartPole 智能体，`enjoy_cartpole.py` 是玩 CartPole 游戏。同理，其他文件也是一样的，但有一点要注意，其中 `train_pong.py` 训练的是 Atari 游戏，所以需要安装 Atari 的 Gym，可以在终端输入下面命令安装：

```
pip install gym[atari]
```

我们分别看看 `train_cartpole.py` 和 `train_pong.py` 是如何使用 deepq 训练的。

下面是 `train_cartpole.py` 的代码，代码说明见注释。

```python
# 引入gym
import gym
# 引入deepq，下面要使用deepq进行训练
from baselines import deepq
```

```python
# 在deepq的训练函数中，有一个参数可以传入callback，在callback中可以得到训练的情况
# lcl（单词是locals）是一个字典变量，保存了很多数据，可以了解训练的情况。lcl字典的
# 所有key如下，这里就不挨个解释了，如果读者对某个key对应的值感兴趣，则可以尝试打印
# 出来看看
# dict_keys(['episode_rewards', 'act', 'obs', 'model_file', 'load_path',
# 'make_obs_ph', 'param_noise', 'new_obs', 'num_episodes', 'exploration',
# 'batch_size', 'td', 'prioritized_replay_eps', 'target_network_update_freq',
# 'act_params', 'prioritized_replay_beta0', 'q_func', 'action', 'network',
# 'exploration_final_eps', 'callback', 'update_eps', 'done', 'lr', 'rew',
# 'network_kwargs', 'seed', 'saved_mean_reward', 'buffer_size',
# 'prioritized_replay', 'replay_buffer', 'sess', 'update_target',
# 'learning_starts', 'total_timesteps', 'model_saved', 'train',
# 'prioritized_replay_alpha', 'checkpoint_path', 'print_freq', 'train_freq',
# '_', 't', 'observation_space', 'exploration_fraction', 'env', 'debug',
# 'update_param_noise_threshold', 'reset', 'checkpoint_freq', 'env_action',
# 'beta_schedule', 'mean_100ep_reward', 'kwargs', 'gamma',
# 'prioritized_replay_beta_iters'])
def callback(lcl, _glb):
    # 如果奖励值超过199就停止训练
    is_solved = lcl['t'] > 100 \
                and sum(lcl['episode_rewards'][-101:-1]) / 100 >= 199
    return is_solved


def main():
    # 首先创建CartPole环境
    env = gym.make("CartPole-v0")
    # 使用deepq进行训练，非常简单，只要调用deepq的learn函数即可
    # 实际上，learn函数能够接收的参数非常多，对于deepq的learn函数的使用，主要
    # 就是调整这些参数
    # 首先传入的是环境env
    # network这里选择的是mlp，就是多层感知机。network参数传入的网络名称必须是
    # baselines.common.models中定义的网络。打开models.py文件，能够看到定义
    # 的网络有mlp、cnn、cnn_small、lstm、cnn_lstm、cnn_lnlstm和conv_only
    # exploration_fraction和exploration_final_eps用来设定探索的参数
    # 每个step都会调用callback，如果callback返回true，则结束训练
```

```python
    act = deepq.learn(
        env,
        network='mlp',
        lr=1e-3,
        total_timesteps=100000,
        buffer_size=50000,
        exploration_fraction=0.1,
        exploration_final_eps=0.02,
        print_freq=10,
        callback=callback
    )
    # 训练完成,保存模型参数
    print("Saving model to cartpole_model.pkl")
    act.save("cartpole_model.pkl")

if __name__ == '__main__':
    main()
```

再来看看 train_pong.py 的代码,代码解释见注释。

```python
from baselines import deepq
from baselines import bench
from baselines import logger
from baselines.common.atari_wrappers import make_atari


def main():
    logger.configure()
    # 这次稍微有些不同,Atari的环境使用make_atari函数来创建
    env = make_atari('PongNoFrameskip-v4')
    # 创建一个监视器,用来将训练情况打印到log文件中,具体看Monitor函数的实现
    env = bench.Monitor(env, logger.get_dir())
    # 下面的函数很重要,大家在训练自定义的环境时,经常会用到这种形式的函数
    # 这是对环境env加了一层包装器,我们可以对环境env加任意多个自定义的包装器
    # 每个包装器负责做不同的事情。关于包装器的解释,稍后进行说明
    env = deepq.wrap_atari_dqn(env)

    # 然后还是调用deepq的learn函数进行训练,不同于训练CartPole,这次传入的network
```

```python
    # 是"conv_only"
    # convs用来定义卷积层的数量和卷积参数
    # dueling设为True，使用dueling DQN，详细内容大家可以看看
    # https://arxiv.org/abs/1511.06581
    # 其他参数应该都容易理解，就不解释了
    model = deepq.learn(
        env,
        "conv_only",
        convs=[(32, 8, 4), (64, 4, 2), (64, 3, 1)],
        hiddens=[256],
        dueling=True,
        lr=1e-4,
        total_timesteps=int(1e7),
        buffer_size=10000,
        exploration_fraction=0.1,
        exploration_final_eps=0.01,
        train_freq=4,
        learning_starts=10000,
        target_network_update_freq=1000,
        gamma=0.99,
    )

    # 训练结束，保存模型参数
    model.save('pong_model.pkl')
    env.close()

if __name__ == '__main__':
    main()
```

现在以 env = deepq.wrap_atari_dqn(env) 为例进行说明，希望大家能够领会包装器的作用。在 wrap_atari_dqn 函数中，根据不同的条件对环境 env 加了很多包装器，这里只介绍其中两个，一个是 env = WarpFrame(env)；另一个是 env = ClipRewardEnv(env)。

先看看 WarpFrame 的实现。

```
# WarpFrame继承了gym.ObservationWrapper，说明这是一个观察包装器
# 可以重载的方法有step、reset和observation，对于观察包装器的主要作用就是对观察的
# 状态做处理
```

```python
class WarpFrame(gym.ObservationWrapper):
    def __init__(self, env, width=84, height=84, grayscale=True):
        # 初始化函数要做的事情需要根据实际情况来决定，在这里是重新定义了观察空间
        # 也就是说，无论使用的环境env是怎样定义观察空间的，你都可以自定义一个观察
        # 包装器再次对观察结果做处理
        gym.ObservationWrapper.__init__(self, env)
        self.width = width
        self.height = height
        self.grayscale = grayscale
        if self.grayscale:
            self.observation_space = spaces.Box(low=0, high=255,
                shape=(self.height, self.width, 1), dtype=np.uint8)
        else:
            self.observation_space = spaces.Box(low=0, high=255,
                shape=(self.height, self.width, 3), dtype=np.uint8)
    # 重载observation函数，参数frame就是环境env自己的观察结果，然后传给我们的包装器
    # 做自定义处理
    # 这里的处理就是将观察的画面重新调整大小并返回，根据条件也可以将3通道的RGB图像
    # 转换成灰度图
    def observation(self, frame):
        if self.grayscale:
            frame = cv2.cvtColor(frame, cv2.COLOR_RGB2GRAY)
        frame = cv2.resize(frame, (self.width, self.height), interpolation=cv2.
        INTER_AREA)
        if self.grayscale:
            frame = np.expand_dims(frame, -1)
        return frame
```

至此，对于观察包装器的作用大家也了解了，你可以根据自己的需求任意实现观察包装器。

再看看 ClipRewardEnv 的实现。

```python
# 通过继承的父类能够看出，这是一个奖励包装器，其作用是处理环境Env的奖励
# 也就是说，对环境Env返回的奖励做第二次处理
class ClipRewardEnv(gym.RewardWrapper):
    def __init__(self, env):
        gym.RewardWrapper.__init__(self, env)
    # 重载reward函数，参数reward就是环境env返回的奖励，在这里进行第二次处理，将奖励
```

```python
    # 值转成{+1, 0, -1}
    def reward(self, reward):
        return np.sign(reward)
```

还有一个可以继承的包装器是gym.ActionWrapper，其作用是对动作空间做处理，这里仅提供一个小例子。

```python
class MyActionSpaceWrapper(gym.ActionWrapper):
    # 假设环境env的动作空间是11个离散量，这里定义了一个字典，用于将动作id转换成序列
    # 笔者特意定义了不同的序列值，这里需要根据实际情况来定义，可以是one-hot编码，也
    # 可以不是，比如这里的10号动作
    mapping = {
        0: [1, 0, 0, 0, 0, 0, 0, 0, 0, 0],  # NOOP
        1: [0, 1, 0, 0, 0, 0, 0, 0, 0, 0],  # Up
        2: [0, 0, 1, 0, 0, 0, 0, 0, 0, 0],  # Down
        3: [0, 0, 0, 1, 0, 0, 0, 0, 0, 0],  # Left
        4: [0, 0, 0, 0, 1, 0, 0, 0, 0, 0],  # Right
        5: [0, 0, 0, 0, 0, 1, 0, 0, 0, 0],  # A
        6: [0, 0, 0, 0, 0, 0, 1, 0, 0, 0],  # B
        7: [0, 0, 0, 0, 0, 0, 0, 1, 0, 0],  # Left + A
        8: [0, 0, 0, 0, 0, 0, 0, 0, 1, 0],  # Right + A
        9: [0, 0, 0, 0, 0, 0, 0, 0, 0, 1],  # A + B
        10: [1, 1, 1, 1, 1, 1],  # Select + A + B
    }

    def __init__(self, env):
        gym.ActionWrapper.__init__(self, env)
        self.action_space = spaces.Discrete(11)

    # 重载action方法，将环境env传来的action转换成序列值返回
    def action(self, action):
        return self.mapping.get(action)

    # 重载reverse_action方法，传入的参数是序列，返回对应的id。也就是说，原本环境env
    # 的动作空间是离散的，而我们的包装器将离散值转换成了序列值，在将动作传给环境
    # 时，必须将动作的序列值再次转换成离散值才行
    def reverse_action(self, action):
```

```
        for k in self.mapping.keys():
            if(self.mapping[k] == action):
                return k
        return 0
```

可以在命令行输入以下命令训练 CartPole，运行上面讲解的 train_cartpole.py 代码：

```
python -m baselines.deepq.experiments.train_cartpole
```

通过本节的讲解，我们已经可以使用 Baselines 训练智能体了，当训练好后，就可以加载所保存的模型参数玩游戏了。

9.5.4 使用训练好的智能体玩游戏

当运行 train_cartpole 训练完成后，可以在命令行输入以下命令加载模型参数玩 CartPole 游戏：

```
python -m baselines.deepq.experiments.enjoy_cartpole
```

我们来看看 enjoy_cartpole.py 是如何实现的。

```python
import gym

from baselines import deepq


def main():
    # 创建CartPole环境
    env = gym.make("CartPole-v0")
    # 加载训练好的模型参数cartpole_model.pkl
    act = deepq.learn(env, network='mlp', total_timesteps=0,
                      load_path="cartpole_model.pkl")

    # 开始游戏循环
    while True:
        # 重置游戏
        obs, done = env.reset(), False
        episode_rew = 0
        # 一局游戏的循环，直到游戏结束
        while not done:
```

```
            # 显示一帧游戏画面
            env.render()
            # 执行动作，这个动作是训练好的模型选择出来的
            obs, rew, done, _ = env.step(act(obs[None])[0])
            # 累计奖励值
            episode_rew += rew
        # 游戏结束，打印总奖励值
        print("Episode reward", episode_rew)

if __name__ == '__main__':
    main()
```

对于训练好的智能体，只要加载模型参数即可，是不是很容易?!

9.5.5 开始训练马里奥游戏智能体

通过前面几节的介绍，我们已经知道如果要训练任意场景（或游戏）的智能体，需要有相应场景的环境。对于《超级马里奥》游戏，我们可以尝试自己实现一个游戏环境类。这里就不实现了，而是给大家介绍一个开源的马里奥环境实现和开源的马里奥训练实现。不过，因为这两个开源实现使用的 Baselines 和 Gym 版本比较旧，因此需要做一些修改才能成功运行起来。接下来逐步介绍如何训练马里奥智能体。

首先安装马里奥环境，执行如下命令安装：

```
pip install git+https://github.com/chris-chris/gym-super-mario
```

当安装好后，执行以下命令查看环境安装的位置：

```
python -c "import ppaquette_gym_super_mario;print(ppaquette_gym_super_mario)"
```

通常位于 /本地的路径/lib/python3.5/site-packages/ppaquette_gym_super_mario/。找这个路径的原因是需要修改两个文件。

第一个文件是 `__init__.py`。在该文件中引入了 gym 已经删除的函数：

```
from gym.scoreboard.registration import add_task, add_group
```

所以这句代码要注释掉。然后在后面代码中调用 add_task 和 add_group 的地方也都要注释掉。代码这里就不贴出来了，删除代码还是比较好改的。

第二个文件是 `nes_env.py`，其第 219 行有这样一句代码：

```
args.extend(['>log/fceux.stdout.log', '2>log/fceux.stderr.log', '&'])
```

这句代码指定了要输出的 log 文件位置，但是笔者在 Mac OS 和 Ubuntu 系统上运行都在这句代码上失败了。修改方法是将 log/fceux.stdout.log 和 log/fceux.stderr.log 的路径修改成本地已经存在的位置，如果不存在就自己创建，并且将文件也一起创建好，这样路径和文件都有了，程序就会找到 log 文件所在的位置。

笔者的修改如下：

```
args.extend(['>/tmp/log/fceux.stdout.log', '2>/tmp/log/fceux.stderr.log', '&'])
```

现在马里奥的游戏环境就准备好了。接下来下载马里奥训练代码，项目地址是：https://github.com/chris-chris/mario-rl-tutorial.git。

当下载好后，还要做一些修改，因为代码中使用的 Baselines 版本较旧。当然，你也可以使用旧版本的 Baselines，这样就不用修改代码了，这里介绍修改代码使用新版本的 Baselines。

（1）下载 OpenAI 的 Baselines 代码，项目地址是：https://github.com/openai/baselines.git。

（2）将 Baselines 文件夹中的同名文件夹（也叫 baselines）复制到 mario-rl-tutorial 文件夹（就是马里奥训练项目）内。

（3）mario-rl-tutorial 文件夹内有 acktr 和 deepq 文件夹，这两个文件夹其实来自 baselines，大家打开 baselines 文件夹就能看到。但是 mario-rl-tutorial 文件夹内的这两个版本很旧，就不要了，删除 acktr 和 deepq 文件夹，然后将 baselines 文件夹中的 acktr 和 deepq 复制到 mario-rl-tutorial 文件夹内即可，这样 mario-rl-tutorial 项目使用的就是最新版本的算法了。这么说有些绕，简单点说，就是用 baselines 文件夹内的 acktr 和 deepq 替换 mario-rl-tutorial 文件夹内原来的文件。但其实我们是不使用 acktr 的，一会还要删除代码中对 acktr 的使用。替换文件如图 9.4 所示。

（4）修改训练代码（train.py）。现在 Baselines 算法的版本已经是新的了，还要修改在训练代码中使用的旧代码，替换成新的调用方法。首先要修改的是对 acktr 的引用，所有跟 acktr 有关的代码都要注释掉，我们只使用 deepq。当然，有兴趣的话，你也可以尝试修改代码使 acktr 的代码也能正常运行。要注释掉的代码不多，只要注释掉 train_acktr 函数和 main 函数最后两句对 acktr 的 if 判断，以及调用 train_acktr 函数的地方即可。接下来要修改 train_dqn 函数，因为最新的 deepq 的 learn 函数发生了变化，不接收 q_func 作为参数了。所以在代码中定义的 model = cnn_to_mlp 用不上了，就是下面这段代码：

```
model = cnn_to_mlp(
    convs=[(32, 8, 4), (64, 4, 2), (64, 3, 1)],
    hiddens=[256],
    dueling=FLAGS.dueling
)
```

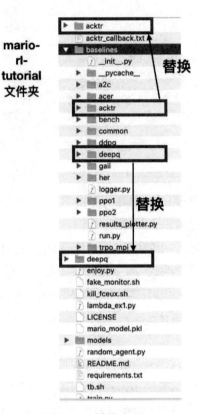

图 9.4　替换文件

但是，这里面对卷积层的设置、隐节点数和 dueling 参数都是可以传给 learn 函数的。修改调用 learn 函数，把这几个参数传给 learn，修改后的代码如下：

```
act = deepq.learn(
    env,
    network="conv_only",
    convs=[(32, 8, 4), (64, 4, 2), (64, 3, 1)],
    hiddens=[256],
    dueling=FLAGS.dueling,
    # q_func=model,
    lr=FLAGS.lr,
    # max_timesteps=FLAGS.timesteps,
    buffer_size=10000,
    exploration_fraction=FLAGS.exploration_fraction,
    exploration_final_eps=0.01,
```

```
    train_freq=4,
    learning_starts=10000,
    target_network_update_freq=1000,
    gamma=0.99,
    prioritized_replay=FLAGS.prioritized,
    callback=deepq_callback
)
```

实际上就是注释掉了 q_func 和 max_timesteps 参数，并且加入了 network、convs、hiddens 和 dueling 参数。convs、hiddens 和 dueling 不用多说了，直接使用 cnn_to_mlp 的即可。新加入的 network 参数，用来指定使用的网络是卷积网络。

（5）修改 wrappers.py 文件。将代码中的 _observation、_action 和 _reverse_action 三个函数的下画线 "_" 前缀删除，变成 observation、action 和 reverse_action。如果不修改的话，运行代码时终端会输出找不到重载函数的警告。然后修改 reverse_action 函数，将倒数第二句代码 return self.mapping[k] 修改成 return k。修改后的 reverse_action 函数如下：

```
def reverse_action(self, action):
    for k in self.mapping.keys():
        if(self.mapping[k] == action):
            return k   # self.mapping[k]
    return 0
```

问题很明显，当 self.mapping[k] == action 时，返回 self.mapping[k]，怎么看都很"违和"。

代码修改到这里，已经可以运行了，在终端输入 python train.py 命令，训练就开始了。在运行过程中，有可能会提示某些软件包没有安装，按照提示自行安装所缺少的软件包即可。训练马里奥智能体的画面如图 9.5 所示。

这样经过漫长的训练之后，会将模型参数保存到 models/deepq 文件夹下。原项目中提供了一个模型参数文件 mario_reward_930.6.pkl，但是因为 Baselines 版本的关系，我们是不能使用的。

在训练好之后，就可以调用 enjoy.py 加载所保存的模型参数玩《超级马里奥》游戏了。同样，对于这个文件也需要做些修改。

首先，将跟 acktr 有关的代码都注释掉。

然后，将 flags.DEFINE_string("file", "mario_reward_930.6.pkl", "Trained model file to use.") 这句代码中的模型参数文件名修改成本地的文件名。

图 9.5 训练马里奥智能体的画面

act = deepq.load("models/deepq/%s" % FLAGS.file) 这句代码已经不能使用了，修改成如下代码：

```
act = deepq.learn(
    env,
    network="conv_only",
    convs=[(32, 8, 4), (64, 4, 2), (64, 3, 1)],
    hiddens=[256],
    load_path="models/deepq/%s" % FLAGS.file,
)
```

接下来，还是调用 deepq.learn 函数，只不过训练用的参数可以不再使用了，传入网络名和与网络相关的参数，并且传入模型参数路径即可。在终端输入 python enjoy.py 命令，便可以开始玩《超级马里奥》游戏了。在过程中如果提示缺少软件包，自行安装即可。

另外，如果提示 "ModuleNotFoundError: No module named 'dill.dill'" 错误，则需要修改 /本地路径/lib/python3.6/pickle.py 文件，在 __import__(module, level=0) 代码前加入 dill.dill 的判断。如果 module == "dill.dill"，则将 module 修改成 "dill._dill"。修改的代码如下：

```
def find_class(self, module, name):
    # Subclasses may override this.
    if self.proto < 3 and self.fix_imports:
        if (module, name) in _compat_pickle.NAME_MAPPING:
```

```
            module, name = _compat_pickle.NAME_MAPPING[(module, name)]
        elif module in _compat_pickle.IMPORT_MAPPING:
            module = _compat_pickle.IMPORT_MAPPING[module]
    if module == "dill.dill":
        module="dill._dill"
    __import__(module, level=0)
    if self.proto >= 4:
        return _getattribute(sys.modules[module], name)[0]
    else:
        return getattr(sys.modules[module], name)
```

这样训练马里奥的例子就介绍完了。其实训练什么游戏并不是重点，希望本节介绍的例子能够让大家熟悉 Baselines 的使用，以及了解自定义 Gym 环境的方法。

9.6 具有好奇心的强化学习算法

除本章介绍的 DQN 算法之外，OpenAI 还提出一个理论，让智能体具有好奇心。这源于 DQN 这类算法的缺点——都是由奖励驱动的。做得好就给奖励，做得不好就给惩罚，就好比马戏团里面训练动物一样，训练的智能体只学会我们想让它学的，没有思想和灵魂。而且这些奖励天生就带有误差在里面，因为都是人为设计的，一旦智能体遇到之前没有遇到过的情况，人类也没有设计相应的奖励，那么智能体的泛化性也不好。最终训练出的智能体只是一味地追求高奖励，没有真正掌握环境的其他要素，并不算真正地会玩某个游戏（假如训练的是游戏）。

比如某个游戏，杀怪会得到高奖励，这样就相当于鼓励智能体不断地杀怪，而没有真正掌握如何通关，甚至永远也不会通关。结果智能体会卡在某个不断杀怪的地方，循环下去，这样的智能体肯定不是我们想要的。

再比如《超级马里奥》这个游戏，是含有隐藏房间的，甚至有各种各样的隐藏要素，这些要素对于 DQN 类的算法是很难发现的。因此，赋予智能体好奇心，就像孩子们一样，带着好奇心不断地去发现游戏中未发现的内容，而不仅仅是追求高奖励，这样的智能体不是更像人一样吗？

那么对算法的改进就是，不仅由奖励驱动，更要鼓励智能体去探索游戏世界。如何鼓励呢？当然是给予智能体探索的行为以更高的奖励，如果某个行为不是探索，那么奖励自然就会很低。这样一来为了追求高奖励，智能体自然就愿意去进行探索了。比如哪些房间可以进，哪些地方有隐藏物品等，没试过当然不知道。OpenAI 提出了一种算法叫作随机网络蒸馏（Random Network Distillation，RND），方法是对于所做的动作事先预测结果，如果动作的结果非常容易预测，则说明这是智能体已经掌握的知识，不算探索，给予低奖励；如果结果很难预测，则说明是未知的情况，这样的动作给予高奖励。像这样把探索的奖励也融入算法当中，算法就提供了两种奖励，一种是我们都熟悉的设计好的奖励，比如游戏分数；另一种就是探索的奖励了。这样训练好的智能体，已经在很多游戏中达到甚至超越了人类的水平，详情可以阅读 OpenAI 发表的博客 *Reinforcement Learning with Prediction-Based Rewards*，地址是：https://blog.openai.com/reinforcement-learning-with-prediction-based-rewards/。开源项目地址是：https://github.com/openai/random-network-distillation。

9.7　本章小结

本章讲解了强化学习算法 DQN 的理论和实现，并且通过例子介绍了 Gym 和 Baselines 的使用方法，相信读者已经可以使用本章介绍的内容训练自己的智能体了。最后介绍了 OpenAI 关于强化学习的最新研究，感兴趣的读者可以阅读与智能体好奇心相关的论文和代码，看看 OpenAI 是如何实现的。

10 生成对抗网络实践：人脸生成

10.1 概述

自从 2014 年 Ian Goodfellow 等人发表了生成对抗网络（Generative Adversarial Networks，GAN）的论文 *Generative Adversarial Networks*（论文地址：https://arxiv.org/abs/1406.2661）之后，GAN 引起了业界的广泛关注，关于 GAN 的越来越多的理论和实践被人们研究出来。在 GitHub 上的 the-gan-zoo 项目中，列出了迄今为止多达 500 个 GAN 相关模型，而且还在持续增加中。那么 GAN 都有哪些用途呢？目前 GAN 的大部分应用都与图像有关，可以用来生成逼真的图像。比如 StackGAN，以一段文字作为输入，可以生成文字内容描述的高清图像。再比如 CycleGAN，可以将输入的一幅图像转换成另一种风格的图像。

理论上，GAN 可以生成任何你想要生成的东西，当然前提是你要设计并训练好模型。比如可以尝试用 GAN 来生成语音或者文本。再比如，既然可以将马转换成斑马，那么是不是可以将图像中任何区域变成其他图像，如人脸的转换，将指定的人脸转换成其他人脸；衣服的转换，将西服变成 T 恤；情绪表情的转换，将严肃的表情变成笑脸。可以说 GAN 是一个非常强大的工具，其更多的应用场景，还需要靠大家的灵感和想象力。

怎么样？是否感受到了 GAN 的强大魅力？本章将分别讲解 GAN、DCGAN、WGAN、WGAN-GP 的理论和实现，最后会简单介绍 PG-GAN 和 TL-GAN，我们一起来看吧。

10.2 GAN

GAN 结构

GAN（生成对抗网络）的全名叫作 Generative Adversarial Networks，是一种生成模型，由生成器（Generator，称作 G）和判别器（Discriminator，称作 D）组成。以生成图像为例，生成器 G 以一组随机噪声向量作为输入，经过生成器处理之后输出一幅逼真的图像。在之前介绍的例子中，所有输出的图像都是由训练好的 G 生成的。判别器 D 是一个二分类的模型，用来判断输入的图像是真实的还是伪造的。

判别器 D 的作用是用来指导 G 如何生成真实图像的，因为 G 不是监督学习，并不会给定一幅图像和一个标签告诉 G 生成的图像是什么（比如给定一幅狗的图像，然后标签是狗）。对于生成器 G 来说，输出是不固定的，在训练时我们给予大量狗的图像让它去学习，训练好后 G 会生成各种不同的狗的图像。

对于真实的图像，这些图像数据的概率分布我们称之为真实分布，判别器 D 就是指导让生成器 G 的输出尽量接近真实图像的概率分布。或者说判别器 D 是高维空间中的二分类超平面（想象一下二维空间的线性分类器），在超平面的一边是真实数据，训练 G 的过程就是逐步让 G 生成的图像在真实数据的一边，即生成真实图像的概率最大化。

判别器 D 输出的是 0~1 之间的值，即输入图像是否是真实图像的概率，我们的最终目标是要 G 生成的图像使判别器 D 输出接近 1 的概率。GAN 的训练其实有些难，如果判别器训练得特别好，那么无论 G 如何生成，都会被识别为假图像，这就意味着 G 生成图像的概率分布与真实分布南辕北辙。因为判别器训练得太好了，将不会提供给生成器梯度来学习如何调整生成真实的图像。而如果判别器训练得不好，又无法提供给生成器足够的信息去学习如何生成真实的图像，因为判别器自己都不知道真实图像的概率分布是什么，无法起到指导生成器的作用。只有在判别器训练得不好又不坏的情况下，因为生成假图像会被判别器识别出来，这就逼迫 G 努力生成更逼真的图像，使得判别器难分真假，这样就可以认为 G 生成的是真实的图像，说明生成图像的概率分布接近真实分布，也就达到训练目的了。生成器和判别器之间的这种对抗和博弈，就是生成与对抗名字的由来。GAN 的结构如图 10.1 所示。

GAN 最主要的两个组件就是 G 和 D，你可以使用任何方式去实现，G（或者 D）可以是一个神经网络或者由多个网络组成，也可以不是神经网络，甚至可以加入标签进行监督学习，这要看你想怎么实现了。很多 GAN 的变种就在于采用何种方式去实现 G 和 D，并且对损失函数做各种演变。

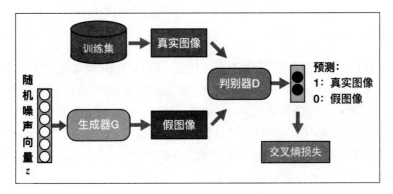

图 10.1　GAN 的结构

我们的任务是训练 G 和 D。G 接收一组低维随机噪声向量 z 作为输入，然后将低维向量映射成高维数据生成出来，即 G 的输出这里称之为假图像（以生成图像为例）。判别器 D 接收任意图像，可以是来自训练集的真实图像，也可以是 G 生成的假图像，并且做出预测，判断输入的图像是真还是假，这样的 D 就是一个二分类分类器，所以损失函数使用交叉熵损失。因为真实图像和假图像我们事先是知道的，此时训练的 D 可以被看成是监督学习。训练的 D 由两个损失组成，一个是对输入的真实图像的交叉熵损失（让 D 的预测接近标签值 1，训练 D 准确识别出真实图像）；另一个是对输入的假图像的交叉熵损失（让 D 的预测接近标签值 0，训练 D 准确识别出假图像）。即训练 D 的预测在真实图像上是 1，在假图像上是 0。

生成器 G 的训练是让 G 生成的图像达到以假乱真的程度，所以对于输入的假图像，让 D 认为是真实图像。这样 G 的损失就是对输入给 D 的假图像，让 D 的预测接近标签值 1 的交叉熵损失，训练让 G 生成的图像被 D 识别成真实图像。

10.3　DCGAN

上一节说过 GAN 由 G 和 D 组成，可以使用任何方式来实现 G 和 D。假如我们的任务是让 G 学会生成真实的彩色图像，比如人脸，那么就需要使用大量人脸数据让 GAN 去学习。让网络能够看懂图像最好的方式就是使用卷积网络，这样使用卷积网络实现的 G 和 D 就叫作 DCGAN（Deep Convolutional GAN）。DCGAN 的论文地址为：https://arxiv.org/pdf/1511.06434.pdf。

根据使用的数据集不同，可能需要改变真实图像的输入占位符的维度，比如下面这样，有图像的宽度、高度和深度。

```
inputs_real = tf.placeholder(tf.float32, (None, image_width, image_height,
image_channels), name='input_real')
```

10.3.1 生成器

上一节介绍的 GAN 网络通常使用的是全连接层，而 DCGAN 要使用卷积层来实现。生成器仍然从一组低维随机噪声向量 z 生成高维图像，在之前的章节中我们使用卷积层都是从浅而宽的层（比如 768×1024×3）逐步变成深而窄的层（比如 4×4×256）的。而生成器需要上采样（Upsample）将比如长度为 100 的向量 z 通过卷积层逐步变成（假设是）768×1024×3 的图像，这样的转换跟之前使用的卷积相反，是从窄而深的层变成宽而浅的层的，我们可以使用转置卷积（Transposed Convolution）来实现。

转置卷积跟之前使用的卷积基本概念都一样，只不过过程是相反的。通过转置卷积生成的新层大小取决于所使用的步幅，假如使用相同填充（Same Padding），步幅为 2，那么经过转置卷积后，输出的层将是输入层的 2 倍。也就是说，对输入层每移动 1 个像素，则在输出层移动 2 个像素。这样就完成了上采样，过程如图 10.2 所示（该图出自 DCGAN 论文）。

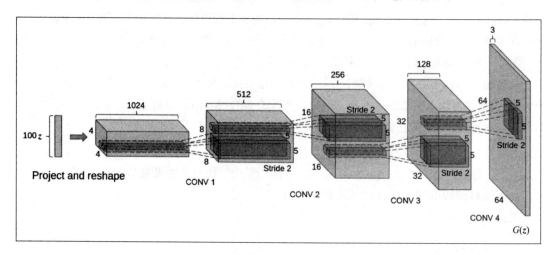

图 10.2　DCGAN 的生成器上采样过程

注意，论文中最终生成的是 64×64×3 的图像，我们的生成器实现要根据实际情况来调整。在 DCGAN 中，我们要在每一层做批次归一化，网络的最后一层除外，这样会加快网络的训练速度。

隐层的激活函数使用 Leaky ReLU，它跟 ReLU 的区别在于负数端的梯度可以继续传递。在 TensorFlow1.x 中可以使用 tf.nn.leaky_relu 函数，参数 alpha 用来设置负数的输出幅度。这样的话，当 x 是正数时，Leaky ReLU 的输出仍然是 x；当 x 是负数时，Leaky ReLU 的输出是 alpha * x。

在 DCGAN 的生成器中，可以使用 ReLU 作为激活函数。研究发现，生成器的输出使用 tanh 效果最好。

10.3.2 判别器

判别器是一个卷积网络，其网络结构跟生成器相反，是逐步进行下采样的。而 DCGAN 的下采样不使用最大池化层，而是使用卷积层将步幅设成 2 来完成。每一层仍然是 [转置卷积 + 批次归一化 +Leaky ReLU] 的结构，第一层和最后一层除外。判别器是二分类，使用 sigmoid 作为输出。

训练 DCGAN 过程中生成的 MNIST 手写数字如图 10.3 所示。

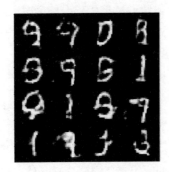

图 10.3　DCGAN 生成的手写数字

10.4　WGAN

如果你试着训练 DCGAN 生成图像，则会发现若不在一段时间内打印出生成器生成的图像，就不知道当前网络训练到什么程度，生成器和判别器的损失也无法指示当前网络的训练情况。另外，你在网上可以搜到很多训练 GAN 的 trick，这是因为原始 GAN 的设计问题会导致训练不稳定、梯度消失、模式崩溃（Collapse Mode）（多样性差）等，所以训练好 GAN 是较困难的。

WGAN（Wasserstein GAN）的出现就是为了解决 GAN 的设计问题的，感兴趣的读者可以看看 WGAN 的第一篇论文 *Towards Principled Methods for Training Generative Adversarial Networks*，该论文从理论上分析了 GAN 的问题点，论文地址是：https://arxiv.org/abs/1701.04862。

总结，判别器既不能训练得太好，也不能训练得太差。如果判别器训练得太好，则生成器梯度会消失。因为生成器的梯度是靠判别器提供的，如果判别器达到了最优，对于真实图像给出的概率是常数 1，生成图像给出的概率是常数 0，这样生成器不会得到任何梯度，造成梯度消失。而如果判别器训练得不好，则生成器的梯度会不准确，导致生成器不收敛，也就学不到要生成图像的特征，即生成器不能生成真实的样本。比如判别器还没有训练好，此时如果生成器生成的假图像并不是很真实，一旦判别器给出了认为是真实图像的预测，就会导致生成器认为自己生成的图像是正确的，那么生成器就会按照这个假图像的模式继续生成，然后判别器仍然会给出正确的评价。结果就是生成器永远不会学到生成真实图像的特征，判别器和生成器都不会训练成功。或者生成器能够生成一些看起来较真实的图像，然后又尝试生成风格不太一样但也属于真实的图像，此时如果判别器没有训练好的话，就会将该图像预测成假图像，并将惩罚梯度输入给生成器。这样生成器为了避免惩罚，则不会再尝试生成多样性的图像了，而是不断生成重复的图像，这就导致了模式崩溃。

这就是训练 GAN 不稳定的直观解释，更理论性的解释大家可以参考论文原文。

这一切都是由于 GAN 的损失函数设计得不够合理导致的，于是 WGAN 的作者推出了第二篇论文 *Wasserstein GAN*，给出了 WGAN 的实现算法，并且提出了新的损失函数，论文地址是：https://arxiv.org/abs/1701.07875。

论文的意思是说原 GAN 的交叉熵损失（JS 散度）不适合衡量生成样本和真实样本之间的距离，因为生成样本和真实样本在高维空间中可能根本不相交（重叠），JS 散度无法反映出两者间的距离，也不能提供有效的梯度，这样就会使得 GAN 不知道该如何优化损失。所以有必要使用新的损失函数代替交叉熵损失，或者换一种方式来衡量生成样本和真实样本之间的距离。

于是，WGAN 引入了 Wasserstein 距离这个概念，其优点在于即使两个样本在高维空间中没有重叠，也可以衡量两者间的距离，并且可以提供有效梯度。但是使用 Wasserteion 距离需要满足连续性条件，称作 Lipschitz 连续。作者将权重限制（截断）在一个范围内，比如 $[-0.01, 0.01]$，以满足 Lipschitz 连续条件。

网络抛弃交叉熵损失，改用 Wasserstein 距离，这样判别器的输出就不再使用 Sigmoid 了。判别器的变化就是删除最后一层的 Sigmoid，由原来的二分类任务变成了 Wasserstein 距离的回归任务，判别器网络的 logits 输出就是 Wasserstein 距离。这样生成器的优化目标就是要最小化 Wasserstein 距离，即让生成样本和真实样本之间的距离最小。每训练一轮判别器之后，都要把判别器的参数截断到一个固定的范围内，比如 $[-0.01, 0.01]$。最后，论文作者建议不要使用基于动量的优化算法（比如 Adam），推荐使用 RMSProp。

论文中给出了一个公式：$\mathbb{E}_{x \sim \mathbb{P}_r}[f_w(x)] - \mathbb{E}_{z \sim p(z)}[f_w(g_\theta(z))]$，其中函数 f_w 是判别器，x 是真实数据，$g_\theta(z)$ 是生成数据。此时判别器输出的是 Wasserstein 距离，所以该公式是计算真实样本与生成样本之间的距离的。训练判别器的目标是让两者间距离最大，从而能够分清楚真实数据和假数据。设该公式为 L，则训练判别器的目标就是最大化 L。将该公式取反，就是最小化 $-L$，得到公式为 $\mathbb{E}_{z \sim p(z)}[f_w(g_\theta(z))] - \mathbb{E}_{x \sim \mathbb{P}_r}[f_w(x)]$，这就是判别器的损失。

训练生成器的目标是最小化 L，即最小化真实样本与生成样本之间的距离，让生成的图像跟真的一样。而第一项 $\mathbb{E}_{x \sim \mathbb{P}_r}[f_w(x)]$ 跟生成器无关，可以删除，最终得到生成器的损失 $-\mathbb{E}_{z \sim p(z)}[f_w(g_\theta(z))]$。

先别着急看代码，尽管理论上很完美，但是 WGAN 的实际训练仍然很难，甚至收敛较慢，这一切都是由于对 Lipschitz 连续的实现有问题导致的。于是，作者又推出一篇论文 *Improved Training of Wasserstein GANs*，提出了改进方案，新的 WGAN 叫作 WGAN-GP，论文地址是：https://arxiv.org/abs/1704.00028。

10.5 WGAN-GP

10.5.1 WGAN-GP 算法

上一节介绍了 WGAN 为满足 Lipschitz 连续的限制，采用了权重裁剪的方式，将判别器的参数限制在一个固定的范围内。然而，这种权重裁剪会带来两个问题，如图 10.4 所示（该图出自 WGAN-GP 论文）。

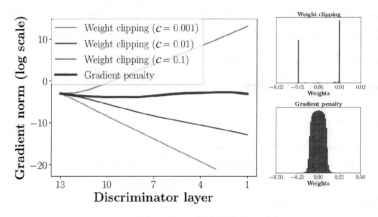

图 10.4　WGAN 权重裁剪带的两个问题

第一个问题，如左图所展示的，权重裁剪会导致梯度消失或梯度爆炸。这是因为在多层网络中，如果把裁剪范围设得很小，比如图中的 $c = 0.001$，那么每经过一层，梯度就会变小一点，经过多层之后梯度消失；相反，如果裁剪范围设得大一些，比如图中的 $c = 0.1$，那么经过多层之后最终梯度爆炸。

第二个问题，如右上图所展示的，权重裁剪过后网络参数分布在了裁剪阈值的最大值和最小值上了，这样的网络参数完全没有发挥出深度神经网络拟合复杂函数的优势。从图中能够看出，经过 WGAN-GP 调整后，能够解决这两个问题。

WGAN-GP 的思想是在判别器的损失上加入一个正则项，这个正则项就是 GP（Gradient Penalty），称之为梯度惩罚。另外，论文中指出不要使用 Batch Normalization（批次归一化），这会破坏梯度惩罚的效果，可以使用 Layer Normalization（层归一化）来替代。

10.5.2 训练 WGAN-GP 生成人脸：基于 TensorFlow 1.x

定义输入占位符，OUTPUT_DIM 是输入图像的维度。

```
OUTPUT_DIM = 28*28*3
input_real = tf.placeholder(tf.float32, [None, OUTPUT_DIM], name="input")
```

然后是实现判别器。需要注意，因为要训练两个网络（G 和 D），所以不同网络的变量名要统一，例如生成器的变量名应该统一成 generator 变量，这样方便后面的训练。可以通过 tf.variable_scope 函数定义变量的命名空间来解决。另外，还要指定 tf.variable_scope 的参数 reuse。因为我们想要重复使用网络的变量，例如对于生成器不仅要训练它，还要用它来生成图像，而判别器在传入真实图像和假图像时使用的判别器变量应该是相同的，也就是创建网络时使用的变量。将 reuse 设成 False，会重新生成网络的变量，将 reuse 设成 True，就是重用变量。

把 batch_norm 改成 layer_norm，将最后的 Sigmoid 删除，只返回最后一个全连接层的 logits 输出即可，此时不再是二分类网络，logits 被认为是 Wasserstein 距离。不要忘了输入图像是 OUTPUT_DIM 维度的向量，我们需要先把输入图像的形状转换成 [28, 28, 3]。

```
def discriminator(images, reuse=False):

    # 先转换输入图像的形状：[28, 28, 3]
    images = tf.reshape(images, [-1, 28, 28, 3])
    # 设置网络的变量命名空间，该网络下的所有变量都是discriminator变量
    with tf.variable_scope("discriminator", reuse=reuse):
        # 输出的形状：[14, 14, 64]
        x1 = tf.layers.conv2d(images, 64, 5, 2, padding='same',
                              kernel_initializer=tf.contrib.layers.
                              xavier_initializer())
        x1 = tf.nn.leaky_relu(x1)

        # 输出的形状：[7, 7, 128]
        x2 = tf.layers.conv2d(x1, 128, 5, 2, padding='same',
kernel_initializer=tf.contrib.layers.xavier_initializer())
        x2 = tf.contrib.layers.layer_norm(x2)
        x2 = tf.nn.leaky_relu(x2)

        # 输出的形状：[4, 4, 256]
        x3 = tf.layers.conv2d(x2, 256, 5, 2, padding='same',
```

```
        kernel_initializer=tf.contrib.layers.xavier_initializer())
    x3 = tf.contrib.layers.layer_norm(x3)
    x3 = tf.nn.leaky_relu(x3)
    
    x_flat = tf.reshape(x3, [-1, 4 * 4 * 256])
    logits = tf.layers.dense(x_flat, 1)

    return logits
```

接下来是实现生成器,同样使用了 layer_norm,激活函数是 ReLU。当然,你也可以使用 Leaky ReLU。最后把输出图像的形状再转换成向量即可。

```
# 噪声向量的维度,当然,也可以作为生成器函数的参数,这里作为全局变量的超参数了
z_dim = 128
# 参数n_samples是生成图像的个数,训练时传入batch_size
def generator(n_samples, reuse=False):
    # 生成随机噪声向量
    z = tf.random_normal([n_samples, z_dim])
    # 设置网络的变量名空间,该网络下的所有变量都是generator变量
    with tf.variable_scope("generator", reuse=reuse):
        x1 = tf.layers.dense(z, 2*2*512)
        x1 = tf.contrib.layers.layer_norm(x1)
        x1 = tf.nn.relu(x1)
        # 输出的形状:[2, 2, 512]
        x1 = tf.reshape(x1, [-1, 2, 2, 512])
        
        # 开始上采样。步幅设成2,使每一层的大小翻倍,深度减少一半
        # 输出的形状:[4, 4, 256]
        x2 = tf.layers.conv2d_transpose(x1, 256, 5, 2, padding='same',
            kernel_initializer=tf.contrib.layers.xavier_initializer())
        x2 = tf.contrib.layers.layer_norm(x2)
        x2 = tf.nn.relu(x2)
        
        # 输出的形状:[8, 8, 128]
        x3 = tf.layers.conv2d_transpose(x2, 128, 5, 2, padding='same',
            kernel_initializer=tf.contrib.layers.xavier_initializer())
        x3 = tf.contrib.layers.layer_norm(x3)
```

```
x3 = tf.nn.relu(x3)

# [8, 8, 128] -> [7, 7, 128]
x3 = x3[:, :7, :7, :] # 8×8只保留7×7

# 输出的形状：[14, 14, 64]
x4 = tf.layers.conv2d_transpose(x3, 64, 5, 2, padding='same',
    kernel_initializer=tf.contrib.layers.xavier_initializer())
x4 = tf.contrib.layers.layer_norm(x4)
x4 = tf.nn.relu(x4)

# 输出的形状：[28, 28, 3]
logits = tf.layers.conv2d_transpose(x4, 3, 5, 2, padding='same',
    kernel_initializer=tf.contrib.layers.xavier_initializer())

output = tf.tanh(logits)
# 转换生成图像的形状
return tf.reshape(output, [-1, OUTPUT_DIM])
```

现在该实现损失函数了，先看看 WGAN-GP 论文中的算法描述，如图 10.5 所示。

Algorithm 1 WGAN with gradient penalty. We use default values of $\lambda = 10$, $n_{\text{critic}} = 5$, $\alpha = 0.0001$, $\beta_1 = 0$, $\beta_2 = 0.9$.

Require: The gradient penalty coefficient λ, the number of critic iterations per generator iteration n_{critic}, the batch size m, Adam hyperparameters α, β_1, β_2.
Require: initial critic parameters w_0, initial generator parameters θ_0.

1: **while** θ has not converged **do**
2:     **for** $t = 1, ..., n_{\text{critic}}$ **do**
3:         **for** $i = 1, ..., m$ **do**
4:             Sample real data $x \sim \mathbb{P}_r$, latent variable $z \sim p(z)$, a random number $\epsilon \sim U[0,1]$.
5:             $\tilde{x} \leftarrow G_\theta(z)$
6:             $\hat{x} \leftarrow \epsilon x + (1-\epsilon)\tilde{x}$
7:             $L^{(i)} \leftarrow D_w(\tilde{x}) - D_w(x) + \lambda(\|\nabla_{\hat{x}} D_w(\hat{x})\|_2 - 1)^2$
8:         **end for**
9:         $w \leftarrow \text{Adam}(\nabla_w \frac{1}{m} \sum_{i=1}^{m} L^{(i)}, w, \alpha, \beta_1, \beta_2)$
10:     **end for**
11:     Sample a batch of latent variables $\{z^{(i)}\}_{i=1}^{m} \sim p(z)$.
12:     $\theta \leftarrow \text{Adam}(\nabla_\theta \frac{1}{m} \sum_{i=1}^{m} -D_w(G_\theta(z)), \theta, \alpha, \beta_1, \beta_2)$
13: **end while**

图 10.5　WGAN-GP 算法描述

行 1~13 的大循环就是整个训练循环，直到网络收敛才结束。

行 2~10 训练判别器，通常循环次数是 5 次，就是 $n_{\text{critic}} = 5$。

行 3 的 m 是一个批次的样本个数（batch_size），行 3~8 处理一个批次的数据。

行 4 从训练集中采样出一个批次的真实数据，然后生成一组随机噪声向量 z，再准备一个随机数 ϵ，ϵ 可以看成是采样的百分比。

行 5 将随机噪声向量 z 传给生成器，得到生成器的输出 $\widetilde{x}$。

行 6 使用随机数 ϵ 进行采样，假设 ϵ 是 0.3，就是采样出 30% 的真实样本（ϵx），再采样出 70% 的生成样本（$(1-\epsilon)\widetilde{x}$），将两者组合起来成为一个新的样本，暂时称之为组合样本 $\hat{x}$。

行 7 有三项，第一项是将生成器的输出 $\widetilde{x}$ 传给判别器得到判别器（生成样本）的输出；第二项是将真实样本输入给判别器得到判别器（真实样本）的输出；最后一项就是梯度惩罚（GP）了。

λ 是梯度惩罚系数，相当于 GP 项的权重，$\lambda = 10$。$(\|\nabla_{\hat{x}} D_w(\hat{x})\|_2 - 1)^2$ 这一项得分开说。首先，$D_w(\hat{x})$ 是将组合样本传给判别器得到判别器（组合样本）的输出，而 $\nabla_{\hat{x}} D_w(\hat{x})$ 是求判别器（组合样本）的输出（可以理解成一个使用组合样本构成的函数）对于组合样本的导数，相当于计算判别器（组合样本）函数对组合样本的梯度。

接下来，$\|\nabla_{\hat{x}} D_w(\hat{x})\|_2$ 是计算梯度值的 L2 范数，一个常用的正则项。最后的 $(\|\nabla_{\hat{x}} D_w(\hat{x})\|_2 - 1)^2$ 是为了让正则化后的梯度离 1 越来越近，1 就是为了保证 Lipschitz 连续的值 K。Lipschitz 连续要求判别器的梯度不能超过 K，在 WGAN 的实现中采用了权重裁剪，但会有副作用。所以在 WGAN-GP 中不再使用权重裁剪，而是通过将判别器的正则化梯度不断接近 K 值即可，这里 K 值设成了 1。

$D_w(\widetilde{x}) - D_w(x)$ 是 WGAN 中定义的判别器损失（就是上一节中提到的 $\mathbb{E}_{z \sim p(z)}[f_w(g_\theta(z))] - \mathbb{E}_{x \sim \mathbb{P}_r}[f_w(x)]$），此时再加上 GP 项，就得到了 WGAN-GP 的判别器损失 L，GP 项是用来替代权重裁剪满足 Lipschitz 连续的。

行 9 使用 Adam 优化器来优化判别器损失，每次迭代都要训练判别器 5 次（$n_{\text{critic}} = 5$）。

行 11 采样出一批随机噪声向量 z。

行 12 就是训练生成器了，生成器的损失是 $D_w(G_\theta(z))$（就是上一节中提到的 $-\mathbb{E}_{z \sim p(z)}[f_w(g_\theta(z))]$），这个损失跟 WGAN 一样没有变化。该损失项是对判别器（生成样本）的输出取负，仍然使用 Adam 优化器来优化损失。

α、β_1 和 β_2 是传给 Adam 优化器的参数。

至此，理论部分解释完了，现在来实现损失函数。

```python
def model_loss(input_real, LAMBDA=10):
    # 传入生成图像个数（batch_size），得到生成样本G_θ(z)，算法中的行5
    # g_output是生成器网络的输出，也就是生成的图像
    # 第一次构建生成器网络，reuse使用False
    g_output = generator(batch_size, reuse=False)
    # 传入真实样本得到判别器(真实样本)的输出f_w(x)
    # 第一次构建判别器网络，reuse使用False
    d_logits_real = discriminator(input_real, reuse=False)
    # 传入生成样本得到判别器(生成样本)的输出f_w(G_θ(z))
    # 此时reuse要设置成True，因为要重用网络的变量
    d_logits_fake = discriminator(g_output, reuse=True)

    # 根据公式-f_w(G_θ(z))得到生成器损失
    gen_cost = -tf.reduce_mean(d_logits_fake)
    # 根据公式f_w(G_θ(z))-f_w(x)得到WGAN的判别器损失
    disc_cost = tf.reduce_mean(d_logits_fake) - tf.reduce_mean(d_logits_real)

    # 接下来开始计算GP项
    # 首先随机得到采样概率ε
    alpha = tf.random_uniform(
        shape=[batch_size, 1],
        minval=0.,
        maxval=1.
    )
    # 这里跟算法有点不太一样，算法中的采样是εx+(1-ε)x̃，将该公式拆开得到：εx+x̃-εx̃
    # 整理得到：x̃ + ε(x - x̃)，其中x - x̃就是生成样本与真实样本之间的差别differences
    differences = g_output - input_real
    # 然后就是x̃ + ε(x - x̃)，你应该注意到了，这里的实现跟公式稍稍有些区别
    # 实际上这里的实现是x + ε(x̃ - x)，得到组合样本，这样就完成了算法中的行6
    interpolates = input_real + (alpha * differences)
    # 开始GP项（算法中的行7）的实现
    # 先计算判别器对于组合样本的梯度，注意要重用判别器的变量
    gradients = tf.gradients(discriminator(interpolates, reuse=True), [interpolates
    ])[0]
    # 然后计算梯度的L2范数
    slopes = tf.sqrt(tf.reduce_sum(tf.square(gradients), reduction_indices=[1]))
```

```python
    # 跟K值1尽量接近，得到GP项
    gradient_penalty = tf.reduce_mean((slopes - 1.) ** 2)
    # 将GP项乘以权重LAMBDA再加上WGAN的判别器损失就得到最终的WGAN-GP的判别器损失
    disc_cost += LAMBDA * gradient_penalty

    tf.summary.scalar('d_loss', disc_cost)
    tf.summary.scalar('g_loss', gen_cost)
    summary_op = tf.summary.merge_all()

    return disc_cost, gen_cost, summary_op
```

优化器部分：

```python
def model_opt(d_loss, g_loss, learning_rate, beta1, beta2, global_step):
    with tf.control_dependencies(tf.get_collection(tf.GraphKeys.UPDATE_OPS)):
        # 之前定义网络时各自定义了变量的命名空间，现在要用命名空间来获得各自网络
        # 训练变量
        var_list = tf.trainable_variables()
        d_var_list = [var for var in var_list if var.name.startswith(
            'discriminator')]
        g_var_list = [var for var in var_list if var.name.startswith('generator')]
        # 使用Adam优化各自的网络
        d_train_opt = tf.train.AdamOptimizer(learning_rate, beta1=beta1,
beta2=beta2).minimize(d_loss, var_list = d_var_list, global_step=global_step)
        g_train_opt = tf.train.AdamOptimizer(learning_rate, beta1=beta1,
beta2=beta2).minimize(g_loss, var_list = g_var_list, global_step=global_step)
    return d_train_opt, g_train_opt
```

接下来实现训练函数。

```python
import time
import numpy as np

def train(epochs, batch_size, learning_rate, beta1, beta2, get_batches,
          Xs, CRITIC_ITERS=5):

    train_times = 0
```

```python
gen = get_batches(Xs, batch_size)
with tf.Session() as sess:
    global_step = tf.train.get_or_create_global_step(sess.graph)

    # 定义输入占位符，调用上面实现的函数，得到损失和优化器
    input_real = tf.placeholder(tf.float32, [None, OUTPUT_DIM], name="input")
    d_loss, g_loss, summary_op = model_loss(input_real)
    d_train_opt, g_train_opt = model_opt(d_loss, g_loss, beta1, beta2,
                                         global_step)

    timestamp = str(int(time.time()))
    out_dir = os.path.abspath(os.path.join(os.path.curdir, "runs", timestamp))
    print("Writing to {}\n".format(out_dir))
    train_summary_dir = os.path.join(out_dir, "summaries", "train")
    train_summary_writer = tf.summary.FileWriter(train_summary_dir, sess.graph)

    # 初始化全局变量
    sess.run(tf.global_variables_initializer())
    saver = tf.train.Saver()
    save_dir = "./models"

    # 如果存在事先保存的模型参数，则加载
    if not os.path.isdir(save_dir):
        os.mkdir(save_dir)
    checkpoint = tf.train.get_checkpoint_state(save_dir)
    if checkpoint and checkpoint.model_checkpoint_path:
        saver.restore(sess, tf.train.latest_checkpoint(save_dir))
        print("Successfully loaded:", tf.train.latest_checkpoint(save_dir))
    else:
        print("Could not find old network weights")

    batch_step = 0

    # try:
    if True:  # 开始训练循环
        for epoch_i in range(epoch_count):
```

行 1~13 的大循环就是整个训练循环，直到网络收敛才结束。

行 2~10 训练判别器，通常循环次数是 5 次，就是 $n_{\text{critic}} = 5$。

行 3 的 m 是一个批次的样本个数（batch_size），行 3~8 处理一个批次的数据。

行 4 从训练集中采样出一个批次的真实数据，然后生成一组随机噪声向量 z，再准备一个随机数 ϵ，ϵ 可以看成是采样的百分比。

行 5 将随机噪声向量 z 传给生成器，得到生成器的输出 $\tilde{x}$。

行 6 使用随机数 ϵ 进行采样，假设 ϵ 是 0.3，就是采样出 30% 的真实样本（ϵx），再采样出 70% 的生成样本（$(1-\epsilon)\tilde{x}$），将两者组合起来成为一个新的样本，暂时称之为组合样本 $\hat{x}$。

行 7 有三项，第一项是将生成器的输出 $\tilde{x}$ 传给判别器得到判别器（生成样本）的输出；第二项是将真实样本输入给判别器得到判别器（真实样本）的输出；最后一项就是梯度惩罚（GP）了。

λ 是梯度惩罚系数，相当于 GP 项的权重，$\lambda = 10$。$(\|\nabla_{\hat{x}} D_w(\hat{x})\|_2 - 1)^2$ 这一项得分开说。首先，$D_w(\hat{x})$ 是将组合样本传给判别器得到判别器（组合样本）的输出，而 $\nabla_{\hat{x}} D_w(\hat{x})$ 是求判别器（组合样本）的输出（可以理解成一个使用组合样本构成的函数）对于组合样本的导数，相当于计算判别器（组合样本）函数对组合样本的梯度。

接下来，$\|\nabla_{\hat{x}} D_w(\hat{x})\|_2$ 是计算梯度值的 L2 范数，一个常用的正则项。最后的 $(\|\nabla_{\hat{x}} D_w(\hat{x})\|_2 - 1)^2$ 是为了让正则化后的梯度离 1 越来越近，1 就是为了保证 Lipschitz 连续的值 K。Lipschitz 连续要求判别器的梯度不能超过 K，在 WGAN 的实现中采用了权重裁剪，但会有副作用。所以在 WGAN-GP 中不再使用权重裁剪，而是通过将判别器的正则化梯度不断接近 K 值即可，这里 K 值设成了 1。

$D_w(\tilde{x}) - D_w(x)$ 是 WGAN 中定义的判别器损失（就是上一节中提到的 $\mathbb{E}_{z \sim p(z)}[f_w(g_\theta(z))] - \mathbb{E}_{x \sim \mathbb{P}_r}[f_w(x)]$），此时再加上 GP 项，就得到了 WGAN-GP 的判别器损失 L，GP 项是用来替代权重裁剪满足 Lipschitz 连续的。

行 9 使用 Adam 优化器来优化判别器损失，每次迭代都要训练判别器 5 次（$n_{\text{critic}} = 5$）。

行 11 采样出一批随机噪声向量 z。

行 12 就是训练生成器了，生成器的损失是 $-D_w(G_\theta(z))$（就是上一节中提到的 $-\mathbb{E}_{z \sim p(z)}[f_w(g_\theta(z))]$），这个损失跟 WGAN 一样没有变化。该损失项是对判别器（生成样本）的输出取负，仍然使用 Adam 优化器来优化损失。

α、β_1 和 β_2 是传给 Adam 优化器的参数。

至此，理论部分解释完了，现在来实现损失函数。

```python
def model_loss(input_real, LAMBDA=10):
    # 传入生成图像个数(batch_size), 得到生成样本G_θ(z), 算法中的行5
    # g_output是生成器网络的输出, 也就是生成的图像
    # 第一次构建生成器网络, reuse使用False
    g_output = generator(batch_size, reuse=False)
    # 传入真实样本得到判别器(真实样本)的输出f_w(x)
    # 第一次构建判别器网络, reuse使用False
    d_logits_real = discriminator(input_real, reuse=False)
    # 传入生成样本得到判别器(生成样本)的输出f_w(G_θ(z))
    # 此时reuse要设置成True, 因为要重用网络的变量
    d_logits_fake = discriminator(g_output, reuse=True)

    # 根据公式-f_w(G_θ(z))得到生成器损失
    gen_cost = -tf.reduce_mean(d_logits_fake)
    # 根据公式f_w(G_θ(z))-f_w(x)得到WGAN的判别器损失
    disc_cost = tf.reduce_mean(d_logits_fake) - tf.reduce_mean(d_logits_real)

    # 接下来开始计算GP项
    # 首先随机得到采样概率ε
    alpha = tf.random_uniform(
        shape=[batch_size, 1],
        minval=0.,
        maxval=1.
    )
    # 这里跟算法有点不太一样, 算法中的采样是εx+(1-ε)x̃, 将该公式拆开得到: εx+x̃-εx̃
    # 整理得到: x̃ + ε(x - x̃), 其中x - x̃就是生成样本与真实样本之间的差别differences
    differences = g_output - input_real
    # 然后就是x̃ + ε(x - x̃), 你应该注意到了, 这里的实现跟公式稍稍有些区别
    # 实际上这里的实现是x + ε(x̃ - x), 得到组合样本, 这样就完成了算法中的行6
    interpolates = input_real + (alpha * differences)
    # 开始GP项(算法中的行7)的实现
    # 先计算判别器对于组合样本的梯度, 注意要重用判别器的变量
    gradients = tf.gradients(discriminator(interpolates, reuse=True), [interpolates
        ])[0]
    # 然后计算梯度的L2范数
    slopes = tf.sqrt(tf.reduce_sum(tf.square(gradients), reduction_indices=[1]))
```

```python
if True:
    start_time = time.time()
    # 训练一次生成器，第一次训练是没有意义的，因为训练生成器需要
    # 判别器来指导
    if epoch_i > 0:
        _, gen_loss = sess.run([g_train_opt, g_loss])

    dev_disc_costs = []

    # 开始CRITIC_ITERS次训练判别器
    for i in range(CRITIC_ITERS):
        # 得到真实样本
        batch_images = next(gen)
        batch_images *= 2
        # 将样本的形状转换成向量再传给网络
        batch_images=batch_images.reshape([-1, OUTPUT_DIM])
        # 训练判别器
        disc_loss, _, step, summary=sess.run([d_loss, d_train_opt,
                                            global_step,
                                            summary_op],
                            feed_dict={input_real: batch_images})
        dev_disc_costs.append(disc_loss)

    disc_loss = np.mean(dev_disc_costs)

    train_summary_writer.add_summary(summary, step)

    batch_step += 1
    train_times += 1
    # 打印训练情况
    if train_times % 10 == 0:
        print(
            "Epochs {}/{} Batch Step {}/{} gen_loss = {}...,
            disc_loss = {}..., train using time = {:.3f}".format(
                epoch_i + 1, epochs, batch_step,
                len(Xs) // batch_size + 1, gen_loss,
```

```
                                    disc_loss, time.time() - start_time))
                        # 保存模型
                        if train_times % 100 == 0:
                            save_path = saver.save(sess, os.path.join(save_dir, '
                            best_model.ckpt'), global_step=step)
                            print("Model saved in file: {}".format(save_path))
                        if batch_step % (data_shape[0] // batch_size) == 0:
                            batch_step = 0
#        except Exception as reason:
#            print("except!!!", type(reason), reason)
#            save_path = saver.save(sess, os.path.join(save_dir,
# 'best_model.ckpt'), global_step=step)
#            print("Model saved in file: {}".format(save_path))
    print("Done!")
```

最后,就可以调用训练函数开始训练了,超参数设置如下。

```
batch_size = 64
z_dim = 128
learning_rate = 1e-4
beta1 = 0
beta2 = 0.9


epochs = 5 * 60000
```

论文作者的代码实现地址是 https://github.com/igul222/improved_wgan_training。

现在可以训练网络来生成人脸了。数据集在 http://mmlab.ie.cuhk.edu.hk/projects/CelebA.html 网站上下载,下载的文件是 img_align_celeba.zip。解压缩后,需要先做预处理,你也可以定义一个处理函数,当读一批次图像时再执行预处理操作也可以。预处理做什么呢?将图像的大小缩放至网络定义的尺寸。例如,判别器接收的图像尺寸是 $28 \times 28 \times 3$,并且生成器生成的图像尺寸也是 $28 \times 28 \times 3$,这样的话,人脸图像就需要缩小到 $28 \times 28 \times 3$ 的尺寸才可以。另外,判别器输入图像的取值范围是 $[-1,1]$,所以不要忘了转换。当然,你也可以使用 MNIST 数据集先验证网络的构建及超参数等设置是否合适。

本节完整代码的实现可以在 GitHub 上找到,地址是:https://github.com/chengstone/face_generation-WGAN-GP,文件名是 face_generation-wgan_gp.ipynb。

当网络训练到 29760 epochs 时给停止了,此时的损失是:

```
Epochs 29760/300000 Batch Step 1275/3166 gen_loss = -71.71266174316406...,
disc_loss = -0.3569445013999939..., train using time = 21.513
```

WGAN 的训练比起传统的 GAN 训练难度要小很多，很关键的一点是损失函数的变化带来了训练效果的可观察性。在训练 DCGAN 时，尽管网络可以打印出判别器和生成器的损失，但是仍然无法知道网络距离收敛到底还差多少。而 WGAN 的训练就很舒服，我们来观察判别器的损失，在训练过程中判别器的损失在逐渐接近 0，因为该损失代表生成样本和真实样本之间的 Wasserstein 距离，距离为 0 说明生成的图像已经可以以假乱真了。再来观察生成器的损失，也是在逐步变小的，这样我们就能时刻知道当前网络的训练状态是否健康。WGAN-GP 在训练过程中生成的人脸如图 10.6 所示。

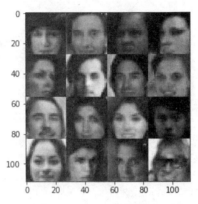

图 10.6　WGAN-GP 生成的人脸

我们来看看 TensorBoard 上的损失曲线，判别器的损失曲线如图 10.7 所示。

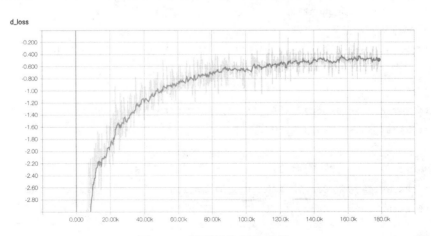

图 10.7　判别器的损失曲线

生成器的损失曲线如图 10.8 所示。

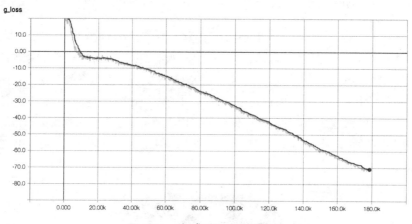

图 10.8 生成器的损失曲线

10.5.3 训练 WGAN-GP 生成人脸：基于 TensorFlow 2.0

本节的代码文件是 face_generation-wgan_gp_tf2.ipynb。

网络结构跟上一节是一样的没有变化，判别器的实现如下：

```
OUTPUT_DIM = 28*28*3

def discriminator():
    # 构建判别器模型
    model = tf.keras.Sequential([
        # 先转换输入图像的形状：[28, 28, 3]
        tf.keras.layers.Reshape([28, 28, 3], input_shape=([OUTPUT_DIM,])),

        # 输出的形状：[14, 14, 64]
        tf.keras.layers.Conv2D(64, (5, 5), strides=(2, 2), padding='same'),
        tf.keras.layers.LeakyReLU(),

        # 输出的形状：[7, 7, 128]
        tf.keras.layers.Conv2D(128, (5, 5), strides=(2, 2), padding='same'),
        tf.keras.layers.LeakyReLU(),

        # 输出的形状：[4, 4, 256]
```

```python
        tf.keras.layers.Conv2D(256, (5, 5), strides=(2, 2), padding='same'),
        tf.keras.layers.LeakyReLU(),

        tf.keras.layers.Flatten(),
        tf.keras.layers.Dense(1)
    ])
    # 返回模型
    return model
```

生成器的实现如下:

```python
# 随机噪声向量的维度
z_dim = 128


def generator():
    # 构建生成器模型
    model = tf.keras.Sequential([
        tf.keras.layers.Dense(2*2*512, activation="relu", input_shape=([z_dim,])),
        # 输出的形状: [2, 2, 512]
        tf.keras.layers.Reshape([2, 2, 512]),
        # 输出的形状: [4, 4, 256]
        tf.keras.layers.Conv2DTranspose(256, (5, 5), strides=(2, 2),
                                        activation="relu", padding='same'),
        # 输出的形状: [8, 8, 128]
        tf.keras.layers.Conv2DTranspose(128, (5, 5), strides=(2, 2),
                                        activation="relu", padding='same'),
        # [8, 8, 128] -> [7, 7, 128]
        tf.keras.layers.Lambda(lambda x:x[:, :7, :7, :]),
        # 输出的形状: [14, 14, 64]
        tf.keras.layers.Conv2DTranspose(64, (5, 5), strides=(2, 2),
                                        activation="relu", padding='same'),
        # 输出的形状: [28, 28, 3]
        tf.keras.layers.Conv2DTranspose(3, (5, 5), strides=(2, 2),
                                        activation="tanh", padding='same'),
        # 转换生成图像的形状
        tf.keras.layers.Reshape([OUTPUT_DIM])
    ])
```

```
# 返回模型
return model
```

实现生成器和判别器的损失函数,跟上一节没有区别。

```python
# 生成器的损失
def generator_loss(d_logits_fake):
    # 根据公式-f_w(G_θ(z))得到生成器的损失
    gen_cost = -tf.reduce_mean(d_logits_fake)
    return gen_cost

batch_size = 64

# 判别器的损失
def discriminator_loss(input_real, g_output, d_logits_real, d_logits_fake):
    LAMBDA=10
    # 根据公式f_w(G_θ(z))-f_w(x)得到WGAN的判别器损失
    disc_cost = tf.reduce_mean(d_logits_fake) - tf.reduce_mean(d_logits_real)

    # 接下来开始计算GP项
    # 首先随机得到采样概率ε
    alpha = tf.random.uniform(
        shape=[batch_size, 1],
        minval=0.,
        maxval=1.
    )
    # 这里跟算法有点不太一样,算法中的采样是εx+(1-ε)x̃,将该公式拆开得到:εx+x̃-εx̃
    # 整理得到:x̃+ε(x-x̃),其中x-x̃就是生成样本与真实样本之间的差别(differences)
    differences = g_output - input_real
    # 然后就是x̃ + ε(x - x̃),你应该注意到了,这里的实现跟公式稍稍有些区别
    # 实际上这里的实现是x + ε(x̃ - x),得到组合样本,这样就完成了算法中的行6
    interpolates = input_real + (alpha * differences)
    # 开始GP项(算法中的行7)的实现
    # 先计算判别器对于组合样本的梯度,注意要重用判别器的变量
    gradients = tf.gradients(discriminator_model(interpolates, training=True),
                             [interpolates])[0]
    # 然后计算梯度的L2范数
```

```python
    slopes = tf.sqrt(tf.reduce_sum(tf.square(gradients), axis=[1]))
    # 跟K值1尽量接近，得到GP项
    gradient_penalty = tf.reduce_mean((slopes - 1.) ** 2)
    # 将GP项乘以权重LAMBDA再加上WGAN的判别器损失就得到最终的WGAN-GP的判别器损失
    disc_cost += LAMBDA * gradient_penalty
    return disc_cost
```

定义优化器，并生成判别器和生成器的模型：

```python
beta1 = 0
beta2 = 0.9
generator_optimizer = tf.keras.optimizers.Adam(1e-4, beta1, beta2)
discriminator_optimizer = tf.keras.optimizers.Adam(1e-4, beta1, beta2)

generator_model = generator()
discriminator_model = discriminator()
```

实现训练函数：

```python
# 判别器的单步训练函数
@tf.function
def d_train_step(input_real):
    # 随机噪声
    noise = tf.random.normal([batch_size, z_dim])
    # 使用磁带记录计算损失的操作
    with tf.GradientTape() as disc_tape:
        # 调用模型，得到生成器生成的样本G_θ(z)（算法中的行5）
        g_output = generator_model(noise, training=True)
        # 传入真实样本得到判别器(真实样本)的输出f_w(x)
        d_logits_real = discriminator_model(input_real, training=True)
        # 传入生成样本得到判别器(生成样本)的输出f_w(G_θ(z))
        d_logits_fake = discriminator_model(g_output, training=True)
        # 计算判别器的损失
        disc_loss = discriminator_loss(input_real, g_output, d_logits_real,
                                      d_logits_fake)
    # 使用磁带计算模型参数的梯度
    gradients_of_discriminator = disc_tape.gradient(disc_loss,
                                                   discriminator_model.variables)
```

```python
        discriminator_optimizer.apply_gradients(zip(gradients_of_discriminator,
                                                   discriminator_model.variables))

    return disc_loss

# 生成器的单步训练函数
@tf.function
def g_train_step():
    # 随机噪声
    noise = tf.random.normal([batch_size, z_dim])
    # 使用磁带记录计算损失的操作
    with tf.GradientTape() as gen_tape:
        # 调用模型,得到生成器生成的样本$G_\theta(z)$(算法中的行5)
        g_output = generator_model(noise, training=True)
        # 传入生成样本得到判别器(生成样本)的输出$f_w(G_\theta(z))$
        d_logits_fake = discriminator_model(g_output, training=True)
        # 计算生成器的损失
        gen_loss = generator_loss(d_logits_fake)
    # 使用磁带计算模型参数的梯度
    gradients_of_generator = gen_tape.gradient(gen_loss, generator_model.variables)

    generator_optimizer.apply_gradients(zip(gradients_of_generator,
                                            generator_model.variables))

    return gen_loss


import time
import numpy as np
from tensorflow.python.ops import summary_ops_v2

# 训练函数
def train(epochs, batch_size, learning_rate, beta1, beta2, get_batches,
          Xs, CRITIC_ITERS=5):

    train_times = 0
    # 获得数据迭代器
```

```python
    gen = get_batches(Xs, batch_size)

if True:
    timestamp = str(int(time.time()))
    save_dir = "./models"
    # 如果模型目录不存在，则创建
    if tf.io.gfile.exists(save_dir):
        pass
    else:
        tf.io.gfile.makedirs(save_dir)

    train_dir = os.path.join(save_dir, 'summaries', 'train')
    # 定义日志记录器，可以不使用
    train_summary_writer = summary_ops_v2.create_file_writer(train_dir,
        flush_millis=10000)
    # 定义保存模型参数检查点
    checkpoint_dir = os.path.join(save_dir, 'checkpoints')
    checkpoint_prefix = os.path.join(checkpoint_dir, 'ckpt')

    checkpoint = tf.train.Checkpoint(generator_optimizer=generator_optimizer,
                                    discriminator_optimizer=
                                    discriminator_optimizer,
                                    generator=generator_model,
                                    discriminator=discriminator_model)

    # 加载存在的参数
    checkpoint.restore(tf.train.latest_checkpoint(checkpoint_dir))

    batch_step = 0
    # 定义平均指标用来计算平均损失
    avg_g_loss = tf.keras.metrics.Mean('g_loss', dtype=tf.float32)
    avg_d_loss = tf.keras.metrics.Mean('d_loss', dtype=tf.float32)
    # try:
    if True:
        # 训练循环
        for epoch_i in range(epochs):
```

```python
        if True:
            start_time = time.time()
            # 训练一次生成器，第一次训练是没有意义的，因为训练生成器需要
            # 判别器来指导
            if epoch_i > 0:
                gen_loss = g_train_step()

            avg_d_loss.reset_states()
            # 开始CRITIC_ITERS次训练判别器
            for i in range(CRITIC_ITERS):
                # 得到真实样本
                batch_images = next(gen)
                batch_images *= 2
                # 将样本的形状转换成向量再传给网络
                batch_images=batch_images.reshape([-1, OUTPUT_DIM])
                # 训练判别器
                disc_loss = d_train_step(batch_images)
                avg_d_loss(disc_loss)

            # summary_ops_v2.scalar('disc_loss', disc_loss)

            batch_step += 1
            train_times += 1
            # 打印训练情况
            if train_times % 10 == 0:
                print(
                    "Epochs {}/{} Batch Step {}/{} gen_loss = {}...,
                    disc_loss = {}..., train using time = {:.3f}".format(
                        epoch_i + 1, epochs, batch_step,
                        len(Xs) // batch_size + 1, gen_loss,
                        avg_d_loss.result(), time.time() - start_time))
            # 保存模型
            if train_times % 100 == 0:
                checkpoint.save(checkpoint_prefix)
            if batch_step % (len(Xs) // batch_size) == 0:
                batch_step = 0
```

```
#         except Exception as reason:
#             print("except!!!", type(reason), reason)
#             save_path = saver.save(sess, os.path.join(save_dir,
# 'best_model.ckpt'), global_step=step)
#             print("Model saved in file: {}".format(save_path))
    print("Done!")
```

10.6　PG-GAN 和 TL-GAN

现在你可以以本章介绍的 GAN 为基础来学习其他的 GAN 模型了，比如 NVIDIA 发表的论文 *Progressive Growing of GANs for Improved Quality, Stability, and Variation* 中提出了名为 PG-GAN 的方法，论文地址是 https://arxiv.org/abs/1710.10196。它可以生成高分辨率的图像，方法是先训练一个生成低分辨率图像的生成器，比如 4×4 的图像。当训练好之后在网络上再加一层，扩大生成图像的尺寸，比如 8×8，然后继续训练，当训练好之后再逐步过渡到生成更高分辨率的图像。这样网络从生成低分辨率的图像逐步成长，最终稳步学会了如何生成高分辨率的图像，过程如图 10.9 所示（该图出自该论文）。

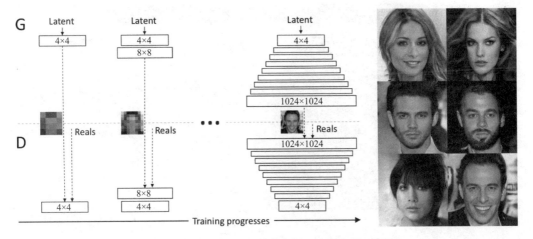

图 10.9　训练 PG-GAN 生成高分辨率图像的过程

最后给大家介绍一个很有意思的模型，叫作 TL-GAN（Transparent Latent-space GAN）。说它很有意思，是因为它利用潜在空间（Latent-space）中的特征轴可以轻松完成人脸特征的定制。大家可以去 GitHub 上看看动图，就知道这是一项超酷的技术，地址是 https://github.com/SummitKwan/transparent_latent_gan，截图如图 10.10 所示。

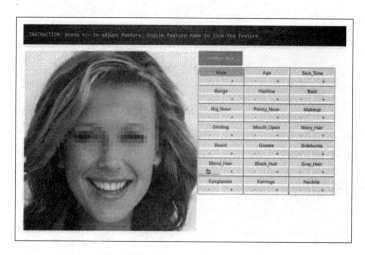

图 10.10　TL-GAN 图例

在图 10.10 右侧是改变人脸的多个特征标签，比如可以改变头发颜色（Blond_Hair、Black_Hair、Gray_Hair）、年龄（Age）、笑脸（Smiling）等。这些特征的修改，可以立即反映到所生成的图像上。稍微修改几个特征，图 10.10 中的人脸就变成了图 10.11 所示的样子。

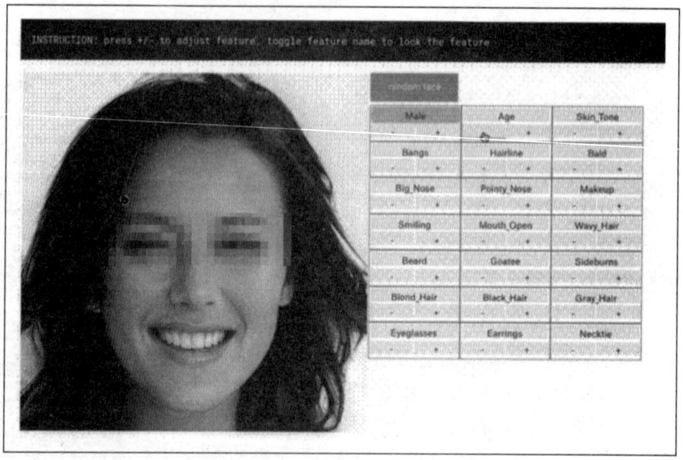

图 10.11　修改特征后的效果

太让人惊奇了！传统的 GAN 使用潜在空间中的随机噪声向量来生成随机图像，如图 10.12 的上半部分所示。本节引用的图片和内容均出自作者的演示文稿，地址是 https://docs.google.com/presentation/d/1OpcYLBVpUF1L-wwPHu_CyKjXqXD0oRwBoGP2peSCrSA/edit#slide=id.p1。

而 TL-GAN 是找到潜在空间中的特征轴（Feature Axes），利用特征轴就可以通过只改变某个特征来控制图像的生成（或修改）。

如何做到呢？

第一步，训练一个生成高质量图像的生成器，TL-GAN 使用了 PG-GAN 训练来生成高分辨率的清晰图像。

其想法是让随机噪声向量变得可以控制，这样就可以控制图像的生成了。作者的说法是，将噪声向量跟可解释的特征标签关联起来，让潜在空间变得透明。通常一个 1024 × 1024 的生成图像，所有的特征均由潜在空间中的 z 维噪声向量决定。如果我们能弄清楚潜在空间代表哪些特征，那么就能定制图像生成了。

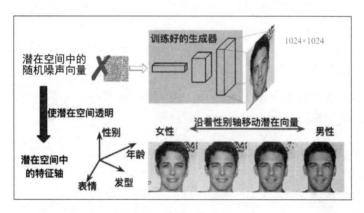

图 10.12　TL-GAN 的特征轴

第二步，训练一个特征提取器（暂时称之为 F），当然少不了准备打上标签的训练集了。这样的网络接收输入的图像，输出图像所具有的特征，比如输入图像的特征是男性、笑脸、发型（短头发）、头发颜色、年龄等，如图 10.13 所示。

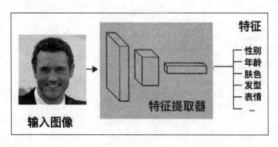

图 10.13　训练特征提取器

为了在潜在空间中找到特征轴，需要使用监督学习方法学习潜在向量 z 和特征标签 y 之间的关系。现在通过生成器 $G(z) = x_{\text{gen}}$ 和特征提取器 $F(x_{\text{gen}}) = y_{\text{pred}}$，就能够在潜在向量 z 和特征标签 y_{pred} 之间建立联系，这样就有了成对的潜在向量和特征，如图 10.14 所示。

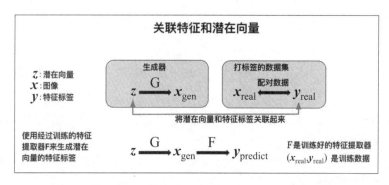

图 10.14 潜在向量与特征标签的关联

第三步，生成大量的随机噪声向量 z，传给生成器生成大量的图像，然后将所生成的图像传给特征提取器得到各生成图像的特征。

第四步，使用广义线性模型（Generalized Linear Model, GLM）在潜在向量和特征之间执行回归训练，回归斜率（Regression Slope）就是特征轴。

第五步，开始探索，找一个潜在向量，修改一个或多个特征轴，观察特征轴的修改对图像生成的影响。

最后给出一张完整的多特征变化的演示图像，如图 10.15 所示。

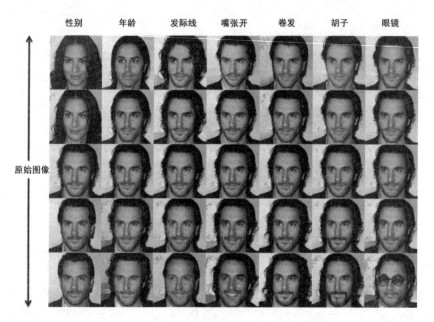

图 10.15　TL-GAN 多特征变化的演示图像

是不是很酷？赶快动手实践一下吧！

10.7 本章小结

本章讨论了 GAN 的基础理论和 GAN 的变种 DCGAN、WGAN、WGAN-GP，通过代码实现了 WGAN-GP，最后简单介绍了 PG-GAN 和 TL-GAN。目前 GAN 仍然是很活跃的研究对象，你现在可以尝试使用 WGAN 来生成一些东西了。比如 *DeepMasterPrints: Generating MasterPrints for Dictionary Attacks via Latent Variable Evolution* 这篇论文中介绍了研究人员使用 WGAN 生成了逼真的指纹，当然，除了 WGAN，他们还使用了其他技术。有兴趣的读者可以读一读并尝试复现，论文地址是 https://arxiv.org/abs/1705.07386，生成的指纹如图 10.16 所示（该图出自该论文）。

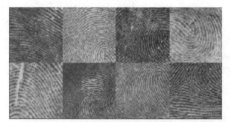
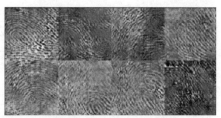

(a) Real (left) and generated (right) samples for the NIST dataset.

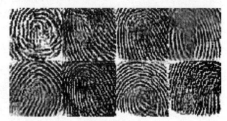

(b) Real (left) and generated (right) samples for the FingerPass capacitive dataset.

图 10.16 生成的指纹